PHYSICS POTENTIAL AND DEVELOPMENT OF $\mu^+\mu^-$ COLLIDERS

PHYSICS POTENTIAL AND DEVELOPMENT OF $\mu^+\mu^-$ COLLIDERS

Fourth International Conference

San Francisco, California   December 1997

EDITOR
David B. Cline
*Center for Advanced Accelerators and the
University of California, Los Angeles, California*

American Institute of Physics

AIP CONFERENCE
PROCEEDINGS 441

Woodbury, New York

Editor:

David B. Cline
Department of Physics and Astronomy
University of California
405 Hilgard Avenue
P.O. Box 951547
Los Angeles, CA 90095-1547
Email: dcline@physics.ucla.edu

Articles on pp. 174–180, pp. 183–208, pp. 282–288, and pp. 289–294 were authored by U. S. Government employees and are not covered by the below mentioned copyright.

Authorization to photocopy items for internal or personal use, beyond the free copying permitted under the 1978 U.S. Copyright Law (see statement below), is granted by the American Institute of Physics for users registered with the Copyright Clearance Center (CCC) Transactional Reporting Service, provided that the base fee of $15.00 per copy is paid directly to CCC, 222 Rosewood Drive, Danvers, MA 01923. For those organizations that have been granted a photocopy license by CCC, a separate system of payment has been arranged. The fee code for users of the Transactional Reporting Service is: 1-56396-723-5/ 98 /$15.00.

© 1998 American Institute of Physics

Individual readers of this volume and nonprofit libraries, acting for them, are permitted to make fair use of the material in it, such as copying an article for use in teaching or research. Permission is granted to quote from this volume in scientific work with the customary acknowledgment of the source. To reprint a figure, table, or other excerpt requires the consent of one of the original authors and notification to AIP. Republication or systematic or multiple reproduction of any material in this volume is permitted only under license from AIP. Address inquiries to Office of Rights and Permissions, 500 Sunnyside Boulevard, Woodbury, NY 11797-2999; phone: 516-576-2268; fax: 516-576-2499; e-mail: rights@aip.org.

L.C. Catalog Card No. 98-86574
ISBN 1-56396-723-5
ISSN 0094-243X
DOE CONF- 971215

Printed in the United States of America

Contents

Preface ix

PARTICLE PHYSICS GOALS OF FUTURE COLLIDERS

Overview of Physics at a Muon Collider 3
    V. Barger
**Workshop on Physics at the First Muon Collider and Front-End
of a Muon Collider: A Brief Summary** 18
    S. Geer
Neutral Higgs Searches at LEP-2: Present Status 31
    P. Bambade
Higgs and Technicolor Goldstone Bosons at a Muon Collider 44
    J. F. Gunion
The Top Quark and Higgs Boson at Hadron Colliders 57
    C. Quigg
Top Quark Physics at a Polarized Muon Collider 72
    S. Parke
Threshold Cross Section Measurements 79
    M. S. Berger
Physics Potential of the CMS/LHC 85
    S. Dasu for the CMS Collaboration
R-Parity Violation and Sneutrino Resonances at Muon Colliders 92
    J. L. Feng
New Particles and Interactions at High Energy Muon Colliders 98
    S. Godfrey
Sparticle Masses from Kinematic Fitting at a Muon Collider 106
    J. D. Lykken
The Physics of Like-Sign Muon Collisions at High Energy 116
    C. A. Heusch
Doubly Charged Particles at a $\mu^\pm\mu^\pm$ Collider 126
    S. Rajpoot
Neutrino Physics at Muon Colliders 132
    B. J. King

LOW ENERGY PHYSICS WITH COLD $\mu^\pm$ BEAMS

A $\mu \to e + \gamma$ Experiment with 10^{-14} Sensitivity? 143
    H.-K. Walter
**The MECO Experiment to Search for $\mu^- N \to e^- N$ with Sensitivity
Below 10^{-16}** 146
    W. R. Molzon

PHYSICS AT HIGGS FACTORY AND PRECISION ELECTROWEAK DATA STUDIES

The Scientific Case for a Higgs Bosons Factory $\mu^+\mu^-$ Collider 159
    D. B. Cline
Precision Electroweak Data: Present Status and Future Prospects 168
    P. B. Renton
Resonant Higgs Enhancement at the First Muon Collider................... 174
    B. Kamal, W. J. Marciano, and Z. Parsa

$\mu^+\mu^-$ COLLIDER STUDIES

Muon Collider Design 183
    R. B. Palmer for the Muon Collaboration
An Isochronous Lattice Design for a 50 on 50 GeV Muon Collider........... 209
    C. Johnstone, A. Drozhdin, N. Mokhov, W. Wan, and A. Garren
The Lattice for the 50-50 GeV Muon Collider............................. 220
    K.-Y. Ng and D. Trbojevic
Calibrating the Energy of a 50×50 GeV Muon Collider Using
Spin Precession.. 228
    R. Raja and A. Tollestrup
Scraping Beam Halo in $\mu^+\mu^-$ Colliders.................................. 242
    A. Drozhdin, N. Mokhov, C. Johnstone, W. Wan, and A. Garren
Towards Ultimate Luminosity Polarized Muon Collider
(Problems and Prospects) 249
    A. Skrinsky
Luminosity Monitoring at μp and $\mu\mu$ Colliders......................... 265
    I. F. Ginzburg
Phase Space Exchange in Thick Wedge Absorbers for Ionization Cooling..... 270
    D. Neuffer
Muon Dynamics in a Toroidal Sector Magnet............................. 282
    J. C. Gallardo, R. C. Fernow, and R. B. Palmer
Ionization Cooling and Muon Dynamics................................. 289
    Z. Parsa
New $\mu^\pm$ Cooling for $\mu^\pm$ Colliders and Possible Realization at JHF/KEK...... 295
    K. Nagamine
Ring Cooler for Muon Collider 310
    V. I. Balbekov and A. Van Ginneken
An AGS Experiment to Test Bunching for the Proton Driver
of the Muon Collider 314
    J. Norem, C. Ankenbrandt, K.-Y. Ng, M. Popvic, Z. Qian, L. A. Ahrens,
    M. Brennan, V. Mane, T. Roser, D. Trbojevic, and W. van Asselt
Pion Production and Targetry at $\mu^+\mu^-$ Colliders 320
    N. Mokhov and A. Van Ginneken
Mesoscopic Quantum Multiplex for Channeling Bunches 325
    J. Shen

μ^-p COLLIDERS

Some Physical Problems for Future μ^-p Colliders........................ 333
    I. F. Ginzburg
Muon-Proton Colliders: Leptoquarks and Contact Interactions.............. 338
    K. Cheung

PHYSICS SUMMARY

The Physics of Muon Colliders: A Perspective 347
    W. J. Marciano

APPENDIX

e^+e^- and e^-p Options for the Very Large Hadron Collider 357
    J. Norem
Suppressing Beam Loss and De-channeling Control via
Modulating Potential by Phononics...................................... 360
    J. Shen
Lane Fuzzy Collision in Channel with Potential Deformation by
Photon-Phonon-Electron Excitation and Sub-atomic Control 367
    J. Shen

Conference Program 373
List of Attendees.. 379
Author Index.. 383

Preface

The 1997 $\mu^+\mu^-$ Collider Conference (the fourth) will stand out as a landmark in the development of a real $\mu^+\mu^-$ collider in the future. There were three key advances:

(1) The first real evidence from the electroweak parameters that the Higgs Boson mass is low, which could justify the construction of a Higgs Factory $\mu^+\mu^-$ Collider. Such a machine was first proposed at the first $\mu^+\mu^-$ collider conference in 1992.

(2) A successful defense of the program to the HEPAP sub-panel, resulting in the first specific funding from the U.S. Department of Energy for the collider study. This is a key development in getting the physics community behind this project.

(3) The first coherent plan for the muon cooling experiment, which will be a key milestone in the program, and an excellent workshop at Fermi National Accelerator Laboratory, which was held at the end of 1997.

All of these were discussed at length at this meeting.

Among many of the highlights of the meeting, the following stand out: (1) The excellent summary of the conference by Bill Marciano; (2) the precise discussion of the electroweak parameters by Peter Renton; (3) the strong motivation talk by Chris Quigg; (4) the summary of the FNAL workshop by Steve Geer; (5) a concise statement of the particle physics by Vernon Barger; (6) a very interesting discussion of the program in Japan by K. Nagamine; (7) a stimulating discussion of the ultimate possible luminosity of the $\mu^+\mu^-$ collider by Alexander Skrinsky; and finally (8) a very informative discussion of the view from the Department of Energy concerning accelerator research and development by David Sutter.

During the conference the weather was great – as always. San Francisco, and specifically the Fairmont Hotel, provides a superb location for future exciting meetings on our search for the ultimate nature of matter.

There are many fine articles in this volume, from which we can all continue to learn.

I wish to thank the Advisory Committee, the U.S. Department of Energy, and the UCLA staff – Jim Kolonko, Joan George, and especially Kevin Lee, for the excellent work they did for the conference and on these Proceedings.

David B. Cline
UCLA and the Center for Advanced Accelerators

PARTICLE PHYSICS GOALS
OF FUTURE COLLIDERS

Overview of Physics at a Muon Collider

V. Barger

Physics Department, University of Wisconsin, Madison, WI 53706, USA

Abstract. Muon colliders offer special opportunities to discover and study new physics. With the high intensity source of muons at the front end, orders of magnitude improvements would be realized in searches for rare muon processes, in deep inelastic muon and neutrino scattering experiments, and in long-baseline neutrino oscillation experiments. At a 100 to 500 GeV muon collider, neutral Higgs boson (or techni-particle) masses, widths and couplings could be precisely measured via s-channel production. Also, threshold cross-section studies of W^+W^-, $t\bar{t}$, Zh and supersymmetric particle pairs would precisely determine the corrresponding masses and test supersymmetric radiative corrections. At the high energy frontier a 3 to 4 TeV muon collider is ideally suited for the study of scalar supersymmetric particles and extra Z-bosons or strong WW scattering.

I INTRODUCTION

The agenda of physics at a muon collider falls into three categories: front end physics with a high-intensity muon source, First Muon Collider (FMC) physics at a machine with center-of-mass energies of 100 to 500 GeV, and Next Muon Collider (NMC) physics at 3–4 TeV center-of-mass energies.

At the front end, a high-intensity muon source will permit searches for rare muon processes at branching sensitivities that are orders of magnitude below present upper limits. Also, a high-energy muon-proton collider can be constructed to probe high Q^2 phenomena beyond the reach of the HERA ep collider. In addition, the decaying muons will provide high-intensity neutrino beams for precision neutrino cross-section measurements and for long-baseline experiments.

The FMC will be a unique facility for neutral Higgs boson (or techni-resonance) studies through s-channel resonance production. Measurements can also be made of the threshold cross sections for W^+W^-, $t\bar{t}$, Zh, $\chi_1^+\chi_1^-$, $\chi_2^0\chi_1^0$, $\tilde{\ell}^+\tilde{\ell}^-$ and $\tilde{\nu}\tilde{\nu}$ production that will determine the corresponding masses to high precision. Chargino, neutralino, slepton and sneutrino pair production cross section measurements would probe the loop corrections to gauge couplings in the supersymmetric sector. A $\mu^+\mu^- \to Z^0$ factory, utilizing the partial polarization of the muons, could allow significant improvements in $\sin^2\theta_\mathrm{w}$ precision and in B-mixing and CP-violating studies.

The NMC will be particularly valuable for reconstructing supersymmetric particles of high mass from their complex cascade decay chains. Also, any Z' resonances within the kinematic reach of the machine would give enormous event rates. The effects of virtual Z' states would be detectable to high mass. If no Higgs bosons exist below $\sim$1 TeV, then the NMC would be the ideal machine for the study of strong WW scattering at TeV energies.

Plus, there are numerous other new physics possibilities for muon facilities that are beyond the scope of the present report [1]. In the following sections the physics opportunities above are discussed in greater detail. The work on physics at muon colliders reported in Sections III, VI and IX is largely based on collaborations with M.S. Berger, J.F. Gunion and T. Han.

II FRONT END PHYSICS

A Rare muon decays

The planned muon flux $\sim 10^{14}$ muons/sec for a muon collider dramatically eclipses the flux $\sim 10^8$ muons/sec of present sources. With an intense source the rare muon processes $\mu \to e\gamma$ (current branching fraction $< 0.49 \times 10^{-12}$), $\mu N \to eN$ conversion, and the muon electric dipole moment can be probed at very interesting levels. A generic prediction of supersymmetric grand unified theories is that these lepton flavor violating or CP violating processes should occur via loops at significant rates, e.g. BF$(\mu \to e\gamma) \sim 10^{-13}$ [2]. Lepton flavor violation can also occur via Z' bosons, lepton quarks, and heavy neutrinos [3].

B μp collider

The possibility of colliding 200 GeV muons with 1000 GeV protons at Fermilab is under study [4]. This collider would reach a maximum $Q^2 \sim 8 \times 10^4$ GeV^2, which is $\sim$90 times the reach of the HERA ep collider, and deliver a luminosity $\sim 10^{33}$ cm^{-2} s^{-1}, which is $\sim$300 times the HERA luminosity. The μp collider would produce $\sim 10^6$ neutral current deep inelastic scattering events per year at $Q^2 > 5000$ GeV^2, which is over a factor 10^3 higher than at HERA. This μp collider would have a sensitivity to probe leptoquarks up to a mass $M_{LQ} \sim 800$ GeV and contact interactions to a scale $\Lambda \sim 6$–9 TeV [5].

C Neutrino flux

Muon decays are the way to make neutrino beams of well defined flavors [4,6]. A muon collider would yield a neutrino flux 1000 times that of the presently available neutrino flux. Then $\sim 10^6$ νN and $\bar{\nu} N$ events per year could be obtained to measure charm production (~6% of the total cross section) and to measure $\sin^2\theta_w$ (and infer the W-mass to an accuracy $\Delta M_W \simeq 30\text{--}50$ MeV in one year) [4].

D Neutrino oscillations

A special purpose muon ring has been proposed by S. Geer [7] to store $\sim 10^{21}$ μ^+ or μ^- per year and obtain $\sim 10^{20}$ neutrinos per year from muon decays along ~ 75 m straight sections of the ring, which would be pointed towards a distant neutrino detector. One would have known neutrino fluxes from $\mu^- \to \nu_\mu \bar{\nu}_e e^-$ or from $\mu^+ \to \bar{\nu}_\mu \nu_e e^+$ decays. Then, for example, from the decays of stored μ^-, the following neutrino oscillation channels could be studied by detection of the charged leptons from the interactions of neutrinos in the detector:

oscillation	detect
$\nu_\mu \to \nu_e$	e^-
$\nu_\mu \to \nu_\tau$	τ^-
$\bar{\nu}_e \to \bar{\nu}_\mu$	μ^+
$\bar{\nu}_e \to \bar{\nu}_\tau$	τ^+

The detected e^- or μ^+ have the "wrong-sign" from the leptons produced by the interactions of the $\bar{\nu}_e$ and ν_μ flux. The known neutrino fluxes from muon decays could be used for long-baseline oscillation experiments at any detector on Earth. The probabilities for vacuum oscillations between two neutrino flavors are given by

$$P(\nu_a \to \nu_b) = \sin^2 2\theta \, \sin^2(1.27 \delta m^2 L/E) \tag{1}$$

with δm^2 in eV^2 and L/E in km/GeV. In a very long baseline experiment from Fermilab to Gran Sasso laboratory ($L = 9900$ km) with ν-energies $E_\nu = 10$ to 50 GeV ($L/E = 1000\text{--}200$ km/GeV), neutrino charged current interaction rates $\sim 10^3$/year would result. Such an experiment would have sensitivity to oscillations down to $\delta m^2 \sim 10^{-5}$ eV^2 for $\sin^2 2\theta = 1$ [7].

III HIGGS PARTICLES

The MSSM has five Higgs bosons: h^0, H^0, A^0, $H^\pm$. The mass of the lightest neutral Higgs boson is bounded from above by $m_h \lesssim 130$ GeV and accordingly is the "jewel in the SUSY crown". Global analyses of precision electroweak data now indicate a preference for a light SM Higgs boson. Davier-Höcker [8] infer $m_h = 129^{+103}_{-62}$ GeV and Erler-Langacker [9] obtain $m_h = 122^{+134}_{-77}$ GeV. Since in the decoupling limit [10] the couplings of the SM and SUSY Higgs bosons are approximately equal, these findings may be the "smoking gun" for the SUSY Higgs boson.

The goals of a muon collider for the Higgs sector are to precisely determine the light Higgs mass, width, and branching fractions, to differentiate the h_{MSSM} from the h_{SM}, and to find and study the heavy neutral Higgs boson H^0 and A^0. The production of Higgs bosons in the s-channel with interesting rates is an unique feature of a muon collider [11,12]. The resonance cross section is

$$\sigma_h(\sqrt{s}) = \frac{4\pi \Gamma(h \to \mu\bar{\mu}) \, \Gamma(h \to X)}{(\hat{s} - m_h^2)^2 + m_h^2 \left(\Gamma_{\text{tot}}^h\right)^2} \quad (2)$$

Gaussian beams with root mean square resolution down to $R = 0.003\%$ are realizable [13]. The corresponding root mean square spread $\sigma_{\sqrt{s}}$ in c.m. energy is

$$\sigma_{\sqrt{s}} = (2 \text{ MeV}) \left(\frac{R}{0.003\%}\right) \left(\frac{\sqrt{s}}{100 \text{ GeV}}\right). \quad (3)$$

The effective s-channel Higgs cross section convolved with a Gaussian spread

$$\bar{\sigma}_h(\sqrt{s}) = \frac{1}{\sqrt{2\pi}\,\sigma_{\sqrt{s}}} \int \sigma_h(\sqrt{\hat{s}}) \exp\left[\frac{-\left(\sqrt{\hat{s}} - \sqrt{s}\right)^2}{2\sigma_{\sqrt{s}}^2}\right] d\sqrt{\hat{s}} \quad (4)$$

is illustrated in Fig. 1 for $m_h = 110$ GeV, $\Gamma_h = 2.5$ MeV, and resolutions $R = 0.01\%$, 0.06% and 0.1% [11,12]. A resolution $\sigma_{\sqrt{s}} \sim \Gamma_h$ is needed to be sensitive to the Higgs width. The light Higgs width is predicted to be [12]

$$\begin{array}{ll} \Gamma \approx 2 \text{ to } 3 \text{ MeV} & \text{if } \tan\beta \sim 1.8 \\ \Gamma \approx 2 \text{ to } 800 \text{ MeV} & \text{if } \tan\beta \sim 20 \end{array} \quad (5)$$

for $80 \text{ GeV} \lesssim m_h \lesssim 120 \text{ GeV}$.

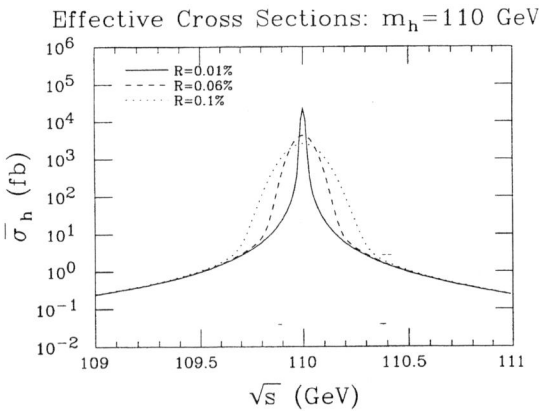

FIGURE 1. Effective s-channel higgs cross section $\bar{\sigma}_h$ obtained by convoluting the Breit-Wigner resonance formula with a Gaussian distribution for resolution R. From Ref. [11].

At $\sqrt{s} = m_h$, the effective s-channel Higgs cross section is [11]

$$\bar{\sigma}_h \simeq \frac{4\pi}{m_h^2} \frac{\mathrm{BF}(h \to \mu\bar{\mu})\,\mathrm{BF}(h \to X)}{\left[1 + \frac{8}{\pi}\left(\frac{\sigma_{\sqrt{s}}}{\Gamma_{\mathrm{tot}}^h}\right)^2\right]^{1/2}}. \quad (6)$$

Note that $\bar{\sigma}_h \propto 1/\sigma_{\sqrt{s}}$ for $\sigma_{\sqrt{s}} > \Gamma_{\mathrm{tot}}^h$. At $\sqrt{s} = m_h \approx 110$ GeV, the $b\bar{b}$ rates are [11,12]

$$\text{signal} \approx 10^4 \text{ events/fb} \quad (7)$$
$$\text{background} \approx 10^4 \text{ events/fb} \quad (8)$$

assuming a b-tagging efficiency $\epsilon \sim 0.5$. The effective on-resonance cross sections for other m_h values and other channels (ZZ^*, WW^*) are shown in Fig. 2 for the SM Higgs. The rates for the MSSM Higgs are nearly the same as the SM rates in the decoupling regime [10], which is relevant at $\tan\beta \sim 1.8$ in mSUGRA, corresponding to the infrared fixed point of the top quark Yukawa coupling [14].

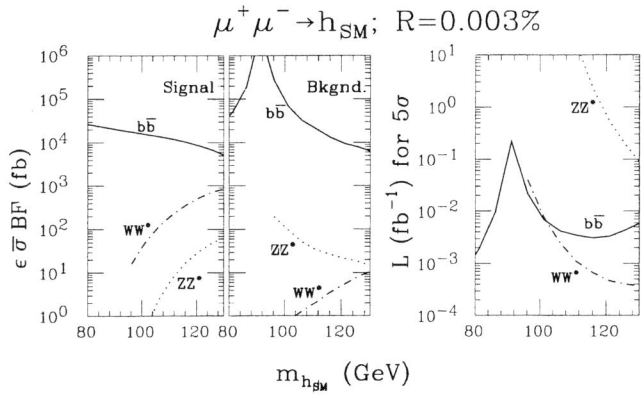

FIGURE 2. The SM Higgs cross sections and backgrounds in $b\bar{b}$, WW^* and ZZ^*. Also shown is the luminosity needed for a 5 standard deviation detection in $b\bar{b}$. From Ref. [11].

The important factors that make s-channel Higgs physics studies possible at a muon collider are energy resolutions $\sigma_{\sqrt{s}}$ of order a few MeV, little bremsstrahlung and no beamstrahlung smearing, and precise tuning of the beam energy to an accuracy $\Delta E \sim 10^{-6} E$ through continuous spin-rotation measurements [15]. As a case study we consider $m_h \approx 110$ GeV. Prior Higgs discovery is assumed at the Tevatron (in $Wh, t\bar{t}h$ production with $h \to b\bar{b}$ decay) or at the LHC (in $gg \to h$ production with $h \to \gamma\gamma, 4\ell$ decays with a mass measurement of $\Delta m_h \sim 100$ MeV for an integrated luminosity of $L = 300$ fb^{-1}) or possibly at a NLC (in $Z^* \to Zh, h \to b\bar{b}$ giving $\Delta m_h \sim 50$ MeV for $L = 200$ fb^{-1}). A muon collider ring design would be optimized to run at energy $\sqrt{s} = m_h$. For an initial Higgs mass uncertainty of $\Delta m_h \sim 100$ MeV, the maximum number of scan points required to locate the s-channel resonance peak at the muon collider is

$$n = 2\Delta m/\sigma_{\sqrt{s}} \approx 100 \qquad (9)$$

for a resolution $\sigma_{\sqrt{s}} \approx 2$ MeV. The necessary luminosity per scan point ($L_{\text{s.p.}}$) to observe or eliminate the h-resonance at a significance level $S/\sqrt{B} = 3$ is $L_{\text{s.p.}} \sim 1.5 \times 10^{-3}$ fb^{-1}. (The scan luminosity requirements increase for m_h closer to M_Z; at $m_h \sim M_Z$ the $L_{\text{s.p.}}$ needed is a factor of 50 higher.) The total luminosity then needed to tune to a Higgs boson with $m_h = 110$ GeV is $L_{\text{tot}} = 0.15$ fb^{-1}. If the machine delivers 1.5×10^{31} cm^{-2} s^{-1} (0.15 fb^{-1}/year), then one year of running would suffice to complete the scan and measure the Higgs mass to an accuracy $\Delta m \sim 1$ MeV. Figure 3 illustrates a simulation of such a scan.

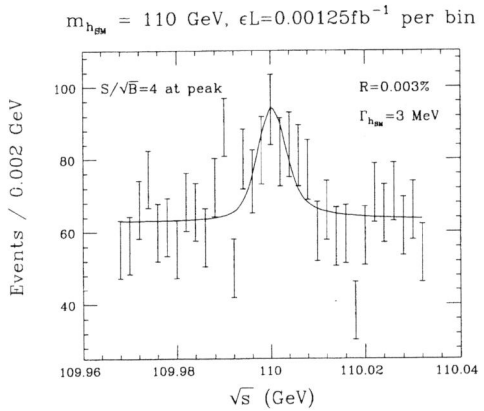

FIGURE 3. Number of events and statistical errors in the $b\bar{b}$ final states as a function of $\sqrt{s}$ in the vicinity of $m_{h_{\text{SM}}} = 100$ GeV, assuming $R = 0.03\%$. From Ref. [11].

Once the h-mass is determined to ~ 1 MeV, a 3-point fine scan [11] can be made across the peak with higher luminosity, distributed with L_1 at the observed peak position in $\sqrt{s}$ and $2.5L_1$ at the wings ($\sqrt{s} = \text{peak} \pm 2\sigma_{\sqrt{s}}$). Then with $L_{\text{tot}} = 0.4$ fb^{-1} the following accuracies would be achievable: 16% for Γ^h_{tot}, 1% for $\sigma\text{BF}(b\bar{b})$ and 5% for $\sigma\text{BF}(WW^*)$. The ratio $r = \text{BF}(WW^*)/\text{BF}(b\bar{b})$ is sensitive to m_A for m_A values below 500 GeV. For example, $r_{\text{MSSM}}/r_{\text{SM}} = 0.3, 0.5, 0.8$ for $m_A = 200, 250, 400$ GeV [11]. Thus, it may be possible to infer m_A from s-channel measurements of h.

The study of the other neutral MSSM Higgs bosons at a muon collider via the s-channel is also of major interest. Finding the H^0 and A^0 may not be easy at other colliders. At the LHC the region $m_A > 200$ GeV is deemed to be inaccessible for $3 \lesssim \tan\beta \lesssim 5$–10 [16]. At an NLC the $e^+e^- \to H^0 A^0$ production process may be kinematically inaccessible if H^0 and A^0 are heavy. At a $\gamma\gamma$ collider, very high luminosity (~ 200 fb^{-1}) would be needed for $\gamma\gamma \to H^0, A^0$ studies.

At a muon collider the resolution requirements for s-channel H^0 and A^0 studies are not as demanding as for the h, because the H^0, A^0 widths are broader; typically $\Gamma \sim 30$ MeV for $m_A < 2m_t$ and $\Gamma \sim 3$ GeV for $m_A > 2m_t$. Consequently $R \sim 0.1\%$ ($\sigma_{\sqrt{s}} \sim 70$ MeV) is adequate for a scan. A luminosity per scan point $L_{\text{s.p.}} \sim 0.1$ fb^{-1} probes the parameter space with $\tan\beta > 2$. The $\sqrt{s}$-range over which the scan should be made depends on other information available to indicate the A^0 and H^0 mass ranges of interest.

In mSUGRA, $m_{A^0} \approx m_{H^0} \approx m_{H^\pm}$ at large m_A, with a very close degeneracy in these masses for large $\tan\beta$. In such a circumstance only an s-channel scan with good resolution may allow separation of the A^0 and H^0 states; see Fig. 4.

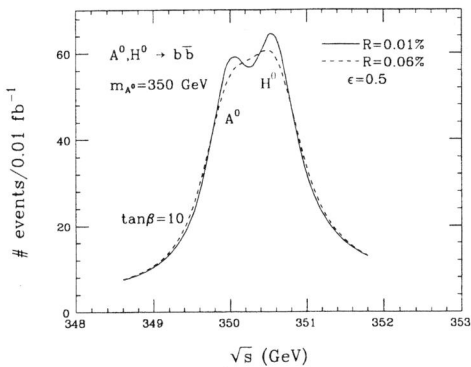

FIGURE 4. Separation of A^0 and H^0 signals for $\tan\beta = 10$. From Ref. [11].

IV TECHNICOLOR PARTICLES

In modern top-assisted technicolor models [17], the lighter neutral technipion resonances are expected to have masses in the range 100 to 500 GeV and widths of order 0.1 to 50 GeV [18]. These resonances would be produced in the s-channel at a muon collider,

$$\mu^+\mu^- \to \pi_T^0, \rho_T^0, \omega_T^0 \tag{10}$$

with high event rates. The peak cross sections for these processes are estimated to be $\approx 10^7$–10^4 fb [18]. The dominant decay modes are [18]

$$\pi_T^0 \to b\bar{b}, \tau\bar{\tau}, c\bar{c}, t\bar{t}, \tag{11}$$

$$\pi_T^{0\prime} \to gg, b\bar{b}, c\bar{c}, t\bar{t}, \tau^+\tau^-, \tag{12}$$

$$\rho_T^0 \to \pi_T\pi_T, W\pi_T, WW, \tag{13}$$

$$\omega_T^0 \to c\bar{c}, b\bar{b}, \tau\bar{\tau}, t\bar{t}, \gamma\pi_T^0, Z\pi_T^0. \tag{14}$$

Such resonances would be easy to find and study at a muon collider.

V Z-FACTORY

A muon collider operating at the Z-boson resonance energy is an interesting option for measurement of polarization asymmetries, $B_s^0 \bar{B}_s^0$ mixing, and of CP violation in the B-meson system [19]. The muon collider advantages are the partial muon beam polarization, the separation of b and $\bar{b}$ in $Z \to b\bar{b}$ events, and the long B-decay length for B-mesons produced at this $\sqrt{s}$. The left-right asymmetry A_{LR} is the most accurate measure of $\sin^2 \theta_w$, since the uncertainty is statistics dominated. The present LEP and SLD polarization measurements show standard deviations of 2.4 in A_{LR}^0, 1.9 in $A_{FB}^{0,b}$ and 1.7 in $A_{FB}^{0,\tau}$ [9]. The CP angle β (see Fig. 5) could be measured from $B^0 \to K_s J/\psi$ decays. To achieve significant improvements over existing measurements and those at future B-facilities, more than 10^7 Z-boson events would be needed, corresponding to a luminosity > 0.15 fb^{-1}, within the domain of muon collider expectations.

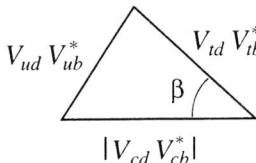

FIGURE 5. Unitarity triangle for 3-generation mixing.

VI THRESHOLD MEASUREMENTS AT A MUON COLLIDER

With 10 fb^{-1} integrated luminosity devoted to a measurement of a threshold cross-section, the following precisions on particle masses may be achievable [20,21]:

$$\begin{aligned} \mu^+\mu^- &\to W^+W^- & \Delta M_W &= 20 \text{ MeV}, \\ \mu^+\mu^- &\to t\bar{t} & \Delta m_t &= 0.2 \text{ GeV}, \\ \mu^+\mu^- &\to Zh & \Delta m_h &= 140 \text{ MeV} \ (\text{if } m_h = 100 \text{ GeV}). \end{aligned} \quad (15)$$

Precision M_W and m_t measurements allow important tests of electroweak radiative corrections through the relation

$$M_W = M_Z \left[1 - \frac{\pi \alpha}{\sqrt{2} G_\mu M_W^2 (1 - \delta r)} \right]^{1/2}, \quad (16)$$

where δr represents loop corrections. In the SM, δr depends on m_t^2 and $\log m_h$. The optimal precision for tests of this relation is $\Delta M_W \approx \frac{1}{140} \Delta m_t$, so the uncertainty on M_W is the most critical. With $\Delta M_W = 20$ MeV the SM Higgs mass could be inferred to an accuracy

$$\Delta m_{h_{\rm SM}} = \pm 30 \text{ GeV} \left(\frac{m_h}{100 \text{ GeV}}\right). \qquad (17)$$

Alternatively, once m_h is known from direct measurements, SUSY loop contributions can be tested.

In top quark production at a muon collider above the threshold region, modest muon polarization would allow sensitive tests of anomalous top quark couplings [22].

One of the important physics opportunities for the First Muon Collider is the production of the lighter chargino, $\tilde{\chi}_1^+$ [23]. Fine tuning arguments in mSUGRA suggest that it should be lighter than 200 GeV [24]. A search at the upgraded Tevatron for the process $q\bar{q} \to \tilde{\chi}_1^+ \tilde{\chi}_2^0$ with $\tilde{\chi}_1^+ \to \tilde{\chi}_1^0 \ell^+ \nu$ and $\tilde{\chi}_2^0 \to \tilde{\chi}_1^0 \ell^+ \ell^-$ decays can potentially reach masses $m_{\tilde{\chi}_1^+} \simeq m_{\tilde{\chi}_2^0} \sim 170$ GeV with 2 fb^{-1} luminosity and ~ 230 GeV with 10 fb^{-1} [25]. The mass difference $M(\tilde{\chi}_2^0) - M(\tilde{\chi}_1^0)$ can be determined from the $\ell^+\ell^-$ mass distribution.

The two contributing diagrams in the chargino pair production process are shown in Fig. 6; the two amplitudes interfere destructively [26]. The $\tilde{\chi}_1^+$ and $\tilde{\nu}_\mu$ masses can be inferred from the shape of the cross section in the threshold region [21]. The chargino decay is $\tilde{\chi}_1^+ \to f\bar{f}'\tilde{\chi}_1^0$. Selective cuts suppress the background from W^+W^- production and leave $\sim 5\%$ signal efficiency for 4 jets + $\not{E}_T$ events. Measurements at two energies in the threshold region with total luminosity $L = 50$ fb^{-1} and resolution $R = 0.1\%$ can give the accuracies listed in Table 1 on the chargino mass for the specified values of $m_{\tilde{\chi}_1^+}$ and $m_{\tilde{\nu}_\mu}$.

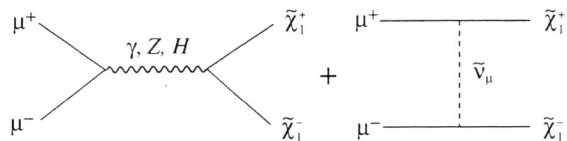

FIGURE 6. Diagrams for production of the lighter chargino.

TABLE 1. Achievable uncertainties with 50 fb^{-1} luminosity on the mass of the lighter chargino for representative $m_{\tilde{\chi}_1^+}$ and $m_{\tilde{\nu}_\mu}$ masses. From Ref. [21].

$\Delta m_{\tilde{\chi}_1^+}$ (MeV)	$m_{\tilde{\chi}_1^+}$ (GeV)	$m_{\tilde{\nu}_\mu}$ (GeV)
35	100	500
45	100	300
150	200	500
300	200	300

VII SUPERSYMMETRIC RADIATIVE CORRECTIONS

In unbroken supersymmetry, the SUSY gaugino couplings h_i to $\tilde{f}f$ are equal to the SM gauge couplings g_i. In broken SUSY a difference in h_i and g_i couplings is induced at the loop level due to different mass scales for squarks and sleptons [27 30]. The differences in the U(1) and SU(2) couplings are [29]

$$\frac{h_1 - g_1}{g_1} \simeq 1.8\% \log_{10}\left(\frac{M_{\tilde{Q}}}{m_{\tilde{\ell}}}\right), \tag{18}$$

$$\frac{h_2 - g_2}{g_2} \simeq 0.7\% \log_{10}\left(\frac{M_{\tilde{Q}}}{m_{\tilde{\ell}}}\right). \tag{19}$$

One-loop amplitudes for SUSY processes are obtained from the tree-level amplitudes by substitution of the modified couplings. The cross-sections of SUSY processes with t-channel exchanges can be enhanced up to $9\% \log_{10}\left(M_{\tilde{Q}}/m_{\tilde{\ell}}\right)$ [28]. Consequently, precision cross-section measurements can be sensitive to squarks of mass $M_{\tilde{Q}} > 1$ TeV. If the first two generations have masses in the 1 to 40 TeV range allowed by naturalness, then precision measurements could provide a way to infer squark masses beyond the kinematic reach of colliders.

Some t-channel exchange processes of interest in this regard at muon colliders are shown in Fig. 7. The technique relies on knowledge of the exchanged particle mass, which must be determined from its production processes. The muon collider advantage in the study of supersymmetric radiative corrections is the accuracy with which mass measurements can be made near thresholds.

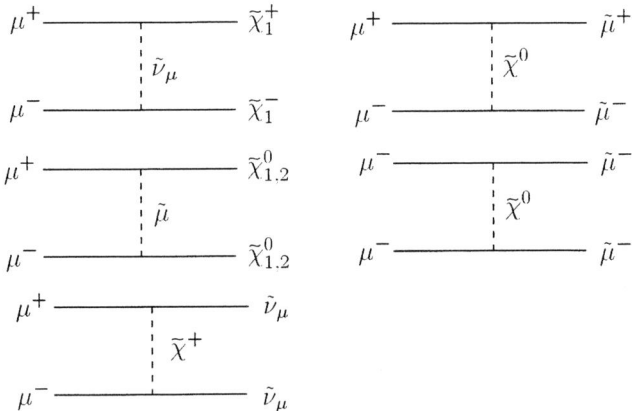

FIGURE 7. t-channel exchange diagrams for processes that can be enhanced by SUSY radiative corrections.

VIII   HEAVY PARTICLES OF SUPERSYMMETRY

The requirements of gauge coupling unification can be used to predict the mean SUSY mass scale, given the value of the strong coupling at the Z-mass scale. Figure 8 shows the SUSY GUT predictions versus $\alpha_s(M_Z)$. For the value $\alpha_s(M_Z) = 0.1214 \pm 0.0031$ from a new global fit to precision electroweak data [9], a mean SUSY mass of order 1 TeV is expected. Thus some SUSY particles will likely have masses at the TeV scale.

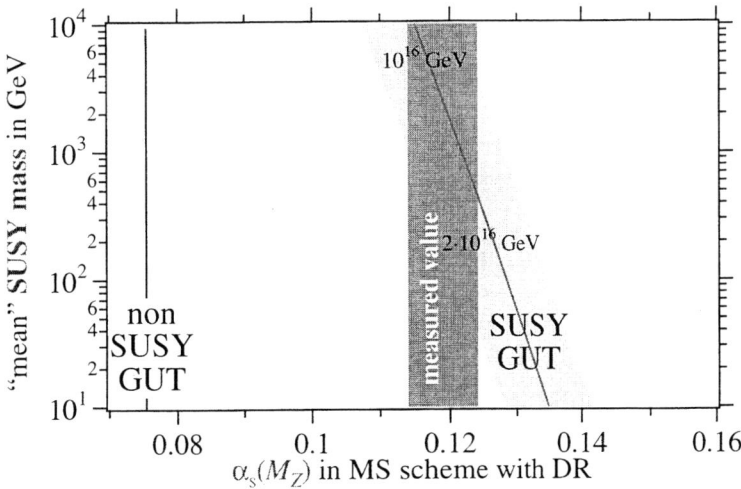

FIGURE 8. α_s prediction in supersymmetric GUT with minimal particle content. From Ref. [31].

At the LHC, mainly squarks and gluinos are produced; these decay to lighter SUSY particles. The LHC is a great SUSY machine, but some sparticle measurements will be very difficult or impossible there [32,33], namely: (i) the determination of the LSP mass (LHC measurements give SUSY mass differences); (ii) study of sleptons of mass $\gtrsim$ 200 GeV because Drell-Yan production becomes too small at these masses; (iii) study of heavy gauginos $\tilde{\chi}_2^\pm$ and $\tilde{\chi}_{3,4}^0$, which are mainly Higgsino and have small direct production rates and small branching fractions to channels usable for detection; (iv) study of heavy Higgs bosons $H^\pm$, H^0, A^0 that have small cross sections and decays to $t\bar{t}$ that are likely dominant (their detection is deemed impossible if SUSY decays dominate).

With supersymmetry there will be many scalar particles. Pair production of scalar particles at a lepton collider is P-wave suppressed. Consequently, energies well above threshold are needed for sufficient production rates; see Fig. 9. A 3 to 4 TeV muon collider with high luminosity ($L \sim 10^2$ to 10^3 fb^{-1}/year) could provide sufficient event rates to reconstruct heavy sparticles from their complex cascade decay chains [33,34].

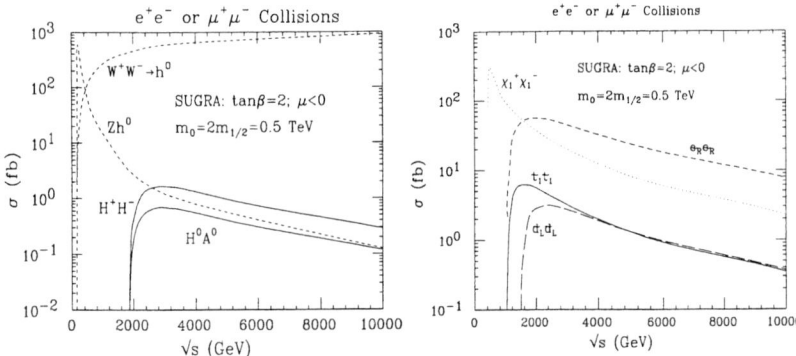

FIGURE 9. Cross sections for pair production of Higgs bosons and scalar particles at a high energy muon collider. From Ref. [35].

Reprinted from *Nuclear Physics B*, **51**, V. Barger, "Particle Physics Opportunities at mu+mu- Colliders," pp 13-31 ©1996 with kind permission from Elsevier Science Ltd., The Boulevard, Langford Lane, Kidlington OX5 1GB, UK.

Extra Z bosons and a low-energy supersymmetry are natural in string models [36]. The s-channel production of a Z' boson at the resonance energy would give enormous event rates at the NMC. Moreover, the s-channel contributions of Z' bosons with mass far above the kinematic reach of the collider could be revealed as contact interactions [37].

IX STRONG SCATTERING OF WEAK BOSONS

The scattering of weak bosons can be studied at a high energy muon collider through the process in Fig. 10. The amplitude for the scattering of longitudinally polarized W-bosons behaves like [38]

$$A(W_L W_L \to W_L W_L) \sim m_H^2/v^2 \qquad (20)$$

if there is a light Higgs boson, and

$$A(W_L W_L \to W_L W_L) \sim s_{WW}/v^2 \qquad (21)$$

if no light Higgs boson exists; here s_{WW} is the square of the WW c.m. energy and $v = 246$ GeV. In the latter scenario, partial wave unitarity of $W_L W_L \to W_L W_L$ requires that strong scattering of weak bosons occurs at the 1 to 2 TeV energy scale. Thus subprocess energies $\sqrt{s_{WW}} \gtrsim 1.5$ TeV are needed to probe strong WW scattering effects.

The nature of the dynamics in the WW sector is unknown. Models for this scattering assume heavy resonant particles (isospin scalar and vector) or a non-resonant amplitude which extrapolates the low-energy theorem behavior $A \sim s_{WW}/v^2$. In all models, impressive signals of strong WW scattering are obtained at the NMC, with cross sections typically of order 50 fb^{-1} [39].

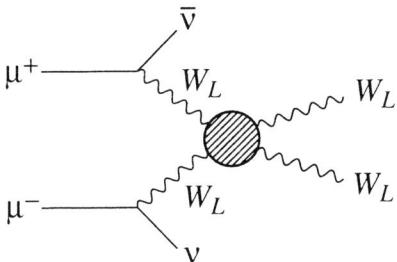

FIGURE 10. Symbolic diagram for strong WW scattering.

X CONCLUSIONS

The Front End of a muon collider offers dramatic improvements in sensitivity for flavor-violating transitions (e.g. $\mu \to e\gamma$), high Q^2 phenomena in deep inelastic muon-proton and neutrino-proton interactions, and neutrino oscillation studies in long-baseline experiments.

The First Muon Collider offers unique probes of supersymmetry (particularly s-channel Higgs boson resonances), high precision threshold measurements of W, t and SUSY particle masses, tests of SUSY radiative corrections that indirectly probe the existence of high mass squarks, and a possible Z^0 factory for improved precision in polarization measurements and for B-physics studies of CP violation and mixing.

The Next Muon Collider guarantees access to heavy SUSY scalar particles and Z' states or to strong WW scattering if there are no Higgs bosons and no supersymmetry.

The bottom line is that muon colliders are robust options for probing new physics that may not be accessible at other colliders.

ACKNOWLEDGMENTS

I would like to thank M.S. Berger, G. Burdman, J.F. Gunion, T. Han and C. Kao for helpful advice in the preparation of this report. This research was supported in part by the U.S. Department of Energy under Grant No. DE-FG02-95ER40896 and in part by the University of Wisconsin Research Committee with funds granted by the Wisconsin Alumni Research Foundation.

REFERENCES

1. See e.g. *Proceedings of the Workshop on Physics at the First Muon Collider and at the Front End of a Muon Collider*, Fermilab, Nov. 1997, ed. by S. Geer and R. Raja, to be published by AIP Press (1998).

2. R. Barbieri, L. Hall, and A. Strumia, Nucl. Phys. **B445**, 219 (1995); J. Hisano, T. Moroi, K. Tobe, M. Yamaguchi, and T. Yanagida, Phys. Lett. **B357**, 579 (1995); J. Hisano, T. Moroi, K. Tobe, M. Yamaguchi, Phys. Rev. **D53** 2442 (1996); J. Hisano, T. Moroi, K. Tobe, M. Yamaguchi, Phys. Lett. **B391**, 341 (1997), Erratum **B397** 357 (1997).
3. W.J. Marciano, in Ref. [1]; W. Molzon, ibid.
4. H. Schellman, these proceedings.
5. Kingman Cheung, hep-ph/9802219, these proceedings.
6. P. Fisher and B. Kayser, in Ref. [1]
7. S. Geer, these proceedings; FERMILAB-PUB-97/389 [hep-ph/9712290]; FERMILAB-CONF-97/417, in Ref. [1].
8. M. Davier and A. Höcker, LAL-97-85 [hep-ph/9711308].
9. J. Erler and P. Langacker, *Electroweak Model and Contraints on New Physics*, in the 1997 partial update for the 1998 edition of the Review of Particle Physics (available from the PDG WWW pages at http://pdg.lbl.gov).
10. H. Haber and Y. Nir, Nucl. Phys. **B335**, 363 (1990).
11. V. Barger, M.S. Berger, J.F. Gunion, and T. Han, Phys. Rep. **286**, no. 1, p. 1 (1997).
12. V. Barger, M.S. Berger, J.F. Gunion and T. Han, Phys. Rev. Lett. **75**, 1462 (1995).
13. R. Palmer, physics/9802005, in Ref. [1].
14. See e.g., V. Barger, M.S. Berger, and P. Ohmann, Phys. Rev. **D47**, 1093 (1993); **D49**, 4908 (1994).
15. R. Raja, talk given at the Workshop on Physics at the First Muon Collider and at the Front End of a Muon Collider, Fermilab, Nov. 1997.
16. E. Richter-Was, D. Froidevaux, F. Gianotti, L. Poggioli, D. Cavalli, S. Resconi, CERN-TH-96-111 (1996).
17. C.T. Hill, Phys. Lett. **B345**, 483 (1995); K. Lane and E. Eichten, Phys. Lett. **352**, 382 (1995); K. Lane, Phys. Rev. **D54**, 2204 (1996).
18. P. Bhat and E. Eichten, FERMILAB-CONF-98/072; K. Lane, BUHEP-98-01 [hep-ph/9810385], in Ref. [1]; J. Womersley, talk presented at Fermilab; D. Dobrescu and C.T. Hill, FERMILAB-PUB-97/409-T [hep-ph/9712319]; R. Casalbuoni et al., hep-ph/9801243, in Ref. [1]; J.F. Gunion, UCD-98-5 [hep-ph/9802258].
19. M. DeMarteau and T. Han, FERMILAB-CONF-98/030 [hep-ph/9801407], in Ref. [1].
20. V. Barger, M.S. Berger, J.F. Gunion and T. Han, Phys. Rev. **D56**, 1714 (1997); Phys. Rev. Lett. **78**, 3991 (1997).
21. V. Barger, M.S. Berger, and T. Han, University of Wisconsin-Madison report MADPH-98-1036 [hep-ph/9801410].
22. S. Parke, hep-ph/9802279.
23. See e.g., M. Carena and S. Protopopescu, in Ref. [1].
24. G.W. Anderson and D.J. Castano, Phys. Rev. **D52**, 1693 (1995); K.L. Chan, U. Chattopadhyay, P. Nath, hep-ph/9710473.
25. H. Baer, Chen, C. Kao, and X. Tata, Phys. Rev. **D52**, 1565 (1995).
26. J.L. Feng and M.J. Strassler, Phys. Rev. **D51**, 4661 (1995); J.L. Feng, M. Peskin, H. Murayama, and X. Tata, Phys. Rev. **D52**, 1418 (1995).
27. P.H. Chankowski, Phys. Rev. **D41**, 2877 (1990); P.H. Chankowski and S. Pokorski, hep-ph/9707497.

28. M.M. Nojiri, K. Fujii, and T. Tsukamoto, Phys. Rev. **D54**, 6756 (1996); M.M. Nojiri, D.M. Pierce, and Y. Yamada, Phys. Rev. **D57**, 1539 (1998); S. Kiyoura, M.M. Nojiri, D.M. Pierce, and Y. Yamada, SLAC-PUB-7754 [hep-ph/9803210].
29. H.-C. Cheng, J.L. Feng, and N. Polonksy, Phys. Rev. **D56**, 6875 (1997); **D57**, 152 (1998).
30. M.A. Diaz, S.F. King, and D.A. Ross, CERN-TH-98-26 [hep-ph/9801373].
31. R. Barbieri, hep-ph/9711232.
32. I. Hinchliffe, in Ref. [1].
33. F. Paige, in Ref. [1]
34. J. Lykken, these proceedings.
35. V. Barger, M.S. Berger, J.F. Gunion, and T. Han, in *Proceedings of the Symposium on Physics Potential and Development of $\mu^+\mu^-$ Colliders*, San Francisco, CA (1995), ed. by D. Cline and D. Sanders, Nucl. Phys. B (proc. suppl.) **51A**, 13 (1996).
36. M. Cvetic and P. Langacker, Mod. Phys. Lett. **A11**, 1247 (1996).
37. S. Godfrey, hep-ph/9802212, these proceedings.
38. B.W. Lee, C. Quigg, and H. Thacker, Phys. Rev. **D16**, 1519 (1977); M. Veltman, Acta Phys. Pol. **B8**, 457 (1977); M.S. Chanowitz and M.K. Gaillard, Nucl. Phys. **B261**, 379 (1985); J. Bagger et al., Phys. Rev. **D52**, 3878 (1995).
39. V. Barger, M.S. Berger, J.F. Gunion and T. Han, Phys. Rev. **D55**, 142 (1997) and in Proceedings of the Institute for Theoretical Physics Symposium *Future High Energy Colliders*, Santa Barbara, California, Oct. 1996, ed. by Z. Parsa (AIP Conference Proceedings 397, New York, 1997) p. 219 [hep-ph/9704290].

Workshop on Physics at the First Muon Collider and Front–End of a Muon Collider: A Brief Summary

S. Geer

Fermi National Accelerator Laboratory, P.O. Box 500, Batavia, Illinois 60510

Abstract. In November 1997 a workshop was held at Fermilab to explore the physics potential of the first muon collider, and the physics potential of the accelerator complex at the "front–end" of the collider. An extensive physics program emerged from the workshop. This paper attempts to summarize this physics program and identify the main conclusions from the workshop.

INTRODUCTION

Over the last couple of years a significant effort has been devoted to exploring the feasibility of designing a high–luminosity high–energy muon collider. This effort has been motivated by the theoretical prejudice that there is new physics beyond the Standard Model (SM) at the TeV energy scale, and that a multi–TeV lepton–lepton collider will be needed to make precision measurements at this new scale. An attractive feature of the muon collider concept is that muon colliders appear to be "stagable" with each stage offering a unique cutting edge physics program. Hence, the path towards a multi–TeV muon collider might be via one or more cheaper "demonstration" stages, each making a significant contribution to our understanding of particle physics. Candidate demonstration stages are (i) part or all of the "front–end" accelerator complex, (ii) a Z^0 factory offering at least an order of magnitude more Z^0s than LEP, (iii) a Higgs factory designed to produce Higgs-like particles in the s-channel if $m_H < 2m_W$, (iv) a WW factory ($\sqrt{s} = 2m_W$), (v) a $t\bar{t}$ factory ($\sqrt{s} = 2m_t$), (vi) a SUSY factory if supersymmetric states are found at LEP2, TEV33, or the LHC, or (vii) a Techni–factory, if Technicolor states are observed.

The Workshop on Physics at the First Muon Collider and Front–end of a Muon Collider was held at Fermilab from 6–9 November 1997. The goal of the workshop was to explore the physics potential of each of the various options for the first muon collider (FMC), including the physics that could be pursued

TABLE 1. Operational parameters of an upgraded Fermilab proton source for a Muon Collider. The right–most column shows parameters for the fully upgraded source, and the other columns for possible intermediate steps in the upgrade.

	Step 1 Scenario 1	Step 1 Scenario 2	Step 2	Step 3
Linac (operating at 15 Hz)				
Kinetic Energy (MeV)	400	1000	1000	1000
Pulse Length (μs)	0.75	0.75	0.75	0.75
H^- per pulse	1×10^{13}	1.5×10^{13}	2.5×10^{13}	1×10^{14}
Pre–Booster (operating at 15 Hz)				
Extraction Kinetic Energy (GeV)				4.5
Momentum Spread (95% FW)				0.5%
Circumference (m)				180.6
Protons per bunch				5×10^{13}
Number of bunches				2
Extracted bunch length (ns)				21
Transverse Emittance (mm–mr)				200π
Longitudinal Emittance (eV-sec)				1.8
Booster (operating at 15 Hz)				
Extraction Kinetic Energy (GeV)	16	8	16	16
Momentum Spread (95% FW)	$< 0.1\%$	$< 0.1\%$	$< 0.1\%$	1.2%
Circumference (m)	474.2	474.2	474.2	474.2
Protons per bunch	1.2×10^{11}	1.8×10^{11}	3×10^{11}	5×10^{13}
Number of bunches	84	84	84	2
Extracted bunch length (ns)	4.9	4.9	4.9	2.3
Transverse Emittance (mm–mr)	50π	30π	50π	240π
Longitudinal Emittance (eV-sec)	2.2	1.8	1.8	4.0

at the accelerator complex at the "front–end" of the collider. The accelerator parameters assumed for the workshop were based on recent studies of how the facilities at Fermilab might evolve towards a high-energy muon collider. A summary of these parameters can be found in Tables 1–3. Figure 1 shows in a schematic how the FMC might fit within the existing accelerator complex at Fermilab.

FRONT–END PARAMETERS AND PHYSICS

The "Front-End" of a muon collider consists of:

(a) A high-intensity proton source. We will assume that the proton source accelerates protons to 16 GeV/c, is cycling at 15 Hz, and produces 2 proton bunches per cycle, each containing 5×10^{13} particles. These parameters are based on the Fermilab summer study summarized in Ref. [1]. This upgrade to the existing proton source at Fermilab would require upgrading the 400 MeV Linac to a 1 GeV Linac, moving the 8 GeV Booster to a new location to overcome radiation limitations, upgrading the Booster energy to 16 GeV, and finally, adding a 4.5 GeV Pre-Booster to enable the protons to be compressed into short (~ 2 ns) long bunches. The up-

TABLE 2. Parameters of muon bunches downstream of the ionization cooling channel.

	Narrow σ_p	Broad σ_p
muons per bunch	5×10^{12}	5×10^{12}
μ^+ bunches per cycle	1	1
μ^- bunches per cycle	1	1
Momentum (MeV/c)	200	200
σ_p/p	5%	10%
Bunch length (cm)	1.5	10
Normalized $\epsilon_\perp$ (mm–mr)	200π	60π
Repetition rate (Hz)	15	15
μ^+ per year (10^7 secs)	7.5×10^{20}	7.5×10^{20}

grade is in principle stagable. Plausible staging steps and the associated proton source parameters are summarized in Table 1.

(b) A pion production and collection system, followed by a pion decay channel. Each incident proton bunch interacts in a target to produce $\sim 3 \times 10^{13}$ charged pions of each sign. The $\pi^\pm$ are confined within a high field solenoid co-axial with the beam direction. At the end of a 20 m long decay channel consisting of a 7 Tesla solenoid with a radius of 25 cm each incident proton results in about 0.2 muons of each charge. If in each accelerator cycle the first incident proton bunch is used to make and collect μ^+s, and the second bunch used for μ^-s, there will be about 10^{13} muons of each charge available at the end of the decay channel per cycle.

(c) A muon cooling channel. The muons exiting the decay channel populate a very diffuse 6-dimensional phase–space. The diffuse muon cloud must be cooled using a new fast cooling technique to form an intense beam before most of the muons have decayed. The cooling method proposed for the muon collider is ionization cooling [2]. Table 2 summarizes the properties of the muons at the end of the cooling channel. Note that the phase-space occupied by the muons can be optimized either to maximize the luminosity of the collider, or alternatively to minimize the beam energy spread at the expense of luminosity. At the end of the cooling channel each muon bunch is expected to contain about 5×10^{12} muons with a momentum of order 200 MeV/c.

(d) A muon acceleration system. A series of recirculating linear accelerators (RLAs) to accelerate the muons up to the colliding beam energy. Each RLA consists of two Linacs connected together by two arcs. Three RLAs with the operational parameters summarized in Table 3 would be able to accelerate the muons up to 250 GeV.

The front–end accelerator complex could be used for a variety of fixed target type physics. Note that the new Fermilab Main Injector can accept a factor of ~ 5 more protons per cycle than can be provided by the existing Fermilab

TABLE 3. Recirculating linear accelerator parameters.

	RLA 1	RLA 2	RLA 3
Input Energy (GeV)	1.0	9.6	70
Output Energy (GeV)	9.6	70	250
No. of turns	9	11	12
Linac Length (m)	100	300	533.3
Arc Length (m)	30	175	520
Bunch Length (ps)	158	43	19
Revolution Time (μs)	0.9	3.1	7.0
Decay Losses	9.0%	5.2%	2.4%
Initial muons per bunch	5×10^{12}	4.6×10^{12}	4.3×10^{12}
μ^+ bunches per sec	15	15	15

proton source. Hence, an upgraded proton source of the type required for a muon collider would directly benefit the foreseen FNAL MI program. In addition, a muon collider front–end offers many other possibilities, some of them quite unique. Four working groups were convened in the workshop to considered the range of possibilities. The main conclusions from these groups are listed in the sub–sections below.

Low Energy Hadron Physics

The proton source required for the FMC would allow a continuation of low and intermediate energy kaon physics with intensities a factor of 20 more than presently available at the AGS, and a factor of a few greater than foreseen at the FNAL MI, an upgraded AGS, or the proposed KEK JHF. Rare kaon decays and precision kaon CP and CPT studies can provide windows on physics beyond the SM and are likely to remain of interest well into the future. As an example consider the rare decays $K^+ \to \pi^+ \nu \bar{\nu}$ and $K_L \to \pi^0 \nu \bar{\nu}$. Precise measurements of these decay modes would enable a precise determination of V_{td} and the CP violation parameter η. The first $K^+ \to \pi^+ \nu \bar{\nu}$ event has recently been reported by the BNL E787 collaboration. The decay $K_L \to \pi^0 \nu \bar{\nu}$ has not yet been observed. Future experiments at the AGS and at the FNAL MI may yield a few of these rare K^+ and K_L decays per year. It has been estimated [3] that at the muon collider proton source of order 100 events per year could be observed in each mode. However, this kaon physics program would require the addition of a stretcher ring to the FMC proton source. Other interesting kaon experiments that might be pursued include muon transverse polarization in $K^+ \to \pi^0 \mu^+ \nu_\mu$ or $K^+ \to \mu^+ \nu_\mu \gamma$, spin–spin correlations in $K^+ \to \pi^+ \mu^+ \mu^-$, and polarization effects in $K^+ \to \mu^+ \mu^-$. Finally, in addition to the kaon physics program there are many other interesting low energy hadron physics experiments that, although they don't require the full potential of future high intensity proton sources, never-the-less are looking for a home. It is therefore likely that the proton source at the FMC would support a healthy low energy hadron physics program.

TABLE 4. Neutrino beam pulses from the straight sections of the Recirculating Linacs.

	1	2	3	4	5	6	7	8	9	10	11	12
RLA 1												
E_μ(start) (GeV)	1.0	1.96	2.92	3.88	4.84	5.8	6.76	7.72	8.68	9.64		
E_μ(end) (GeV)	1.48	2.44	3.4	4.36	5.32	6.28	7.24	8.2	9.16			
$<E_\mu>$ (GeV)	1.24	2.2	3.16	4.12	5.08	6.04	7.0	7.96	8.92			
$\gamma c\tau$ (km)	7.72	13.7	19.7	25.7	31.7	37.8	43.8	49.6	55.7			
$f_{decay} = 100m/\gamma c\tau$(%)	1.3	0.73	0.51	0.39	0.32	0.26	0.23	0.20	0.18			
N_{decay}/bunch ($\times 10^{10}$)	6.5	3.7	2.6	2.0	1.6	1.3	1.2	1.0	0.9			
N_{decay}/year ($\times 10^{18}$)	9.8	5.5	3.8	2.9	2.4	2.0	1.7	1.5	1.4			
RLA 2												
E_μ(start) (GeV)	9.6	15.1	20.6	26.1	31.6	37.1	42.6	48.1	53.6	59.1	64.6	70.1
E_μ(end) (GeV)	12.4	17.9	29.4	28.9	34.4	39.9	45.4	50.9	56.4	61.9	67.4	
$<E_\mu>$ (GeV)	11.0	16.5	22.0	27.5	33.0	38.5	44.0	49.5	55.0	60.5	66.0	
$\gamma c\tau$ (km)	68.7	100	140	170	210	240	270	310	340	380	410	
$f_{decay} = 300m/\gamma c\tau$(%)	0.44	0.30	0.21	0.18	0.14	0.13	0.11	0.097	0.088	0.079	0.073	
N_{decay}/bunch ($\times 10^{10}$)	2.0	1.4	0.97	0.83	0.64	0.60	0.51	0.45	0.40	0.36	0.34	
N_{decay}/year ($\times 10^{18}$)	3.0	2.1	1.5	1.2	0.96	0.90	0.77	0.68	0.60	0.54	0.51	
RLA 3												
E_μ(start) (GeV)	70	85	100	115	130	145	160	175	190	205	220	235
E_μ(end) (GeV)	77.5	92.5	108	123	138	153	168	183	198	213	228	243
$<E_\mu>$ (GeV)	73.8	88.8	104	119	134	149	164	179	194	209	224	239
$\gamma c\tau$ (km)	460	550	650	740	840	930	1000	1100	1200	1300	1400	1500
$f_{decay} = 533m/\gamma c\tau$(%)	0.12	0.10	0.08	0.07	0.06	0.06	0.05	0.05	0.04	0.04	0.04	0.04
N_{decay}/bunch ($\times 10^{10}$)	0.52	0.42	0.35	0.31	0.27	0.25	0.23	0.21	0.19	0.18	0.16	0.15
N_{decay}/year ($\times 10^{18}$)	0.78	0.63	0.53	0.46	0.41	0.37	0.34	0.31	0.28	0.26	0.25	0.23

Neutrino Physics

Conventional neutrino beams are made by allowing pions and kaons to decay in a decay channel. This produces a ν_μ beam with a small ν_e component from $K^+ \to e^+ \pi^0 \nu_e$ decays, and if the primary proton beam energy is sufficient, a small ν_τ component from D_S decays. The uncertainties on the fluxes and flavor content of the resulting neutrino beam introduce significant systematic uncertainties for many neutrino experiments. A muon collider accelerator complex offers the very attractive possibility of making intense neutrino beams using a muon decay channel. The resulting beam would have a precisely calculable flux and, for μ^- decays, would be a mixture of 50% ν_μ and 50% $\bar{\nu}_e$. This would provide a uniquely "clean" tool for neutrino physics.

The characteristics of the neutrino pulses downstream of the RLAs are summarized in Table 4. The resulting "accidental" neutrino beam 600 m downstream of RLA3 is sufficiently intense to produce [4] 7.5×10^5 events per year in a 1 m long 10 cm radius liquid hydrogen target ! Indeed, the neutrino beam intensity downstream of RLA3 is about a factor of 1000 greater than the intensity of existing neutrino beams. These high neutrino fluxes would enable compact highly instrumented detectors to be used with active fine-grained targets, micro-vertexing, and good particle identification ... a "quantum leap" in the design of neutrino detectors ! It has been proposed [5] to optimize the neutrino physics potential at a muon collider accelerator complex by building muon storage rings with straight sections pointing in the desired direction. Low energy storage rings could be built for long–baseline neutrino oscillation experiments with the plane of the storage ring tilted downwards so that the neutrino beam traverses the Earth. The neutrino beam intensity from a 20 GeV/c muon storage ring is sufficient to produce hundreds of CC events per year in a 10 kT detector on the other side of the Earth [5]. Neutrino beams at the front end of a muon collider would clearly enable significant improvements in the sensitivity of experiments searching for, and perhaps eventually measuring, neutrino flavor oscillations. For example, for large mixing angles, values of Δm^2 approaching 10^{-5} eV^2 might be observable for ν_e–ν_μ oscillations, and 10^{-4} eV^2 for ν_e–ν_τ oscillations [5]. Finally, it has been pointed out [6] that if neutrino oscillations are observed, the fluxes and characteristics of the neutrino beams at the front end of the muon collider would facilitate very interesting tests of Lorentz invariance, CPT invariance, and the equivalence principle.

Deep Inelastic Scattering

Deep inelastic scattering measurements at a muon collider facility could be pursued at fixed target experiments using intense muon and neutrino beams, or at a μp collider if the muon collider ring was located near a proton storage ring (Fig. 1). At the workshop there was little enthusiasm for the fixed target muon option. However, there was extensive enthusiasm for exploiting the intense neutrino beams, using light targets in general and liquid hydrogen targets in particular. This would enable the proton structure function to be measured directly without the need for nuclear corrections. In addition, Shiltsev [7] has calculated the parameters of a μp collider using 1 TeV protons stored in the Tevatron and 250 GeV muons stored in a muon-collider type ring. The average luminosity of this machine would be $\sim 10^{33}$ $cm^{-2}s^{-1}$. A 10 fb^{-1} data sample would contain 10^6 events with $Q^2 > 5000$ GeV^2 [4] (the present ZEUS sample, which corresponds to 34 nb^{-1}, contains 326 events with $Q^2 > 5000$ GeV^2), and would allow the discovery of leptoquarks with standard couplings up to 800 GeV/c^2. Hence, the deep inelastic scattering group at the workshop concluded that the high-Q^2 physics at a 200 GeV $\times$ 1 TeV μp collider would be very interesting. However, due to a large background [8] flux in the forward muon direction, it is suspected that small angle scattering measurements would be difficult, and hence the low–x physics program would be limited.

Slow/Stopped Muon Physics

Current low energy muon beam facilities produce typically 10^7–10^8 μ per second. The muon source at a muon collider would provide muon beams with intensities approaching 10^{14} μ per second (Table 2). A small fraction of the available muons could be used to support a broad range of low energy muon experiments. However, it should be noted that in general the bunch structure at a muon collider accelerator facility is not ideal for low energy muon experiments that tend to require either a DC muon beam to minimize instantaneous rates, or a CW beam with $\sim 2\mu s$ between bunches. Hence, either experiments have to be designed to match the bunch structure in Table 2, or the muon source has to be designed so that it can also provide DC and/or CW muon beams. Neither of these options is straightforward, and both deserve detailed study.

Perhaps the best motivated particle physics experiments using low energy muons are searches for muon–number violation in rare muon decays ($\mu \to e\gamma$, $\mu \to eee$), muonium–antimuonium oscillation, or $\mu \to e$ conversion. The detection of muon number non-conservation would be a spectacular signal for physics beyond the SM. Many extensions to the SM predict lepton flavor violation at levels that may be detectable in the next few years. As an example consider $\mu \to e$ conversion, for which the current experimental bound is $< 7 \times 10^{-13}$. Ongoing experiments are expected to improve the sensitivity by a factor of about 3. In the longer term a recently approved BNL experiment proposes to achieve a sensitivity of 10^{-16}. Several models of physics beyond the SM predict signals at this level. For example, due to slepton and gaugino mixing, some supersymmetry (SUSY) models predict $\mu \to e$ conversion at rates that would be observable. At the front end of a muon collider it may be possible [9] to achieve a sensitivity of 10^{-18}–10^{-19} !

Finally, there are many other muon experiments that might be profitably pursued at the front end of a muon collider. Some examples are (i) precision measurements (muon anomalous magnetic moment, muon electric dipole moment, muon lifetime, muonium hyperfine splitting, etc), (ii) ν_μ mass constraints, (iii) searches for parity and CP violation in muonic atoms, (iv) condensed matter physics using μSR, and (v) μ^- catalyzed fusion research.

THE FIRST MUON COLLIDER

The workshop parameters for the FMC are shown in Table 5. Note that the assumptions that went into computing the luminosities were somewhat conservative. To obtain a more aggressive but still reasonable set of goals for the FMC these luminosities can be multiplied by a factor of three. In addition to specific conclusions that emerged from the workshop for each physics sub-topic, there were also some more general conclusions:

TABLE 5. Parameters for (going from left to right) a narrowband low–energy, broadband low–energy, medium–energy, top factory, and higher–energy FMC.

$\sqrt{s}$	100	100	200	350	500
σ_p/p	3×10^{-5}	1×10^{-3}	1×10^{-3}	1×10^{-3}	1×10^{-3}
Muons per bunch	3×10^{12}	3×10^{12}	2×10^{12}	2×10^{12}	2×10^{12}
Number of bunches	1	1	2	2	2
Repetition rate (Hz)	15	15	15	15	15
Norm. $\epsilon_\perp$ (mm-mr)	297π	85π	67π	56π	50π
Collider circum. (m)	380	380	700	864	1000
f_{rev} (Hz)	7.9×10^5	7.9×10^5	4.3×10^5	3.5×10^5	3.0×10^5
turns/lifetime	820	820	890	1260	1560
β^* (cm)	13	4	3	2.6	2.3
σ_z (cm)	13	4	3	2.6	2.3
σ_r (μm)	286	85	47	30	22
L_{peak} ($cm^{-2}s^{-1}$)	6×10^{32}	7×10^{33}	6×10^{33}	1×10^{34}	2×10^{34}
L_{av} ($cm^{-2}s^{-1}$)	5×10^{30}	6×10^{31}	1×10^{32}	3×10^{32}	7×10^{32}

(i) The luminosities in Table 5 are at the threshold of physics interest. A factor of 3–or–more luminosity would be very desirable, and should be the goal of the FMC design. With this increase in luminosity there seems to be a tremendous potential physics program.

(ii) Initial studies [10] show that the absolute energy calibration of the FMC could be at the level of $\delta E/E \sim 10^{-5}$, with a beam energy spread $\sigma_p/p \sim 3 \times 10^{-5}$. This would give the FMC a unique capability as a precision tool to scan and measure the parameters of any resonant states produced in the s–channel.

(iii) It is unclear with what precision the luminosity can be measured at a muon collider. Precise measurements of muon Bhabha scattering may not be possible because of the large backgrounds induced by showering high–energy electrons from muon decay, and the necessity of having shielding cones of 10°–20° half–angle in the forward/backward directions. More work needs to be done on understanding how best to measure the luminosity, and how well it needs to be measured.

(iv) Significant muon polarization would increase the physics potential of the FMC.

Higgs Physics

Current theoretical prejudice suggests that if one or more Higgs bosons exists, the lightest Higgs boson has a mass $m_h < 150$ GeV/c^2. If this is true, the FMC could be designed to be an s–channel Higgs factory. This would be a unique tool for studying the Higgs boson. Consider a specific example. Suppose a Higgs boson has been observed at TEV33 with a mass $m_h = 110$ GeV/c^2, which is then confirmed and pinned down at the LHC with a precision $\sigma_m = 0.1$ MeV/c^2. It has been shown [11] that if the FMC beam energy spread is 0.003% (2 MeV), and the luminosity is a factor of 3 greater than in Table 5, it will take 1 operational year (10^7 secs) to make a first rough scan to determine m_h to 2 MeV/c^2. A precise 3 point scan, taking 3 years, would then determine m_h to ~ 0.1 MeV/c^2. If the width $\Gamma_h = 3$ MeV, it would be determined with a precision $\Delta\Gamma_h/\Gamma_h = 16\%$. At the same time the dominant decay channels would be measured with good precision (3% for $\sigma.B(b\bar{b})$ and 15% for $\sigma.B(WW)$). This "tour de force" in Higgs measurements is a unique capability amongst all currently imagined futuristic colliders. Not only is the width measurement sufficiently precise to distinguish between a SM and MSSM Higgs boson over a large region of SUSY parameter space, but from the ratio of branching ratios B(WW)/B($b\bar{b}$) one should be able to infer the presence of a heavy Higgs (A^0) up to masses $M_A \sim 400$ GeV/c^2. At a higher energy muon collider the direct discovery and measurement of the A^0 would then be possible in the s–channel for a large region of SUSY parameter space.

WW and Z^0 Physics

The LEP era of Z^0–pole physics is over. It is likely that the 2.8σ discrepancy between the $\sin^2\theta_W$ values determined from the SLD left–right asymmetry measurement and from the LEP forward–backward asymmetry measurements will remain. A muon collider Z^0 factory producing 10^8 Z^0 events per year would push beyond the statistical reach of LEP by an order of magnitude. This would require a luminosity of 2×10^{32} cm^{-2} s^{-1}, consistent with a factor of 3 more than in Table 5. A Z^0 factory with this capability might be of interest if there was no known Higgs, SUSY, Technicolor, or other type of new particle to scan and measure in the s–channel. In this case it may become important to resolve the $\sin^2\theta_W$ discrepancy using an FMC with polarized muon beams, and obtain further experimental guidance from precision FMC Z^0 measurements. Anticipated precisions for the various Z^0 measurements at the FMC are discussed in [11]. Ultimately the precision with which m_W is known may limit the sensitivity to new physics obtained from the overall consistency of the measured SM parameters. In this case an FMC with $\sqrt{s} \sim 2m_W + 0.5$ GeV may be desirable to obtain a precision $\delta m_W \sim 6$ MeV/c^2 for an integrated luminosity of 100 pb^{-1}.

SUSY Searches and Measurements

In supersymmetric extensions to the SM each fermion (boson) has a boson (fermion) superpartner. SUSY is broken by introducing soft masses and couplings that do not result in quadratic divergences. In the MSSM, this scheme results in over 100 SUSY-breaking parameters. Hence, if SUSY has something to do with electroweak symmetry breaking there are many SUSY particles to discover, and a large number of measurements at a variety of high energy colliders will be required to pin down the model dependent details. Although much of the SUSY phenomenology is model dependent, the prediction that the lightest Higgs boson has a mass $m_h < 150$ GeV/c^2 is more general. A muon collider Higgs factory would play a unique role in providing precise measurements of the lightest Higgs boson properties. Furthermore, finding the MSSM heavier Higgs particles (H^0 and A^0) may not be easy in futuristic e^+e^- or high energy hadron colliders. Hence, once the parameters of the lightest Higgs boson have been precisely determined, a muon collider at higher energies scanning in the region of the heavier Higgs bosons might make a crucial contribution to our understanding of the details of the emerging underlying SUSY theory.

Muon colliders can also contribute to unraveling the SUSY zoo in other ways. For example, fine tuning arguments in mSUGRA models suggest the lightest chargino is lighter than 200 GeV/c^2, in which case this chargino may well be discovered at TEV33. The energy of the FMC could then be chosen to pair produce charginos, a process that proceeds at lowest order via s–channel production with an intermediate γ, Z, or H, and via t–channel muon–sneutrino exchange. The amplitudes from these diagrams interfere destructively, and the threshold dependence of the cross-section at a muon collider is sensitive to the mass of the muon sneutrino up to masses of a few hundred GeV/c^2. Other t–channel sparticle exchanges in sparticle pair production process at a muon collider are also of interest, and can probe for example the presence of heavy squarks via t–channel enhancements. Finally, it is likely that of the many sparticles, at least some would have masses in the TeV range, and hence ultimately a multi-TeV muon collider would be required. A more detailed discussion of the strength of the muon collider physics program in a SUSY world can be found in Ref. [12].

Strong Dynamics

Although supersymmetry is very appealing, we may not be living in a SUSY–world. Technicolor is perhaps the most actively pursued alternative to a SUSY solution to electroweak symmetry breaking. Modern technicolor models predict narrow neutral technihadrons (π_T, ρ_T, and ω_T) which would appear as spectacular narrow resonances at an FMC with $\sqrt{s} = 100$–200 GeV and beam energy spread $\sigma_E/E < 10^{-4}$. For example, technipions are expected to couple to $\mu^+\mu^-$ with a strength proportional to m_μ. Furthermore the technipion coupling is enhanced with respect to the equivalent Higgs boson coupling. Hence, an FMC at the appropriate energy would be a superb technipion factory. Modern technicolor ideas also suggest that eventually there will be a compelling need for a multi-TeV muon collider to search for a TeV-scale Z', and to search for and measure higher mass techni–particles. A more detailed discussion of the strong dynamics physics potential of muon colliders can be found in Ref. [13].

Top Factory

The shape of the $\mu^+\mu^- \to t\bar{t}$ cross–section over the threshold region is sensitive to $m_t, \Gamma_t, V_{tb}, m_h,$ and α_s. A precise scan over the threshold region can therefore be used to improve our knowledge of some or all of these parameters. These measurements could also be performed at an e^+e^- collider. However, with a smaller beam energy spread and less initial state radiation, for a given integrated luminosity the measurements at a muon collider would be more precise. As an example, if we assume a factor of 3 more luminosity than in Table 5, a 1 year scan would determine m_t with a precision of 200 MeV/c^2, and a 10 year scan would improve the mass determination to 70 MeV/c^2. Further discussion of the top-factory physics at a muon collider can be found in Ref. [14].

CONCLUSIONS

The development of high luminosity muon colliders is motivated primarily by the desire to build a multi-TeV lepton collider. However, before achieving this goal it will probably be necessary to advance along the learning curve by first constructing and operating a more modest facility. The workshop has demonstrated that there are world class cutting edge physics programs that could be pursued at both the front end of a muon collider and at a "low" energy FMC. We do not yet know whether a muon collider is technically feasible, but given the amount of interest in the muon collider and its front end that was manifest at the workshop, and the strength of the physics program that could be pursued at a muon collider facility, I believe that there is a compelling case to vigorously pursue a muon collider R&D program.

Acknowledgments

This work was performed at the Fermi National Accelerator Laboratory, which is operated by Universities Research Association, under contract DE-AC02-76CH03000 with the U.S. Department of Energy.

REFERENCES

1. S. Holmes et al., "A Development Plan for the Fermilab Proton Source", FERMILAB-TM-2021, September 1997, unpublished.
2. A.N. Skrinsky and V.V. Parkhomchuk, Sov. J. Part. Nucl. **12**, 223 (1981).

3. L. Littenberg, "Can BNL-Style Studies of $K \to \pi \nu \bar{\nu}$ be Pushed at the FEMC?" *Proc. Workshop on Physics at the First Muon Collider and at the Front End of a Muon Collider*, edited by S. Geer and R. Raja, New York, AIP #435, 1998, pp. 299-307.
4. H. Schellman, "Deep Inelastic Scattering at a Muon Collider - Neutrino Physics," *Proc. Workshop on Physics at the First Muon Collider and at the Front End of a Muon Collider*, edited by S. Geer and R. Raja, New York, AIP #435, 1998, pp.166-176.
5. S. Geer, "Neutrino Beams from Muon Storage Rings: Characteristics and Physics Potential," hep-ph/9712290, submitted to PRD.
6. R. Mohapatra, "Neutrino Physics in a Muon Collider," hep-ph/9711444, *Proc. Workshop on Physics at the First Muon Collider and at the Front End of a Muon Collider*, edited by S. Geer and R. Raja, New York, AIP #435, 1998, pp. 358-369.
7. V. Shiltsev, "An Asymmetric Muon-Proton Collider: Luminosity Consideration," Fermilab-Conf-97/114.
8. S. Geer, "Backgrounds and Detector Issues at a Muon Collider," Fermilab-Conf-96-313, *Proc. of 1996 DPF/DPB Summer Study on New Directions for High-Energy Physics (Snowmass 96)*, Snowmass, CO, 25 June - 12 July 1996.
9. W. Marciano, "Low Energy Physics and the First Muon Collider," *Proc. of Workshop on Physics at the First Muon Collider and at the Front End of Muon Collider*, edited by S. Geer and R. Raja, New York, AIP #435, 1998, pp. 58-65.
10. R. Raja and A. Tollestrup, "Calibrating the Energy of a 50 x 50 GeV Muon Collider Using Spin Precession," hep-ex/9801004, submitted to PRD.
11. M. Demarteau and T. Han, "Higgs Boson and Z Physics at the First Muon Collider," hep-ph/9801407, *Proc. Workshop on Physics at the First Muon Collider and at the Front End of a Muon Collider*, edited by S. Geer and R. Raja, New York, AIP #435, 1998, pp. 177-192.
12. V. Barger, "Supersymmetry via-a-vis Muon Colliders," *Proc. Workshop on Physics at the First Muon Collider and at the Front End of a Muon Collider*, edited by S. Geer and R. Raja, New York, AIP #435, 1998, pp. 107-120.
13. K. Lane, "Technicolor and the First Muon Collider," hep-ph/9801385, *Proc. Workshop on Physics at the First Muon Collider and at the Front End of a Muon Collider*, edited by S. Geer and R. Raja, New York, AIP #435, 1998, pp. 711-722.
14. M. S. Berger, "The Top-Antitop Threshold at Muon Colliders," hep-ph/9712486, *Proc. Workshop on Physics at the First Muon Collider and at the Front End of a Muon Collider*, edited by S. Geer and R. Raja, New York, AIP #435, 1998, pp. 797-802.

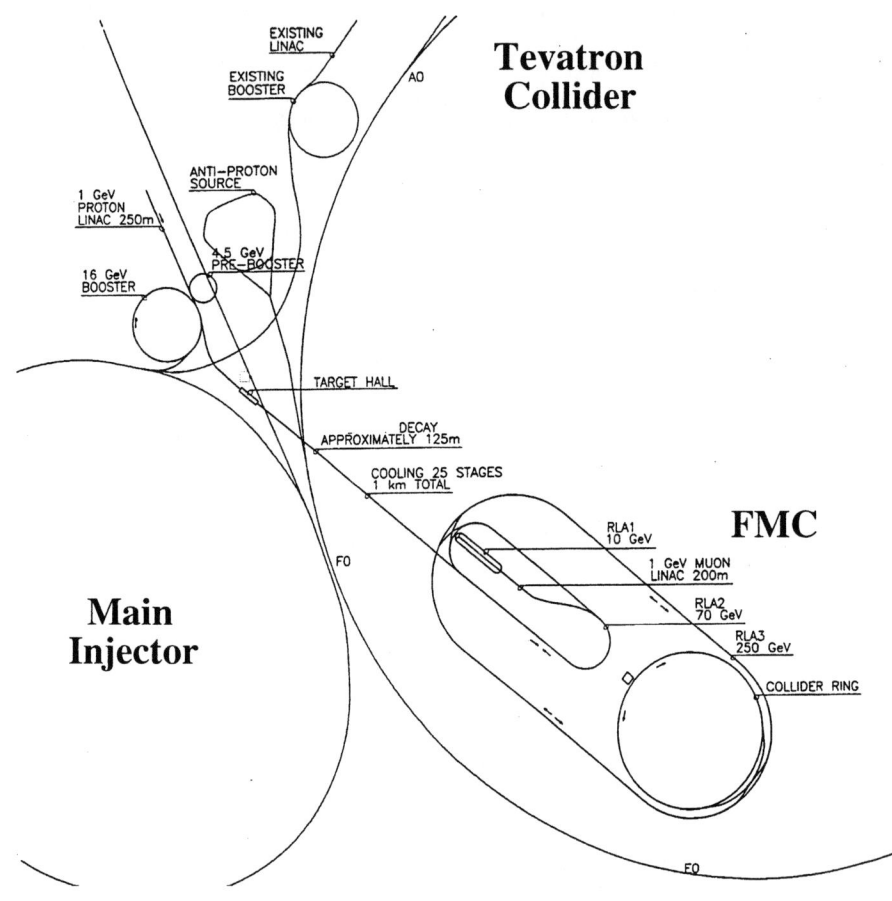

FIGURE 1. Schematic showing a plausible location for the First Muon Collider at Fermilab.

Neutral Higgs Searches at LEP-2: Present Status

Philip Bambade

Laboratoire de l'Accélérateur Linéaire
Bâtiment 200, Centre d'Orsay
91405 ORSAY CEDEX

Abstract.
The present status of neutral Higgs boson searches at LEP-2 is reviewed. Final results using 1996 data at 161 and 172 GeV are reported, along with preliminary ones from the 1997 run at 183 GeV. An overview of the main tools required to isolate the signal (beauty tagging and mass reconstruction) is presented. Prospects for the upcoming 1998 run at 189 GeV, and from higher energies expected to be reached in the following years are also given.

INTRODUCTION

Higgs bosons are scalar particles predicted as a consequence of the mechanism for spontaneous gauge symmetry breaking [1] introduced to explain the generation of masses in the Standard Model (SM) of electroweak interactions [2], and in its supersymmetric extensions [3]. In spite of the general success of the SM in reproducing basically all present electroweak data [4], finding experimental evidence for their existence remains a central issue in elementary particle science.

In this talk we concentrate on the status of neutral Higgs searches at LEP-2 from the recent high energy runs (161-172 and 183 GeV)[1]. After a brief review of the theoretical framework and mass limits available, we summarize the experimental strategies used to separate neutral Higgs bosons from physical background sources in the accessible mass range. The performance of two main tools used to isolate the signal (b quark tagging and mass reconstruction) is also illustrated. We then report on the final result from the four LEP collaborations at 161-172 GeV, using combined statistical procedures, and mention some preliminary results obtained at 183 GeV. Finally we conclude on the outlook for the next few years, with runs planned at energies up to 200 GeV.

---

[1] The case of charged Higgs bosons (predicted in the framework of supersymmetry) is not covered in this talk, since it is likely that their masses will exceed the range of LEP-2 (see Eq. (2)).

Reports of recent experimental results on these topics include [5,6]. An extensive review of the motivations for searching Higgs bosons and on the potential at LEP-2 can be found in [7] and in the references therein.

THEORETICAL FRAMEWORK AND MASS BOUNDS

Standard Model

In the Standard Model, after the spontaneous breakdown of the electroweak symmetry through the self-interactions of a complex isodoublet scalar field and the generation of masses for gauge bosons and fermions, only one scalar field remains, manifesting itself as the physical Higgs particle (H).

Although indirect experimental constraints arising from precision measurements of electroweak observables tend to slightly favor a relatively light mass for the Higgs boson [4], the exact value is not specified in the SM. However it is customary to consider that it must be constrained in two ways as a consequence of the indefinite growing with energy of the quartic self-coupling of the Higgs field in the presence of radiative corrections [7]. From above one needs to maintain a perturbative character to the theory up to a high energy scale, and from below it is important to require that the electroweak vacuum be stable in the presence of a heavy *top* quark. Quantitatively, the SM Higgs boson mass should be heavier than 100 GeV if the electroweak theory is to hold in its present form up to an energy scale of 10^5 GeV (for a *top* quark mass of 175 GeV) [7].

Supersymmetry

Severe fine-tuning problems appear when trying to embed the SM in Grand Unified Theories (GUTs), because of the very different mass scales at which electroweak and GUT symmetries are broken, and because of the quadratically divergent radiative corrections affecting the Higgs boson mass. These so-called *naturalness* – or *hierarchy* – problems are resolved elegantly by introducing a new boson-fermion symmetry (supersymmetry) which effectively removes these divergences [8].

The Higgs sector must in supersymmetric theories consist of two complex isodoublet scalar fields, each one giving masses to *up* and *down* quarks and leptons. In the minimal version (MSSM), the physical Higgs spectrum contains two CP-even and one CP-odd neutral Higgs bosons, h, H and A, respectively, and a charged Higgs boson pair $H^\pm$ [9]. The tree level Higgs spectrum is entirely specified by $m_{W,Z}$, the weak gauge boson masses, m_A, the CP-odd Higgs mass, and $\tan\beta$, the ratio of Higgs vacuum expectation values, giving

$$m_{h,H}^2 = \frac{1}{2}\left(m_A^2 + m_Z^2 \mp \sqrt{(m_A^2 + m_Z^2)^2 - 4m_Z^2 m_A^2 \cos^2 2\beta}\right), \quad (1)$$

and

$$m_{H^\pm}^2 = m_A^2 + m_W^2. \tag{2}$$

It can be shown from (1) that the lightest MSSM neutral Higgs particle is bounded to be smaller than the Z boson mass. However, radiative corrections, mainly involving *top* quarks, weaken this fairly stringent bound to $m_h \leq 130\ GeV$ for a *top* quark mass of $175\ GeV$ [2] [7].

On the other hand, one of the motivations for supersymmetry is the development of GUTs. It has been shown that supersymmetric GUTs are a well suited framework for successfull unification of the gauge couplings of the weak, electromagnetic and strong interactions at some very high mass scale [11]. In the same spirit, it is tempting to also try to unify Yukawa couplings of fermions at the GUT mass scale. The particular case of b and τ Yukawa coupling unification has been pursued extensively and has led to further constrained versions of the minimal supersymmetric model (CMSSM). In such models [12], relatively small values of $\tan\beta$ are favored and upper mass bounds around $105\ GeV$ are obtained for the lightest neutral Higgs particle.

Discussion

The rule of thumb for the Higgs boson mass which can be probed at LEP-2 is approximatively given by $\sqrt{s} - 95\ GeV$. This arises because the Higgs boson is mostly produced in association with a Z boson. Hence if LEP-2 reaches about 200 GeV in coming years as expected, even if it can - and should - be argued that none of the above mass bounds should be taken too literally (for example they depend on the exact value of the *top* quark mass), we will be in a good position to study a good part of the mass range most favored by present theoretical prejudice.

HIGGS DETECTION AT LEP-2

Higgs Production

The standard and lightest supersymmetric Higgs bosons are mainly produced at LEP-2 via bremstrahlung off a Z boson[3], as shown in Figure 1. Contrary to LEP-1, where this process is also available, the Z boson in the final state is predominantly on-shell.

The expected cross-section is shown in Figure 2 (taken from [5]) for the standard Higgs boson. In the MSSM, the cross-section for the lightest Higgs boson is obtained

---

[2] This upper bound on the mass in the MSSM can be weakened even further in the framework of non minimal extensions of the MSSM. For instance in the Next to Minimal Supersymmetric Model (NMSSM), where an extra scalar gauge singlet is introduced [10,7], it can be shown that the upper bound is closer to $150\ GeV$.

[3] Some small contributions to the cross-section also arise from fusion diagrams involving W and Z bosons (at the 10% and 1% levels respectively). These diagrams are also shown in Figure 1.

by simply multiplying the SM results by $\sin^2(\beta - \alpha)$, where α is the mixing angle between the two SUSY Higgs doublets. The case of the CP-odd Higgs boson is more difficult. It can be pair-produced along with the lightest Higgs particle as shown in Figure 1. This process is affected by a factor $\cos^2(\beta - \alpha)$ and is hence complementary to the bremstrahlung process. However not only is the mass for the CP-odd state expected to be typically higher than that of the lightest Higgs boson, but in addition the cross-section is depressed kinematically near threshold as $(p/E)^3$ because of the scalar nature of the two particles produced.

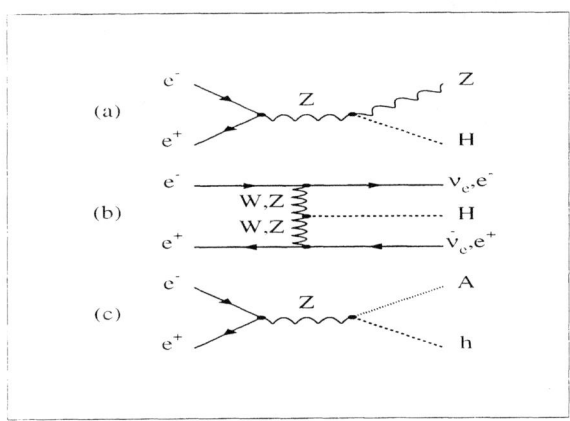

FIGURE 1. Production diagrams for Higgs bosons. In (a), bremstrahlung off a Z boson, in (b), fusion processes, and in (c), hA pair-production.

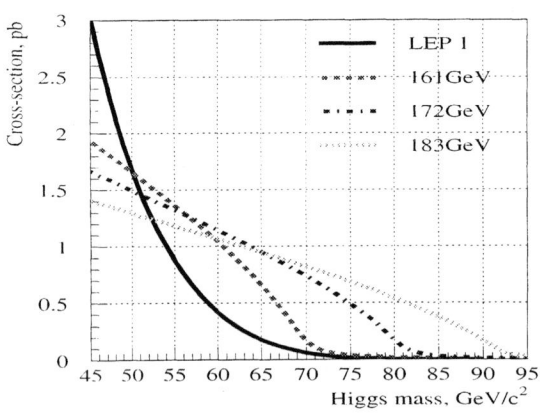

FIGURE 2. Cross-section for the production of Higgs bosons at LEP-1 and LEP-2 energies.

TABLE 1. Decay branching ratios of SM Higgs (for $m_H = 70$ GeV).

Decay Channels	Branching Ratios (%)
$b\bar{b}$	86
$c\bar{c}$	4
gluons	2
$\tau^+\tau^-$	8

Higgs Decays and Final State Topologies

From the proportionality of the Yukawa couplings to fermion masses, and more generally, of the Higgs couplings to particles masses, we know that Higgs bosons decay predominantly into the heaviest particles accessible for any given value of its mass. In the mass range available at LEP-2, this means essentially b quark pairs, as shown in Table 1.

Hence the possible final state event topologies which can be used are determined mostly by the decay of the accompanying Z. Consisting generically of a four jet channel, a missing energy channel and of channels with lepton pairs, we shall see that they correspond each to specific experimental situations, with both different backgrounds and identifying signatures.

Physical Background

The cross-sections for standard processes at LEP-2 energies are shown in Figure 3 (taken from [7]). The main process involves e^+e^- annihilation to $q\bar{q}$ pairs. Because of the proximity and large width of the Z resonance, an important fraction of such events consist of single, and sometimes multiple, radiative returns to the Z pole. Processes contributing to missing energy channels also include $\gamma\gamma$ fusion processes (not shown on Figure 3). Finally, above their respective thresholds, the rates of WW – and particularly ZZ – pair-production become critical.

Search Methods

The Four Jet Channel

Four jet topologies result from both HZ and hA production. Two main backgrounds are WW and ZZ pairs, once above their respective thresholds. Exploiting the high decay probability of Higgs bosons to b quarks, resulting most of the time in $b\bar{b}q\bar{q}$ or $b\bar{b}b\bar{b}$ final states, b-tagging can be efficiently used against WW pairs, and also helps to reduce ZZ pairs. A third important source of background arises from $q\bar{q}$ pair production accompanied by hard gluon radiation, which can also result in multijet event structures. Accurate mass reconstruction is here required for discrimination.

Although detection efficiency, around 30% and 50%, respectively for HZ and hA, is limited by background, this channel is the main one simply because decay branching ratios of Z and Higgs bosons into quarks are dominant.

The Missing Energy Channel

The next most important Higgs detection channel is the missing energy channel, with the Z decaying in $\nu\bar{\nu}$ pairs. The main signatures for signal identification are the presence of two acoplanar b-tagged jets recoiling against an unseen Z boson. Precise energy flow and good hermeticity of the detector are here essential to control backgrounds from radiative returns on the Z in which the high energy photon (or photons) escape detection. Although not quite as essential as in the four jet channels, b-tagging is still important to reduce backgrounds from ZZ pairs, and from semi-leptonic WW pairs in which the lepton is not identified. Detection efficiency typically reaches about 40% in this channel.

The Semi-Leptonic Channels

The cleanest channels are certainly the semi-leptonic channels in which the Z boson decays either into e^+e^- or $\mu^+\mu^-$. Backgrounds are here efficiently controlled simply by requiring the presence of a pair of identified leptons with an invariant mass close to the Z mass. Efficiency ranges between 50 and 80%, the lower values

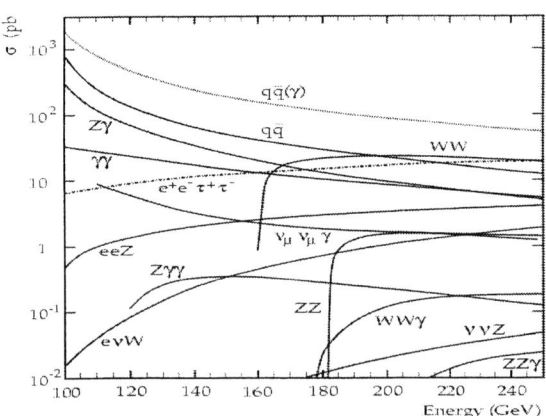

FIGURE 3. Cross-sections for Standard Model processes composing the physical background to the Higgs searches.

corresponding to the electron channels, where the lepton identification is typically more difficult. The b-tagging tools are here only needed to reduce the irreducible ZZ background.

The situation is more difficult in the case of channels with τ-lepton pairs. These topologies arise from both HZ and hA production. In spite of much lower efficiencies (typically 15 to 25%) resulting from the need to achieve good reconstruction of the different τ lepton decays, they nicely complement the main topologies of the Higgs searches, and are therefore in development in several of the LEP experiments.

ANALYSIS TOOLS

Beauty Tagging

The LEP-2 searches benefit from the highly performant b-tagging tools developed for b-physics during the LEP-1 period. Because of the long life-time, large mass and relatively hard spectrum of B hadrons, the tracks of their decay products carry impact parameters large enough to be measured using silicon strip vertex detectors (VD) installed close to the vaccuum chamber. In order to optimally isolate events - and jets within events - containing B hadrons, a likehood ratio is constructed based on the expected probability distributions for the track impact parameters arising from B hadrons and from the other possible sources. At LEP-2 efficiencies as large as 50% for HZ events are obtained, for WW rejection factors of about 100 (see Figure 4 (taken from [13])). In addition, in order to help discriminate against charm, the properties of reconstructed secondary vertices, such as the masses and multiplicities of the corresponding tracks, have been introduced with success into the likelihood. Also lepton tags have been combined with the lifetime ones to maximize performance. Finally, LEP-2 events being more spherical than the typical

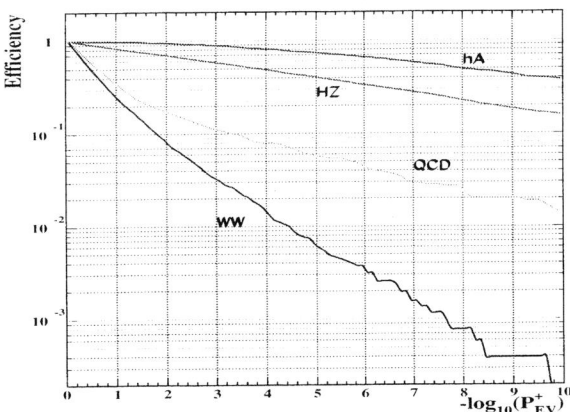

FIGURE 4. Efficiencies to tag LEP-2 events using track impact parameter information.

back to back Z decays produced at LEP-1, extensions of tagging capabilities to the largest possible solid angle have been sought. This is illustrated for the DELPHI experiment in Figure 5 (taken from [13]), where the improvements in the tagging efficiency for $Z \to b\bar{b}$ events is shown following an extension of the VD angular coverage from $\theta = 42\,\text{deg}$ down to $\theta = 25\,\text{deg}$.

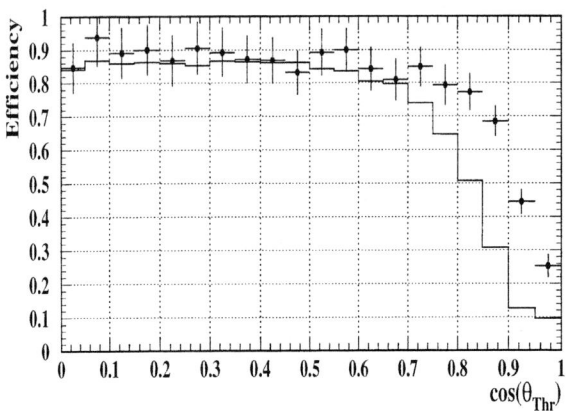

FIGURE 5. Efficiencies to tag $Z \to b\bar{b}$ events using the old version of the DELPHI VD, limited to $\theta = 42\,\text{deg}$ (solid line), and using the new one extended to $\theta = 25\,\text{deg}$ (full points), as a function of the cosine of the event thrust angle.

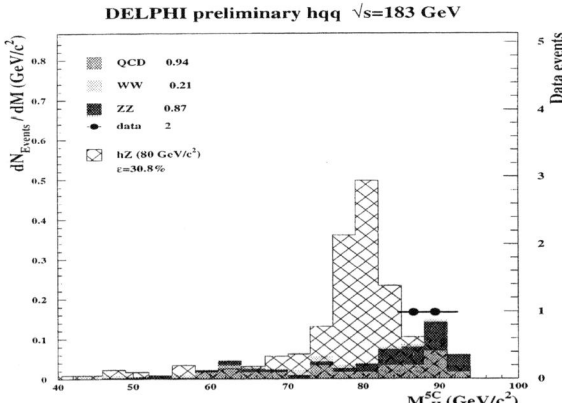

FIGURE 6. Mass distribution obtained in the preliminary HZ analysis performed by DELPHI at 183 GeV.

Mass Reconstruction

Estimating the invariant masses of the two decaying heavy bosons is important to separate a possible Higgs signal from the different background hypotheses. This is particularly true for the QCD background in the four jet topology. Mass reconstruction is in this case achieved via a constrained kinematic fit taking into account energy and momentum conservation in the event as well as the known mass of the Z boson for two of the four jets, in the case of the HZ channel, or some relation between boson masses (equality in first approximation) for the hA channel. In order to best solve the ambiguity in the pairing of the four jets into dijets, a likelihood is constructed for each hypothesis which combines the fit probabilities with the information from the jet b-tagging (for jets appropriately contained in the VD acceptance, and in the case of events not satisfying a $4b$-hypothesis). As an example, the mass distribution achieved by DELPHI in the preliminary HZ analysis performed at 183 GeV is shown in Figure 6 (taken from [14]). Resolutions of about 3 GeV are attained in the core of the signal distribution, although significant tails remain at lower masses, due in part to limitations inherent to the jet clustering algorithm. These tails can be reduced to some extent by separately treating events which tend to cluster into five or more jets, rather than into four (because of hard gluon radiation), and by introducing additional information into the likelihood, such as jet angular variables and estimators representating the dijet charges.

RESULTS

LEP Performance

The LEP-2 period began in the fall of 1995 with short runs at 130 and 136 GeV centre of mass energy, mainly dedicated to test the operation of the new super-conductive radiofrequency cavities installed as part of the energy upgrade program. About 6 pb^{-1} of luminosity were accumulated by each experiment. Since then, the schedule for new cavity installation has enabled to reach 161 and 172 GeV during the 1996 running, and 183 GeV during the 1997 running, with total accumulated luminosities of respectively about 20 and 55 pb^{-1}.

Summary of Final 1996 Results

Final results from the 1996 runs where presented at the Jerusalem EPS Conference, and have been published by the four LEP experiments [15]. In addition, the LEP Higgs Working Group has produced a combined LEP limit in the case of the SM Higgs boson [5,16]. In short, none of the experiments had candidates in excess of background expectations.

HZ Search

Aleph had no candidates in any of the topologies, a both relatively low combined signal efficiency (29 %) and very low expected background (0.5 events).

Delphi had two candidates, one in the four jet search, with an estimated mass of 59 GeV for the dijet assigned to the Higgs boson hypothesis, and one in the missing energy topology, with visible mass of 65 GeV. The combined efficiency and expected background events were 30 % and 2.6 respectively.

Opal had also two candidates, one in the 4 jet channel, with a mass of 75 GeV for the supposed Higgs boson, and one in the missing energy channel, with a visible mass of 39 GeV. The combined efficiency and expected background events were 33 % and 2 respectively.

L3, rather than devise an analysis with specific selection cuts, selected a large number of candidate events (33) and used a weight technique to measure the Higgs-likeness of its signal.

hA Search

No hA candidates were retained by any of the four LEP experiments, using selection algorithms with efficiencies ranging from 37 to 55 % and expected backgrounds of typically one event at 172 GeV.

Combined Mass Limit from 1996 Data

In the absence of a clearly detected signal, in the form of an excess of events as compared to the background expectation, a limit on the mass of the SM Higgs boson

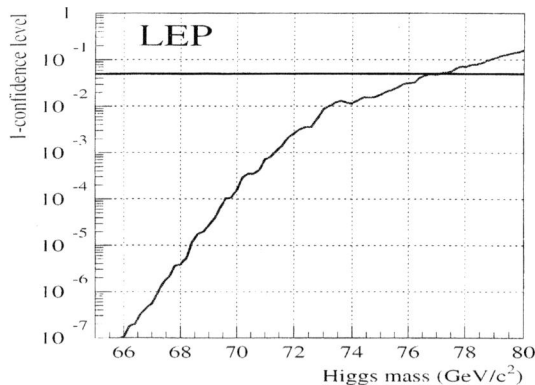

FIGURE 7. Combined signal probability from LEP in the case of the SM HZ search. A 95 % CL limit of 77 GeV on the mass of the Higgs boson is placed.

was estimated, combining all search channels, in all of the four LEP experiments. The method, at the conceptual level, consists in estimating, as a function of the mass of the Higgs boson, the Poisson probability to realize the actual observation of N obtained events with their specific reconstructed invariant masses and values of other discriminant variables, given the expected numbers of signal and background, and given the respective shapes of their mass distributions and of the other variables considered. The Higgs boson mass corresponding to the 95 % CL limit is the smallest possible mass for which the probability that the found candidates - thought to be background - really contained the searched signal, is less than 5 %.

The actual practical methods for computing the combined probability are involved, and different receipes have been devised in each LEP experiment [16], which can however be shown to all be different ways of approximating the fully general product of likelihood ratio technique already mentioned.

Fortunately, the different techniques applied have resulted in nearly equal limits. An example is shown in Figure 7 (taken from [5]). The final 95 % CL limit on the SM Higgs boson mass is 77 GeV.

Preliminary Results from 1997

Only very preliminary and necessarily brief results are presented here from the 183 GeV data collected in 1997, based on the reports given to the LEPC Committee in November [14]. Having crossed the threshold for ZZ pair production for the first time at this new energy, some learning is still in progress relative to dealing with this most signal-like of all backgrounds.

The L3 experiment presented a 95 % CL limit of 82.2 GeV on the SM Higgs boson mass, based on the analysis of only 60 % of the new data. Candidates were still under study, particularly in relation to the newly installed 3-dimensional b-tagging capabilities.

The Opal experiment, having analysed 80 % of the new data, obtained 7 candidates, while expecting 11 from background sources. A new 82 GeV 95 % CL limit was placed on the SM Higgs boson mass. One particularly nice on-shell ZZ candidate was found (which could in principle also arise from HZ, (see Table 1)), in the $e^+e^-\tau^+\tau^-$ final state.

The Delphi and Aleph experiments, having both analysed all of the new data, found respectively 7 and 3 candidates while expecting 6 and 5. The 95 % CL limits obtained for the SM Higgs boson mass were 83.6 and 88.6 GeV. The quite a bit higher limit found in Aleph results directly from the low number of candidates obtained in comparison with the expected background. It has been pointed out that although the limit statistically expected from the Aleph analysis was as low as 83 GeV, the higher obtained limit only corresponds to a one sigma fluctuation, and hence was not so unlikely to have been achieved [17].

Interpreting the above result on HZ in the framework of the MSSM and combining with preliminary results from its hA search, the Delphi experiment has in addition produced preliminary 95 % CL limits on the masses of h and A at respectively 73 and 75 GeV[4].

CONCLUSIONS AND OUTLOOK FOR 1998

Using the data accumulated in 1996 at 161-172 GeV, the four LEP experiments have achieved a combined 77 GeV 95 % CL limit on the SM Higgs boson.

Using the new 1997 data at 183 GeV, the best individual (preliminary) 95 % CL limit was obtained by Aleph, at 88 GeV. It may be speculated that by combining results from the four LEP experiments a limit at 90 GeV will be reached, although to this end it will be necessary to adequately monitor the cross-section of the ZZ process, from which very Higgs-like background candidates can arise.

With the upcoming runs in 1998, at 189 GeV, and then in 1999 (as well as possibly in 2000[5]), at energies which could reach up to 200 GeV[6], it is expected that Higgs boson masses close to 105 GeV will be probed, if sufficient luminosity can be collected. This may be enough to fully cover the low $\tan\beta$ scenario favored in the constrained versions of the MSSM mentioned above.

The Higgs search at LEP-2 is one of the hottest topics in elementary particle phycics for the next couple of years !

REFERENCES

1. P.W. Higgs, *Phys. Rev. Lett.* **12**, (1964) 132; and *Phys. Rev.* **145** (1966) 1156; F. Englert and R. Brout, *Phys. Rev. Lett.* **13**, (1964) 321; G.S. Guralnik, C.R. Hagen, and T.W. Kibble, *Phys. Rev. Lett.* **13**, (1964) 585.
2. S. Glashow, *Nucl. Phys.* **20**, (1961) 579; A. Salam, in *Elementary Particle Theory*, ed. N. Svartholm, (1968); S. Weinberg, *Phys. Rev. Lett.* **19** (1967) 1264.
3. J. Wess and B. Zumino, *Nucl. Phys.* **B70**, (1974) 39.
4. D. Ward, *Test of the Standard Model W mass and WWZ coupling*, talk at the HEP97 Conference, Jerusalem, Aug. 19-26, 1997, to be published in the proceedings.
5. W. Murray, *Search for the Standard Model Higgs Boson at LEP*, talk at the HEP97 Conference, Jerusalem, Aug. 19-26, 1997, to be published in the proceedings.
6. Y. Pan, *Searches for Higgs Bosons beyond the Standard Model at LEP*, talk at the HEP97 Conference, Jerusalem, Aug. 19-26, 1997, to be published in the proceedings, and *Review of recent LEP-2 Results*, talk at this conference.
7. CERN Report 96-01, *Physics at LEP-2*, Vol. 1.

---

[4] For any value of $\tan\beta$, and assuming a *top* quark mass of 175 GeV and a SUSY mass scale at 1 TeV.

[5] Formal decision is still pending regarding running in year 2000.

[6] An ongoing R & D program has shown that close to 200 GeV energies can be reached in the course of year 1999 by raising the voltages of most of the installed superconductive cavities from the design value of 6 MV/m to 7 MV/m, and by upgrading the cryogenic plant as part of the LHC program.

8. J. Wess and B. Zumino, *Phys. Lett.* **49B**, (1974) 74; J. Iliopoulos and B. Zumino, *Nucl. Phys.* **B76** (1974) 310.
9. See, e.g., J.F. Gunion, H.E. Haber, G.L. Kane and S. Dawson, *The Higgs Hunter's Guide*, Addison-Wesley 1990.
10. P. Fayet, *Nucl. Phys.* **B90**, (1975) 104.
11. U. Amaldi, W. de Boer, P.H. Frampton, H. Furstenau, and J.T. Liu, *Phys. Lett.* **B281**, (1992) 374 V. Barger, M.S. Berger, and P. Ohmann, *Phys. Rev.* **D47**, (1993) 1093.
12. M. Diaz and H. Haber, *Phys. Rev.* **D46** (1992) 3086; V. Barger, M.S. Berger, P. Ohmann, and R.J.N. Phillips, *Phys. Lett.* **314B**, (1993) 351; M. Carena,M. Olechowski, S. Pokorski, and C.E.M. Wagner, *Nucl. Phys.* **B419**, (1994) 213.
13. A. Zalewska, *The Silicon Tracker of the DELPHI Experiment at LEP-2*, talk at the HEP97 Conference, Jerusalem, Aug. 19-26, 1997, to be published in the proceedings.
14. P.J. Dornan, *Aleph Status Report*, P. Charpentier, *DELPHI Status Report*, M. Pohl, *L3 Report to LEPC*, A. Honma, *Opal Status 1997*, talks given at the LEPC, at CERN, November 11, 1997.
15. Aleph Collaboration, R. Barate et al., *Search for the Standard Model Higgs Boson in e^+e^- Collisions at $\sqrt{s} = 161$, 170 and 172 GeV*, CERN PPE/97-070
    Delphi Collaboration, P. Abreu et al., *Search for Neutral and Charged Higgs Bosons in e^+e^- Collisions at $\sqrt{s} = 161$ and 172 GeV*, CERN PPE/97-85
    L3 Collaboration, M. Acciarri et al., *Search for the Standard Model Higgs Boson in e^+e^- interactions at $161 \leq \sqrt{s} \leq 172$ GeV*, CERN PPE/97-81
    Opal Collaboration, K. Ackerstaff et al., *Search for the Standard Model Higgs Boson in e^+e^- Collisions at $\sqrt{s} = 161$, 170 and 172 GeV*, CERN PPE/97-115.
16. CERN Report LEPC 97-11, *Lower Bound for the SM Higgs boson mass: combined result from the four LEP experiments*.
17. P. Janot, *Private Communication*.

Higgs and Technicolor Goldstone Bosons at a Muon Collider [1]

John F. Gunion

Davis Institute for High Energy Physics, Department of Physics, University of California, Davis, CA 95616, USA

Abstract. I discuss the exciting prospects for Higgs and technicolor Goldstone boson physics at a muon collider.

INTRODUCTION

The prospects for Higgs and Goldstone boson physics at a muon collider depend crucially upon the instantaneous luminosity, $\mathcal{L}$, possible for $\mu^+\mu^-$ collisions as a function of E_{beam} and on the percentage Gaussian spread in the beam energy, denoted by R. The small level of bremsstrahlung and absence of beamstrahlung at muon collider implies that very small R can be achieved. The (conservative) luminosity assumptions for the recent Fermilab-97 workshop were: [2]

- $\mathcal{L} \sim (0.5, 1, 6) \cdot 10^{31} \text{cm}^{-2}\text{s}^{-1}$ for $R = (0.003, 0.01, 0.1)\%$ at $\sqrt{s} \sim 100$ GeV;
- $\mathcal{L} \sim (1, 3, 7) \cdot 10^{32} \text{cm}^{-2}\text{s}^{-1}$, at $\sqrt{s} \sim (200, 350, 400)$ GeV, $R \sim 0.1\%$.

With modest success in the collider design, at least a factor of 2 better can be anticipated. Note that for $R \sim 0.003\%$ the Gaussian spread in $\sqrt{s}$, given by $\sigma_{\sqrt{s}} \sim 2$ MeV $\left(\frac{R}{0.003\%}\right)\left(\frac{\sqrt{s}}{100 \text{ GeV}}\right)$, can be comparable to the few MeV widths of very narrow resonances such as a light SM-like Higgs boson or a (pseudo-Nambu-Goldstone) technicolor boson. This is critical since the effective resonance cross section $\overline{\sigma}$ is obtained by convoluting a Gaussian $\sqrt{s}$ distribution of

---

[1] Work supported in part by U.S. Department of Energy grant No. DE-FG03-91ER40674.
[2] For yearly integrated luminosities, we use the standard convention of $\mathcal{L} = 10^{32} \text{cm}^{-2}\text{s}^{-1} \Rightarrow L = 1 \text{ fb}^{-1}/\text{yr}$.

width $\sigma_{\sqrt{s}}$ with the standard s-channel Breit Wigner resonance cross section, $\sigma(\sqrt{\hat{s}}) = 4\pi\Gamma(\mu\mu)\Gamma(X)/([\hat{s} - M^2]^2 + [M\Gamma^{tot}]^2)$. For $\sqrt{s} = M$, the result, [3]

$$\bar{\sigma} \simeq \frac{\pi\sqrt{2\pi}\Gamma(\mu\mu)\,B(X)}{M^2\sigma_{\sqrt{s}}} \times \left(1 + \frac{\pi}{8}\left[\frac{\Gamma^{tot}}{\sigma_{\sqrt{s}}}\right]^2\right)^{-1/2}, \qquad (1)$$

will be maximal if Γ^{tot} is small and $\sigma_{\sqrt{s}} \sim \Gamma^{tot}$. [4] Also critical to scanning a narrow resonance is the ability [2] to tune the beam energy to one part in 10^6.

HIGGS PHYSICS

The potential of the muon collider for Higgs physics is truly outstanding. First, it should be emphasized that away from the s-channel Higgs pole, $\mu^+\mu^-$ and e^+e^- colliders have similar capabilities for the same $\sqrt{s}$ and $\mathcal{L}$ (barring unexpected detector backgrounds at the muon collider). At $\sqrt{s} = 500$ GeV, the design goal for a e^+e^- linear collider (eC) is $L = 50$ fb^{-1} per year. The conservative $\mathcal{L}$ estimates given earlier suggest that at $\sqrt{s} = 500$ GeV the μC will accumulate *at least* $L = 10$ fb^{-1} per year. If this can be improved somewhat, the μC would be fully competitive with the eC in high energy ($\sqrt{s} \sim 500$ GeV) running. (We will use the notation of ℓC for either a eC or μC operating at moderate to high $\sqrt{s}$.)

The totally unique feature of the μC is the dramatic peak in the cross section $\bar{\sigma}_h$ for production of a narrow-width Higgs boson in the s-channel that occurs when $\sqrt{s} = m_h$ and R is small enough that $\sigma_{\sqrt{s}}$ is smaller than or comparable to Γ_h^{tot} [1]. The peaking is illustrated below in Fig. 1 for a SM Higgs (h_{SM}) with $m_{h_{SM}} = 110$ GeV ($\Gamma_{h_{SM}}^{tot} \sim 3$ MeV).

A Standard Model-Like Higgs Boson

For SM-like $h \to WW, ZZ$ couplings, Γ_h^{tot} becomes big if $m_h \gtrsim 2m_W$, and $\bar{\sigma}_h \propto B(h \to \mu^+\mu^-)$ [Eq. (1)] will be small; s-channel production will not be useful. But, as shown in Fig. 1, $\bar{\sigma}_h$ is enormous for small R when the h is light, as is very relevant in supersymmetric models where the light SM-like h^0 has $m_{h^0} \lesssim 150$ GeV. In order to make use of this large cross section, we must first center on $\sqrt{s} \sim m_h$ and then proceed to the precision measurement of the Higgs boson's properties.

---

[3] In actual numerical calculations, bremsstrahlung smearing is also included (see Ref. [1]).
[4] Although smaller $\sigma_{\sqrt{s}}$ (*i.e.* smaller R) implies smaller $\mathcal{L}$, the $\mathcal{L}$'s given earlier are such that when Γ^{tot} is in the MeV range it is best to use the smallest R that can be achieved.

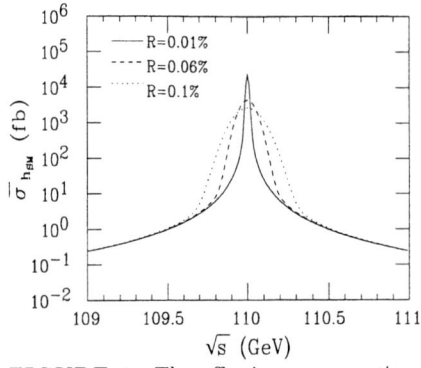 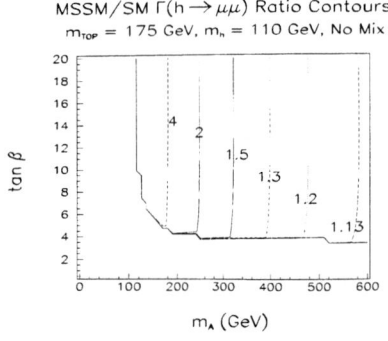

FIGURE 1. The effective cross section, $\bar{\sigma}_{h_{SM}}$, for $R = 0.01\%$, $R = 0.06\%$, and $R = 0.1\%$ vs. $\sqrt{s}$ for $m_{h_{SM}} = 110$ GeV.

FIGURE 2. We give $(m_{A^0}, \tan\beta)$ parameter space contours for $\frac{\Gamma(h^0 \to \mu^+\mu^-)}{\Gamma(h_{SM} \to \mu^+\mu^-)}$: no-squark-mixing, $m_{h^0}, m_{h_{SM}} = 110$ GeV.

For a SM-like Higgs with $m_h \lesssim 2m_W$ one expects [3] to determine m_h to within $\Delta m_h \sim 100$ MeV from LHC data ($L = 300$ fb^{-1}) (the uncertainty Δm_h will be even smaller if ℓC data is available). Thus, a final ring that is fully optimized for $\sqrt{s} \sim m_h$ can be built. Once it is operating, we scan over the appropriate Δm_h interval so as to center on $\sqrt{s} \simeq m_h$ within a fraction of $\sigma_{\sqrt{s}}$. Consider first the "typical" case of $m_h \sim 110$ GeV. For m_h of order 100 GeV, $R = 0.003\%$ implies $\sigma_{\sqrt{s}} \sim 2$ MeV. $\Delta m_h \sim 100$ MeV implies that $\Delta m_h/\sigma_{\sqrt{s}} \sim 50$ points are needed to center within $\lesssim \sigma_{\sqrt{s}}$. At this mass, each point requires $L \sim 0.0015$ fb^{-1} in order to observe or eliminate the h at the 3σ level, implying a total of $L_{\rm tot} \leq 0.075$ fb^{-1} is needed for centering. (Plots as a function of $m_{h_{SM}}$ of the luminosity required for a 5σ observation of the SM Higgs boson when $\sqrt{s} = m_{h_{SM}}$ can be found in Ref. [1].) Thus, for the anticipated $L \sim 0.05 - 0.1$ fb^{-1}/yr, centering would take no more than a year. However, for $m_h \simeq m_Z$ a factor of 50 more $L_{\rm tot}$ is required just for centering because of the large $Z \to b\bar{b}$ background. Thus, for the anticipated $\mathcal{L}$ the μC is not useful if the Higgs boson mass is too close to m_Z.

Once centered, we will wish to measure with precision: (i) the very tiny Higgs width — $\Gamma_h^{\rm tot} = 1-10$ MeV for a SM-like Higgs with $m_h \lesssim 140$ GeV; (ii) $\sigma(\mu^+\mu^- \to h \to X)$ for $X = \tau^+\tau^-, b\bar{b}, c\bar{c}, WW^\star, ZZ^\star$. The accuracy achievable was studied in Ref. [1]. The three-point scan of the Higgs resonance described there is the optimal procedure for performing both measurements simultaneously. We summarize the resulting statistical errors in the case of a SM-like h with $m_h = 110$ GeV, assuming

$R = 0.003\%$ and an integrated (4 to 5 year) $L_{\rm tot} = 0.4 \text{ fb}^{-1}$.[5] One finds 1σ errors for $\sigma B(X)$ of $8, 3, 22, 15, 190\%$ for the $X = \tau^+\tau^-, b\bar{b}, c\bar{c}, WW^\star, ZZ^\star$ channels, respectively, and a $\Gamma_h^{\rm tot}$ error of 16%. The individual channel X results assume the τ, b, c tagging efficiencies described in Ref. [4]. We now consider how useful measurements at these accuracy levels will be.

If only s-channel Higgs factory μC data are available (i.e. no Zh data from an eC or μC), then the σB ratios (equivalently squared-coupling ratios) that will be most effective for discriminating between the SM Higgs boson and a SM-like Higgs boson such as the h^0 of supersymmetry are $\frac{(WW^\star h)^2}{(b\bar{b}h)^2}$, $\frac{(c\bar{c}h)^2}{(b\bar{b}h)^2}$, $\frac{(WW^\star h)^2}{(\tau^+\tau^-h)^2}$, and $\frac{(c\bar{c}h)^2}{(\tau^+\tau^-h)^2}$. The 1σ errors (assuming $L_{\rm tot} = 0.4 \text{ fb}^{-1}$ in the three-point scan centered on $m_h = 110$ GeV, or $L_{\rm tot} = 0.2 \text{ fb}^{-1}$ with $\sqrt{s} = m_h = 110$ GeV) for these four ratios are $15\%, 20\%, 18\%$ and 22%, respectively. Systematic errors for $(c\bar{c}h)^2$ and $(b\bar{b}h)^2$ of order $5\% - 10\%$ from uncertainty in the c and b quark mass will also enter. In order to interpret these errors one must compute the amount by which the above ratios differ in the minimal supersymmetric model (MSSM) vs. the SM for $m_{h^0} = m_{h_{SM}}$. The percentage difference turns out to be essentially identical for all the above ratios and is a function almost only of the MSSM Higgs sector parameter m_{A^0}, with very little dependence on $\tan\beta$ or top-squark mixing. At $m_{A^0} = 250$ GeV (420 GeV) one finds MSSM/SM ~ 0.5 (~ 0.8). Combining the four independent ratio measurements and including the systematic errors, one concludes that a $> 2\sigma$ deviation from the SM predictions would be found if the observed 110 GeV Higgs is the MSSM h^0 and $m_{A^0} < 400$ GeV. Note that the magnitude of the deviation would provide a determination of m_{A^0}.

If, in addition to the s-channel measurements we also have ℓC $\sqrt{s} = 500$ GeV, $L_{\rm tot} = 200 \text{ fb}^{-1}$ data, it will be possible to discriminate at an even more accurate level between the h^0 and the h_{SM}. The most powerful technique for doing so employs the four determinations of $\Gamma(h \to \mu^+\mu^-)$ below:

$$\frac{[\Gamma(h \to \mu^+\mu^-)B(h \to b\bar{b})]_{\mu C}}{B(h \to b\bar{b})_{\ell C}}; \quad \frac{[\Gamma(h \to \mu^+\mu^-)B(h \to WW^\star)]_{\mu C}}{B(h \to WW^\star)_{\ell C}};$$
$$\frac{[\Gamma(h \to \mu^+\mu^-)B(h \to ZZ^\star)]_{\mu C}[\Gamma_h^{\rm tot}]_{\mu C + \ell C}}{\Gamma(h \to ZZ^\star)_{\ell C}}; \quad \frac{[\Gamma(h \to \mu^+\mu^-)B(h \to WW^\star)\Gamma_h^{\rm tot}]_{\mu C}}{\Gamma(h \to WW^\star)_{\ell C}}. \quad (2)$$

The resulting 1σ error for $\Gamma(h \to \mu^+\mu^-)$ is $\lesssim 5\%$. Fig. 2, which plots the ratio of the h^0 to h_{SM} partial width in $(m_{A^0}, \tan\beta)$ parameter space for $m_{h^0} = m_{h_{SM}} =$

---

[5] For σB measurements, $L_{\rm tot}$ devoted to the optimized three-point scan is equivalent to $\sim L_{\rm tot}/2$ at the $\sqrt{s} = m_h$ peak.

110 GeV, shows that this level of error allows one to distinguish between the h^0 and h_{SM} at the 3σ level out to $m_{A^0} \gtrsim 600$ GeV. Additional advantages of a $\Gamma(h \to \mu^+\mu^-)$ measurement are: (i) there are no systematic uncertainties arising from uncertainty in the muon mass; (ii) the error on $\Gamma(h \to \mu^+\mu^-)$ increases only very slowly as the s-channel L_{tot} decreases, [6] in contrast to the errors for the previously discussed ratios of branching ratios from the μC s-channel data which scale as $1/\sqrt{L_{\text{tot}}}$. Finally, we note that Γ_h^{tot} alone cannot be used to distinguish between the MSSM h^0 and SM h_{SM} in a model-independent way. Not only is the error substantial ($\sim 12\%$ if we combine μC, $L = 0.4$ fb^{-1} s-channel data with ℓC, $L = 200$ fb^{-1} data) but also Γ_h^{tot} depends on many things, including (in the MSSM) the squark-mixing model. Still, deviations from SM predictions are generally substantial if $m_{A^0} \lesssim 500$ GeV implying that the measurement of Γ_h^{tot} could be very revealing.

We note that the above errors and results hold approximately for all $m_h \lesssim 150$ GeV so long as m_h is not too close to m_Z.

Precise measurements of the couplings of the SM-like Higgs boson could reveal many other types of new physics. For example, if a significant fraction of a fermion's mass is generated radiatively (as opposed to arising at tree-level), then the $hf\bar{f}$ coupling and associated partial width will deviate from SM expectations [5]. Deviations of order 5% to 10% (or more) in $\Gamma(h \to \mu^+\mu^-)$ are quite possible and, as discussed above, potentially detectable.

The MSSM H^0, A^0 and $H^\pm$

We begin by recalling [3] that the possibilities for H^0, A^0 discovery are limited at other machines. (i) Discovery of H^0, A^0 is not possible at the LHC for all $(m_{A^0}, \tan\beta)$: e.g. if $m_{\tilde{t}} = 1$ TeV, consistency with the observed value of $B(b \to s\gamma)$ requires $m_{A^0} > 350$ GeV, in which case the LHC might not be able to detect the H^0, A^0 at all, and certainly not for all $\tan\beta$ values. If $\tan\beta \lesssim 3$, detection might be possible in the $H^0, A^0 \to t\bar{t}$ final state, but would require $\lesssim 10\%$ systematic uncertainty in understanding the absolute normalization of the $t\bar{t}$ background. Otherwise, and certainly for $\tan\beta \gtrsim 3$, one must employ $b\bar{b}A^0, b\bar{b}H^0$ associated production, first analyzed in Refs. [6,7] and recently explored further in [8,9]. There is currently considerable debate as to what portion of $(m_{A^0}, \tan\beta)$ parameter space can be covered using the associated production modes. In the update of [7], it

---

[6] This is because the $\Gamma(h \to \mu^+\mu^-)$ error is dominated by the $\sqrt{s} = 500$ GeV measurement errors.

is claimed that $\tan\beta \gtrsim 5$ ($\gtrsim 15$) is required for $m_{A^0} \sim 200$ GeV (~ 500 GeV). Ref. [8] claims that still higher $\tan\beta$ values are required, $\tan\beta \gtrsim 20$ ($\tan\beta \gtrsim 30$), whereas Ref. [9] claims $\tan\beta \gtrsim 2$ ($\gtrsim 4$) will be adequate. (ii) At $\sqrt{s} = 500$ GeV, $e^+e^- \to H^0 A^0$ pair production probes only to $m_{A^0} \sim m_{H^0} \lesssim 230-240$ GeV. (iii) A $\gamma\gamma$ collider could potentially probe up to $m_{A^0} \sim m_{H^0} \sim 0.8\sqrt{s} \sim 400$ GeV, but only for $L_{\text{tot}} \gtrsim 150-200$ fb^{-1} [10].

Thus, it is noteworthy that $\mu^+\mu^- \to H^0, A^0$ in the s-channel potentially allows production and study of the H^0, A^0 up to $m_{A^0} \sim m_{H^0} \lesssim \sqrt{s}$. To assess the potential, let us (optimistically) assume that a total of $L_{\text{tot}} = 50$ fb^{-1} (5 yrs running at $<\mathcal{L}>= 1 \times 10^{33}$) can be accumulated for $\sqrt{s}$ in the $250-500$ GeV range. (We note that $\Gamma_{A^0}^{\text{tot}}$ and $\Gamma_{H^0}^{\text{tot}}$, although not big, are of a size such that resolution of $R \gtrsim 0.1\%$ will be adequate to maximize the s-channel cross section, thus allowing for substantial $\mathcal{L}$.)

There are then several possible scenarios. (a) If we have some preknowledge or restrictions on m_{A^0} from LHC discovery or from s-channel measurements of h^0 properties, then $\mu^+\mu^- \to H^0$ and $\mu^+\mu^- \to A^0$ can be studied with precision for all $\tan\beta \gtrsim 1-2$. (b) If we have no knowledge of m_{A^0} other than $m_{A^0} \gtrsim 250-300$ GeV from LHC, then we might wish to search for the A^0, H^0 in $\mu^+\mu^- \to H^0, A^0$ by scanning over $\sqrt{s} = 250-500$ GeV. If their masses lie in this mass range, then their discovery by scanning will be possible for most of (m_{A^0}, $\tan\beta$) parameter space such that they cannot be discovered at the LHC (in particular, if $m_{A^0} \gtrsim 250$ GeV and $\tan\beta \gtrsim 4-5$). (c) Alternatively, if the μC is simply run at $\sqrt{s} = 500$ GeV and $L_{\text{tot}} \sim 50$ fb^{-1} is accumulated, then H^0, A^0 in the $250-500$ GeV mass range can be discovered in the $\sqrt{s}$ bremsstrahlung tail if the $b\bar{b}$ mass resolution (either by direct reconstruction or hard photon recoil) is of order ± 5 GeV and if $\tan\beta \gtrsim 6-7$ (depending on m_{A^0}). Typical peaks are illustrated in Fig. 3. [7]

Finally, once the closely degenerate A^0, H^0 are discovered, it will be extremely interesting to be able to separate the resonance peaks. This will probably only be possible at a muon collider with small $R \lesssim 0.01\%$ if $\tan\beta$ is large, as illustrated in Fig. 4.

We note that the above results assume that SUSY decays of the H^0 and A^0 do not have a large net branching ratio for $m_{A^0}, m_{H^0} \lesssim 500$ GeV. If SUSY decays are significant, the possibilities and strategies for H^0, A^0 discovery at all machines would have to be re-evaluated.

---

[7] SUSY decays are assumed to be absent in this and the following figure.

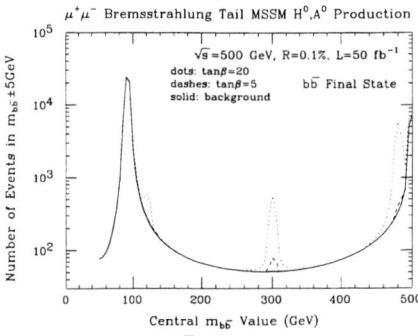
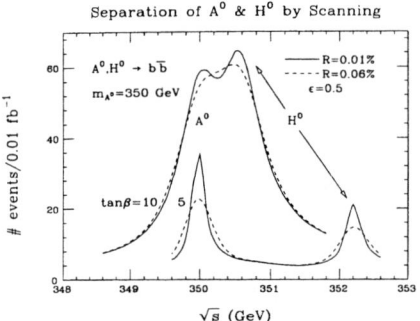

FIGURE 3. $N(b\bar{b})$ in the $m_{b\bar{b}} \pm 5$ GeV interval vs. $m_{b\bar{b}}$ for $\sqrt{s} = 500$ GeV, $L_{\text{tot}} = 50$ fb^{-1}, and $R = 0.1\%$: peaks are shown for $m_{A^0} = 120, 300$ or 480 GeV, with $\tan\beta = 5$ and 20 in each case.

FIGURE 4. $N(b\bar{b})$ (for 0.01 fb^{-1}) vs. $\sqrt{s}$, for $m_{A^0} = 350$ GeV H^0, A^0 resonance (with $\tan\beta = 5$ and 10), including the $b\bar{b}$ continuum background.

We end this sub-section with just a few remarks on the possibilities for production of $H^0 A^0$ and $H^+ H^-$ pairs at a high energy μC (or eC). Since $m_{A^0} \gtrsim 1$ TeV cannot be ruled out simply on the basis of hierarchy and naturalness (although fine-tuning is stretched), it is possible that energies of $\sqrt{s} > 2$ TeV could be required for pair production. If available, then it has been shown [11,12] that discovery of $H^0 A^0$ in their $b\bar{b}$ or $t\bar{t}$ decay modes and $H^+ H^-$ in their $t\bar{b}$ and $b\bar{t}$ decays will be easy for expected luminosities, even if SUSY decays are present. As a by-product, the masses will be measured with reasonable accuracy.

Regardless of whether we see the H^0, A^0 in s-channel production or via pair production, one can measure branching ratios to other channels, including supersymmetric pair decay channels with good accuracy. In fact, the ratios of branching ratios and the value of $m_{A^0} \sim m_{H^0} \sim m_{H^\pm}$ will be measured with sufficient accuracy that, in combination with one gaugino mass, say the chargino mass (which will also presumably be well-measured) it will be possible [11] to discriminate with incredible statistical significance between different closely similar GUT scenarios for the GUT-scale soft-supersymmetry-breaking masses. Thus, Higgs pair production could be very valuable in the ultimate goal of determining all the soft-SUSY-breaking parameters.

Finally, entirely unexpected decays of the heavy Higgs bosons of SUSY (or other extended Higgs sector) could be present. For example, non-negligible branching ratios for $H^0, A^0 \to t\bar{c} + c\bar{t}$ FCNC decays are not inconsistent with current theoretical model-building ideas and existing constraints [13]. The muon collider s-channel $\mu^+\mu^- \to H^0, A^0$ event rate is sufficient to probe rather small values for such FCNC branching ratios.

Verifying Higgs CP Properties

Once a neutral Higgs boson is discovered, determination of its CP nature will be of great interest. For example, direct verification that the SM Higgs is CP-even would be highly desirable. Indeed, if a neutral Higgs boson is found to have a mixed CP nature (implying CP violation in the Higgs sector), then neither the SM nor the MSSM can be correct. In the case of the SM, one must have a multi-doublet (or more complicated) Higgs sector. In the case of the MSSM, at least a singlet Higgs boson (as in the NMSSM) would be required to be present in addition to the standard two doublets.

One finds that the $\gamma\gamma$ and $\mu^+\mu^-$ single Higgs production modes provide the most elegant and reliable techniques for CP determination. In $\gamma\gamma$ collisions at the eC (a $\gamma\gamma$ collider is not possible at the μC), one establishes definite polarizations $\vec{e}_{1,2}$ for the two colliding photons in the photon-photon center of mass. Since $\mathcal{L}_{\gamma\gamma h} = \vec{e}_1 \cdot \vec{e}_2 \mathcal{E} + (\vec{e}_1 \times \vec{e}_2)_z \mathcal{O}$, where $\mathcal{E}$ and $\mathcal{O}$ are of similar size if the CP-even and CP-odd (respectively) components of the h are comparable. There are two important types of measurement. The first [14] is the difference in rates for photons colliding with $++$ vs. $--$ helicities, which is non-zero only if CP violation is present. Experimentally, this difference can be measured by simultaneously flipping the helicities of both of the initiating back-scattered laser beams. The second [14–16] is the dependence of the h production rate on the relative orientation of transverse polarizations $\vec{e}_1$ and $\vec{e}_2$ for the two colliding photons. In the case of a CP-conserving Higgs sector, the production rate is maximum for a CP-even (CP-odd) Higgs boson when $\vec{e}_1$ is parallel (perpendicular) to $\vec{e}_2$. The limited transverse polarization that can be achieved at a $\gamma\gamma$ collider implies that very high luminosity is needed for such a study.

In the end, a $\mu^+\mu^-$ collider might well prove to be the best machine for directly probing the CP properties of a Higgs boson that can be produced and detected in the s-channel mode [17,18]. Consider transversely polarized muon beams. For 100% transverse polarization and an angle ϕ between the μ^+ transverse polarization and the μ^- transverse polarization, one finds

$$\sigma(\phi) \propto 1 - \frac{a^2 - b^2}{a^2 + b^2} \cos\phi + \frac{2ab}{a^2 + b^2} \sin\phi, \tag{3}$$

where the coupling of the h to muons is given by $h\bar{\mu}(a + ib\gamma_5)\mu$, a and b being the CP-even and CP-odd couplings, respectively. If the h is a CP mixture, both a and b are non-zero and the asymmetry

$$A_1 \equiv \frac{\sigma(\pi/2) - \sigma(-\pi/2)}{\sigma(\pi/2) + \sigma(-\pi/2)} = \frac{2ab}{a^2 + b^2} \quad (4)$$

will be large. For a pure CP eigenstate the cross section difference

$$A_2 \equiv \frac{\sigma(\pi) - \sigma(0)}{\sigma(\pi) + \sigma(0)} = \frac{a^2 - b^2}{a^2 + b^2} \quad (5)$$

is $+1$ or -1 for a CP-even or CP-odd h, respectively. Since background processes and partial transverse polarization will dilute the statistics, further study will be needed to fully assess the statistical level of CP determination that can be achieved in various cases.

Exotic Higgs Bosons

If there are doubly-charged Higgs bosons, $e^-e^- \to \Delta^{--}$ probes λ_{ee} and $\mu^-\mu^- \to \Delta^{--}$ probes $\lambda_{\mu\mu}$, where the λ's are the strengths of the Majorana-like couplings [19-21]. Current $\lambda_{ee,\mu\mu}$ limits are such that factory-like production of a Δ^{--} is possible if $\Gamma^{tot}_{\Delta^{--}}$ is small. Further, a Δ^{--} with $m_{\Delta^{--}} \lesssim 500 - 1000$ GeV will be seen previously at the LHC (for $m_{\Delta^{--}} \lesssim 200 - 250$ GeV at TeV33) [22]. For small $\lambda_{ee,\mu\mu,\tau\tau}$ in the range that would be appropriate, for example, for the Δ^{--} in the left-right symmetric model see-saw neutrino mass generation context, it may be that $\Gamma^{tot}_{\Delta^{--}} \ll \sigma_{\sqrt{s}}$,[8] leading to $\bar{\sigma}_{\ell^-\ell^- \to \Delta^{--}} \propto \lambda_{\ell\ell}^2/\sigma_{\sqrt{s}}$. Note that the absolute rate for $\ell^-\ell^- \to \Delta^{--}$ yields a direct determination of $\lambda_{\ell\ell}^2$, which, for a Δ^{--} with very small $\Gamma^{tot}_{\Delta^{--}}$, will be impossible to determine by any other means. The relative branching ratios for $\Delta^{--} \to e^-e^-, \mu^-\mu^-, \tau^-\tau^-$ will then yield values for the remaining $\lambda_{\ell\ell}^2$'s. Because of the very small $R = 0.003\% - 0.01\%$ achievable at a muon collider, $\mu^-\mu^-$ collisions will probe weaker $\lambda_{\mu\mu}$ coupling than the λ_{ee} coupling that can be probed in e^-e^- collisions. In addition, it is natural to anticipate that $\lambda_{\mu\mu}^2 \gg \lambda_{ee}^2$. A more complete review of this topic is given in Ref. [23].

PROBES OF NARROW TECHNICOLOR RESONANCES

In this section, I briefly summarize the ability of a low-energy muon collider to observe the pseudo-Nambu-Goldstone bosons (PNGB's) of an extended technicolor

---

[8] For small $\lambda_{ee,\mu\mu,\tau\tau}$, $\Gamma^{tot}_{\Delta^{--}}$ is very small if the $\Delta^{--} \to W^-W^-$ coupling strength is very small or zero, as required to avoid naturalness problems for $\rho = m_W^2/[\cos^2\theta_w m_Z]^2$.

theory. These narrow states need not have appeared at an observable level in Z decays at LEP. Some of the PNGB's have substantial $\mu^+\mu^-$ couplings. Thus, a muon collider search for them will bear a close resemblance to the light Higgs case discussed already. The main difference is that, assuming they have not been detected ahead of time, we must search over the full expected mass range.

The first results for PNGB's at a muon collider appear in Refs. [24] and [25]. Here I summarize the results for the lightest P^0 PNGB as given in Ref. [24]. Although the specific P^0 properties employed are those predicted by the extended BESS model [24], they will be representative of what would be found in any extended technicolor model for a strongly interacting electroweak sector. The first point is that m_{P^0} is expected to be small; $m_{P^0} \lesssim 80$ GeV is preferred in the BESS model. Second, the Yukawa couplings and branching ratios of the P^0 are easily determined. In the BESS model, $\mathcal{L}_Y = -i \sum_f \lambda_f \bar{f} \gamma_5 f P^0$ with $\lambda_b = \sqrt{\frac{2}{3}} \frac{m_b}{v}$, $\lambda_\tau = -\sqrt{6} \frac{m_\tau}{v}$, $\lambda_\mu = -\sqrt{6} \frac{m_\mu}{v}$. Note the sizeable $\mu^+\mu^-$ coupling. The P^0 couplings to $\gamma\gamma$ and gg from the ABJ anomaly are also important. Overall, these couplings are not unlike those of a light Higgs boson. Not surprisingly, therefore, $\Gamma^{\rm tot}_{P^0}$ is very tiny: $\Gamma^{\rm tot}_{P^0} = 0.2, 4, 10$ MeV for $m_{P^0} = 10, 80, 150$ GeV, respectively, for $N_{TC} = 4$ technicolor flavors. For such narrow widths, it will be best to use $R = 0.003\%$ beam energy resolution.

For the detailed tagging efficiencies *etc.* described in [24], the $L_{\rm tot}$ required to achieve $\sum_k S_k / \sqrt{\sum_k B_k} = 5$ at $\sqrt{s} = m_{P^0}$, after summing over the optimal selection

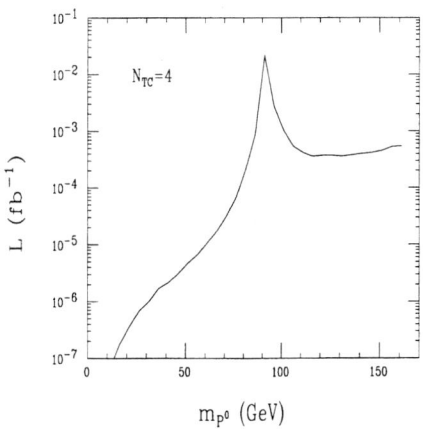
FIGURE 5. $L_{\rm tot}$ required for a 5σ P^0 signal at $\sqrt{s} = m_{P^0}$.

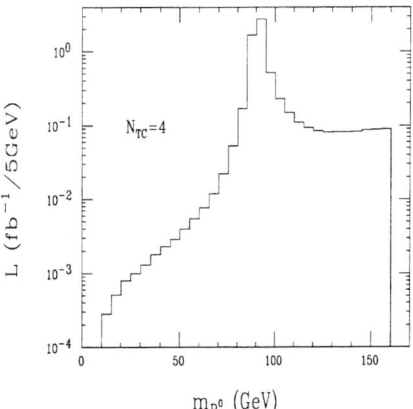
FIGURE 6. $L_{\rm tot}$ required to scan indicated 5 GeV intervals and either discover or eliminate the P^0 at the 3σ level.

of the $k = b\bar{b}$, $\tau^+\tau^-$, $c\bar{c}$, and gg channels (as defined after tagging), is plotted in Fig. 5. Very modest L_{tot} is needed unless $m_{P^0} \sim m_Z$. Of course, if we do not have any information regarding the P^0 mass, we must scan for the resonance. The (very conservative, see [24] for details) estimate for the luminosity required for scanning a given 5 GeV interval and either discovering or eliminating the P^0 in that interval at the 3σ level is plotted in Fig. 6. If the P^0 is as light as expected in the extended BESS model, then the prospects for discovery by scanning would be excellent. For example, a P^0 lying in the ~ 10 GeV to ~ 75 GeV mass interval can be either discovered or eliminated at the 3σ level with just 0.11 fb^{-1} of total luminosity, distributed in proportion to the luminosities plotted in Fig. 6. The $\mathcal{L}$ that could be achieved at these low masses is being studied [26]. A P^0 with $m_{P^0} \sim m_Z$ would be much more difficult to discover unless its mass was approximately known. A 3σ scan of the mass interval from ~ 105 GeV to 160 GeV would require about 1 fb^{-1} of integrated luminosity, which is more than could be comfortably achieved for the conservative $R = 0.003\%$ $\mathcal{L}$ values assumed at the Fermilab-97 workshop.

DISCUSSION AND CONCLUSIONS

There is little doubt that a variety of accelerators will be needed to explore all aspects of the physics that lies beyond the Standard Model and accumulate adequate luminosity for this purpose in a timely fashion. For any conceivable new-physics scenario, a muon collider would be a very valuable machine, both for discovery and detailed studies. Here we have reviewed the tremendous value of a muon collider for studying any narrow resonance with $\mu^+\mu^-$ (or $\mu^-\mu^-$) couplings, focusing on neutral light Higgs bosons and the Higgs-like pseudo-Nambu-Goldstone bosons that would be present in almost any technicolor model. A muon collider could well provide the highest statistics determinations of many important Higgs or PNGB fundamental couplings. In particular, it might provide the only direct measurement of the very important $\mu^+\mu^-$ coupling. Measurement of this coupling will very possibly allow discrimination between a SM Higgs boson and its light h^0 SUSY counterpart. Comparison of the $\mu^+\mu^-$ coupling to the $\tau^+\tau^-$ coupling (one may be able to approximately determine the latter from branching ratios) will also be of extreme interest. For Higgs physics, developing machine designs that yield the highest possible luminosity at low energies, while maintaining excellent beam energy resolution, should be a priority.

REFERENCES

1. V. Barger, M. Berger, J. Gunion and T. Han, Phys. Rev. Lett. **75**, 1462 (1995); *Phys. Rep.* **286**, (1997) 1.
2. R. Raja, to appear in Proceedings of *Workshop on Physics at the First Muon Collider and at the Front End of a Muon Collider*, Fermilab, Chicago, November 6–9, 1997, editors S. Geer and R. Raja, AIP Press, hereafter denoted FNAL$\mu\mu$97.
3. J.F. Gunion, L. Poggioli, R. Van Kooten, C. Kao and P. Rowson, in *New Directions for High-Energy Physics*, Proceedings of the 1996 DPF/DPB Summer Study on High-Energy Physics, June 25—July 12, 1996, Snowmass, CO, edited by D.G. Cassel, L.T. Gennari, and R.H. Siemann (Stanford Linear Accelerator Center, 1997) pp. 541–587.
4. B. King, to appear in FNAL$\mu\mu$97.
5. F. Borzumati, G. Farrar, N. Polonsky and S. Thomas, hep-ph/9712428, to appear in FNAL$\mu\mu$97.
6. J. Dai, J.F. Gunion and R. Vega, Phys. Rev. Lett. **71**, 2699 (1993); Phys. Lett. **B315**, 355 (1993); Phys. Lett. **B345**, 29 (1995); Phys. Lett. **B371**, 71 (1996).
7. J. Dai, J.F. Gunion and R. Vega, Phys. Lett. **B387**, 801 (1996).
8. E. Richter-Was and D. Froidevaux, Z. Phys. **C76**, 665 (1997).
9. J. Lorenzo Diaz-Cruz, H.-J. He, T. Tait and C.-P. Yuan, hep-ph/9802294.
10. J.F. Gunion and H.E. Haber, Phys. Rev. **D48**, 5109 (1993); and in *Research Directions for the Decade*, Proceedings of the 1990 DPF Summer Study on High Energy Physics, Snowmass, CO, 25 June–13 July 1990, edited by E. Berger (World Scientific, Singapore, 1992) p. 469–472.
11. J.F. Gunion and J. Kelly, Phys. Rev. **D46**, 1730 (1997.)
12. J. Feng and T. Moroi, Phys. Rev. **D56**, 5962 (1997).
13. L. Reina, hep-ph/9712426, to appear in FNAL$\mu\mu$97.
14. J.F. Gunion and B. Grzadkowski, Phys. Lett. **B294**, 361 (1992).
15. J.F. Gunion and J. Kelly, Phys. Lett. **B333**, 110 (1994).
16. M. Kramer, J. Kuhn, M. Stong, and P. Zerwas, Z. Phys. **C64**, 21 (1994).
17. D. Atwood and A. Soni, Phys. Rev. **D52**, 6271 (1995).
18. J.F. Gunion, A. Stange, and S. Willenbrock, *Weakly-Coupled Higgs Bosons*, in *Elec-*

troweak Physics and Beyond the Standard Model (World Scientific 1996), eds. T. Barklow, S. Dawson, H. Haber, and J. Siegrist, pp. 23–145.

19. J.F. Gunion, Int. J. Mod. Phys. **A11**, 1551 (1996).
20. P. Frampton, Int. J. Mod. Phys. **A11**, 1621 (1996).
21. F. Cuypers, Nucl. Phys. **B510**, 3 (1997).
22. J.F. Gunion, C. Loomis and K. Pitts, in *New Directions for High-Energy Physics*, Proceedings of the 1996 DPF/DPB Summer Study on High Energy Physics, Snowmass '96, edited by D.G. Cassel, L.T. Gennari and R.H. Siemann (Stanford Linear Accelerator Center, Stanford, CA, 1997), p. 603.
23. J.F. Gunion, hep-ph/9803222, to appear in e^-e^- *1997: Proceedings of the Electron-Electron Linear Collider Workshop*, Santa Cruz, California, September, 1997, edited by C. Heusch, to be published in Int. J. Mod. Phys. **A**.
24. R. Casalbuoni, S. De Curtis, D. Dominici, A. Deandrea, R. Gatto and J. F. Gunion, hep-ph/9801243, to appear in FNAL$\mu\mu$97.
25. K. Lane, hep-ph/9801385, to appear in FNAL$\mu\mu$97.
26. R. Palmer, private communication.

The Top Quark and Higgs Boson at Hadron Colliders

Chris Quigg

Fermi National Accelerator Laboratory[1]
P.O. Box 500, Batavia, Illinois 60510 USA

Abstract. To provide context for discussions of experiments at future muon colliders, I survey what is known and what will be known about the top quark and the Higgs boson from experiments at hadron colliders.

INTRODUCTION

When we discuss whether there should be muon colliders in our future, we must answer a number of important questions.
What machines are possible? When? At what cost?
What are the physics opportunities?
Can we do physics in the environment? (What does it take?)
How will these experiments add to existing knowledge *when they are done?*
The aim of this talk is to provide a survey of what we might expect to know about the top quark and the Higgs boson before a $\mu^+\mu^-$ collider operates [1].

THE HADRON COLLIDERS

Let us take a moment to recall the characteristics of the hadron colliders that will contribute to the study of the top quark and Higgs boson. The combination of the Fermilab Tevatron and the new Main Injector with the CDF and DØ detectors will in the future bring us $\bar{p}p$ collisions at 2 TeV. In Fermilab parlance, the data now under analysis come from Run I: 100 pb^{-1} at 1.8 TeV, recorded in 1994–1996. We look forward to the first 2-TeV data. The approved quantum of data is Run II: 2 fb^{-1} in 2000–2002. Beyond the approved running, we are enthusiastic about the physics prospects for another high-luminosity run while the Tevatron defines the energy frontier. Although

---

[1] Fermilab is operated by Universities Research Association Inc. under Contract No. DE-AC02-76CH03000 with the United States Department of Energy.

the laboratory hasn't taken a position, we refer to this possibility as Run III: 30 fb^{-1} by the year 2006.

On that time scale, the Large Hadron Collider at CERN will open the study of pp collisions at 14 TeV in the ATLAS and CMS detectors. A modest goal for the beginning of the LHC era is to accumulate $\int \mathcal{L}dt = 100$ fb^{-1} in 2005–2009.

Three elements inform the way we think about experiments in these high-energy hadron colliders. First, they promise high sensitivity from high integrated luminosity. Second, the success of b-tagging in the hadron-collider environment encourages the hope that heavy-flavor tags, and perhaps even triggers, can make future experiments more sensitive to the exotic events that may signal new physics. I have in mind here both the CDF Silicon Microvertex Detector (SVX), with resolution $\sim 11\mu$m, and the "soft"-lepton tag used by CDF and DØ to identify the transition $b \to c\ell\nu$. Third, both the physics and the experimental approach to the new energy regime are colored by the great mass of the top quark.

THE TOP QUARK

The top quark has been observed at the Tevatron in the reaction [2]

$$\bar{p}p \to t\,\bar{t} + \cdots$$
$$\hookrightarrow W^-\bar{b}$$
$$\hookrightarrow W^+b$$

In the Tevatron experiments, the b-quarks are identified as displaced vertices or through soft-lepton tags. The channels studied to date are dileptons (including $\tau + (e,\mu)$), lepton + jets, and all jets.

Top Mass

The top mass has already been determined to impressive precision. An "unofficial" average including the latest data from CDF and DØ is [3]

$$m_t = 174.3 \pm 5.3 \text{ GeV}/c^2 \,.$$

Within the electroweak theory, fermion masses are set by the scale of electroweak symmetry breaking v and by apparently arbitrary Yukawa couplings,

$$m_f = \frac{\zeta_f v}{\sqrt{2}} \approx (176 \text{ GeV}/c^2) \cdot \zeta_f \,.$$

It is striking that the top quark's Yukawa coupling $\zeta_t \approx 1$. Does this mean that top is special, or might top be the only "normal" fermion, with a mass close to the electroweak scale?

Top Lifetime

The top-quark lifetime is governed by the semiweak decay $t \to bW^+$; the decay width is given by [4]

$$\Gamma(t \to bW^+) = \frac{G_F m_t^3}{8\pi\sqrt{2}} |V_{tb}|^2 \left(1 - \frac{M_W^2}{m_t^2}\right)^2 \left(1 + \frac{2M_W^2}{m_t^2}\right).$$

If there are three generations of quarks, so that we can use 3×3 unitarity to determine $|V_{tb}| = 0.9991 \pm 0.0002 \approx 1$, then $\Gamma(t \to bW^+) \approx 1.55$ GeV. This corresponds to a top lifetime,

$$\tau_t \approx 0.4 \times 10^{-24} \text{ s},$$

that is very short compared with the time-scale for confinement,

$$1/\Lambda_{\text{QCD}} \approx \text{few} \times 10^{-24} \text{ s}.$$

As a consequence, the top quark decays before it can be hadronized. No discrete lines will be observed in the $t\bar{t}$ spectrum, and there will be no dressed hadronic states containing top. This freedom from the confining effects of the strong interaction means that the characteristics of top production and the hadronic environment near top in phase space should be calculable in perturbative QCD. The fact that top is, in this sense, the purest, freest quark we have to study will have important consequences for future experiments.

Top Production

It is useful to summarize some important characteristics of top pair production. At the Tevatron, at 1.8 TeV, the top-pair production cross section is $\sigma \approx 6$ pb [5]. Approximately 90% arises from the reaction $q\bar{q} \to t\bar{t}$, and only about 10% from the reaction $gg \to t\bar{t}$. Top is a heavy particle for the Tevatron, and this is reflected in the dominance of $q\bar{q}$ collisions. The measured cross sections are in reasonable accord with this estimate. CDF measures $7.6^{+1.8}_{-1.5}$ pb [6], while DØ has determined 5.5 ± 1.8 pb [7].

At the LHC, the pair-production cross section rises to $\sigma \approx 800$ pb. The origin of the top events is markedly different. In 14-TeV pp collisions, the reaction $q\bar{q} \to t\bar{t}$ accounts for only about 10% of the rate, whereas $gg \to t\bar{t}$ accounts for 90%. At the LHC, top will be a moderately light particle.

Future Top Yields

For Run II, the Tevatron energy will increase to 2 TeV. Accordingly, the top-pair production cross section will rise by about 40%. In a run of 30 fb^{-1}

TABLE 1. Anticipated top-quark yields in future Tevatron runs

Mode	2 fb^{-1}	30 fb^{-1}	S/B
Dilepton	80	1200	5:1
ℓ + 3jets/1b	1300	20000	3:1
ℓ + 4jets/2b	600	9000	12:1
Single top (all)	170	2500	1:2.2
Single top ($W^\star$)	20	300	1:1.3

at 2 TeV, approximately 225K $t\bar{t}$ pairs will be produced. I show in Table 1 a *Snowmass '96* projection of the number of top events available for study in the Tevatron's Run II and Run III [8]. The LHC is a veritable fountain of tops: it will produce 8×10^6 $t\bar{t}$ pairs in a modest-luminosity exposure of only 10 fb^{-1}.

It seems reasonable to expect that experiments at the Tevatron and LHC will determine the top-quark mass within $\delta m_t = (1\text{-}2)$ GeV/c^2.

Measuring $|V_{tb}|$

By studying the number of top events in which they register 0, 1, or 2 b tags, CDF measures [9] the fraction of top decays that lead to b quarks in the final state as

$$B_b \equiv \frac{\Gamma(t \to bW)}{\Gamma(t \to qW)} = \frac{|V_{tb}|^2}{|V_{td}|^2 + |V_{ts}|^2 + |V_{tb}|^2} = 0.99 \pm 0.29 \;.$$

If there are three generations, so that $|V_{td}|^2 + |V_{ts}|^2 + |V_{tb}|^2 = 1$, this measurement leads to a lower bound on the strength of the $t\bar{b}W$ coupling,

$$|V_{tb}| > 0.76 \; (95\% \text{ CL}).$$

Without the three-generation unitarity constraint, we learn only that

$$|V_{tb}| \gg |V_{td}|, |V_{ts}|.$$

Increased sensitivity in the forthcoming runs should lead to significant improvements in δB_b. For Run II, we anticipate ±10%, and for Run III, ± a few percent. At the LHC, it should be possible to reduce the uncertainty to about ±1%.

Direct measurement of the coupling $|V_{tb}|$ will become possible in single-top production through the reactions $\bar{q}q \to W^\star \to t\bar{b}$ and $gW \to t\bar{b}$ [10]. The cross sections for both reactions are $\propto |V_{tb}|^2$. We can expect to measure the coupling with an uncertainty $\delta|V_{tb}| = \pm(10\%, 5\%)$ in Run II and III, using both the virtual-W^* channel and gW fusion. I am not aware of any detailed studies for the LHC environment, but the fact that the gW fusion cross section is a hundred times larger than at the Tevatron means that there will be a very large sample of single-top events.

Searches for new physics

Top decay is an excellent source of longitudinally polarized gauge bosons. In the decay of a massive top, W-bosons with |helicity| = 1 occur with weight = 1, while longitudinally polarized W-bosons with helicity = 0 occur with weight = m_t^2/M_W^2. If the decays of top proceed by the standard $V-A$ interaction, we therefore expect that the longitudinal fraction $f_0 = (m_t^2/M_W^2)/(1 + m_t^2/M_W^2) \approx 70\%$. The polarization of the W-boson is reflected in the decay angular distribution of leptons from its subsequent decay:

$$\frac{d\Gamma(W^+ \to \ell^+\nu_\ell)}{d(\cos\theta)} = \tfrac{3}{8}(1-f_0)(1-\cos\theta)^2 + \tfrac{3}{4}f_0\sin^2\theta \ .$$

In experiments at the Tevatron, it should be possible to determine the longitudinal fraction to $\delta f_0 = \pm 3\%$ in Run II. The LHC experiments will improve the measurement to $\pm 1\%$. Departures from the canonical expectation would give a hint of unexpected structure at the tbW vertex.

The flavor-changing–neutral-current decays

$$t \to \begin{pmatrix} g \\ Z \\ \gamma \end{pmatrix} + \begin{pmatrix} c \\ u \end{pmatrix}$$

are unobservably small ($\ll 10^{-10}$) in the standard model [11], but the present indirect constraints on the $Zt\bar{c}$ couplings would permit branching fractions as large a a few percent [12]. The ultimate sensitivity at the Tevatron might reach about 1% for these decays, while the LHC experiments could reach a level of $\sim 10^{-4}$.

It is possible that the rare decay $t \to bWZ$, with a branching fraction $\sim 10^{-6}$ in the standard model, might be detectable at the LHC.

Because top is so massive, top decays may surprise by providing a conduit to final states that would otherwise be reached with difficulty. One of the favorite targets is the search for a charged scalar or pseudoscalar P^+ in the semiweak decay $t \to bP^+$. Such charged scalars may occur in multi-Higgs models, supersymmetry, and technicolor. Both CDF and DØ have reported searches [13].

Resonances in $t\bar{t}$ Production?

We have noted that the top quark decays before it can be incorporated into a color-singlet hadron. That fact does not exclude the possibility that some new object might include tops among its decay products. Because objects associated with the breaking of electroweak symmetry tend to couple to fermion mass, the discovery of top opens a new window on electroweak symmetry breaking. Indeed, top-condensate models and multiscale technicolor both imply the existence of color-octet resonances with masses of several hundred GeV/c^2 that decay into $t\bar{t}$. In technicolor models [14], the prime candidate is a colored pseudoscalar produced in the elementary reactions

$$gg \to \eta_T \to (t\bar{t}, gg).$$

Topcolor models [15] typically include a colored vector state that would appear in the reactions

$$q\bar{q} \to V_8 \to (t\bar{t}, b\bar{b}).$$

The first hint for such objects would come from the observation of structure in the $t\bar{t}$ invariant mass spectrum. A first look from CDF, based on a small sample, resembles the conventional spectrum.

Top-Quark Measurements: Summary

Until the LHC operates, top-quark measurements will only be possible at the Tevatron. The LHC will, in time, be a prodigious source of tops. We expect that the top-quark mass will be determined within $\delta m_t \approx$ 1-2 GeV/c^2 at both the Tevatron and the LHC. The production cross section should be measured to $\pm 5\%$ at the Tevatron, and to $\pm$ a few % at the LHC. The branching fraction $\delta\Gamma(t \to bW)/\Gamma(t \to qW)$ will improve to $\pm 10\%$ in Run II, $\pm$ a few percent in Run III, and $\pm 1\%$ at the LHC. Studies of single-top production should yield $\delta|V_{tb}| \approx \pm 10\%$ in Run II, and $\pm 5\%$ in Run III at the Tevatron. In the current Tevatron experiments, searches are under way for $t\bar{t}$ resonances, rare decays, and other signs of new physics.

THE HIGGS BOSON

The central challenge in particle physics is to explore the 1-TeV scale and elucidate the nature of electroweak symmetry breaking. A key element in this quest is the search for the Higgs boson, the agent of electroweak symmetry breaking in the standard electroweak theory. The unique opportunity offered by a muon collider to construct a "Higgs factory" using the formation reaction $\mu^+\mu^- \to H$ calls attention to a not-too-heavy Higgs boson, as favored in supersymmetric models. In such models, it is plausible that the mass of the lightest Higgs boson—which has much in common with the standard-model Higgs boson—is no more than ~ 130 GeV/c^2. It is important to bear in mind that a heavy Higgs boson remains a logical possibility, as we shall see momentarily. I will abbreviate to the search for the standard-model Higgs boson in what follows.

Constraints on the Higgs Mass

One of the shortcomings of the electroweak theory is that it fails to make a prediction for the mass of the Higgs boson. Perhaps the most general statement that can be made is the upper bound derived [16] from the requirement of perturbative unitarity,

$$M_H \lesssim \left(\frac{8\pi\sqrt{2}}{3G_F}\right)^{1/2} \approx 1 \text{ TeV}/c^2 \ .$$

This condition is the most straightforward way to expose the importance of the 1-TeV scale.

We can obtain sharper constraints, in the form of upper and lower bounds, at the price of assuming that no new physics intervenes up to a cutoff scale Λ. The so-called "triviality" bound says that, for a given value of M_H, the electroweak theory makes sense up to a scale [17]

$$\Lambda < M_H \exp\left(\frac{4\pi^2 v^2}{3M_H^2}\right) \ .$$

Read the other way, if we regard the electroweak theory as an effective theory, apt up to some scale Λ, the triviality bound gives an upper limit on M_H. If, for example, we demand that the electroweak theory apply up to the Planck scale, the Higgs-boson mass must not exceed 175 GeV/c^2.

The requirement that the electroweak vacuum correspond to an absolute minimum of the Higgs potential in the face of quantum corrections leads to a lower bound,

$$M_H^2 > \frac{3G_F\sqrt{2}}{16\pi^2}(2M_W^4 + M_Z^4 - 4m_t^4) \cdots ,$$

that also depends on the scale of new physics [18]. If we exclude any new physics up to the Planck scale, then $M_H \gtrsim 130$ GeV/c^2.

These are informative constraints—given the assumptions that lead to them—but they do not really narrow the search. Crucial guidance comes from the direct searches for the standard-model Higgs boson, specifically from the study of the reaction $e^+e^- \to HZ$ at $161+172+183$ GeV in experiments at LEP2. The four LEP experiments examine the $qqbb$, $\nu\nu qq$, $\tau\tau qq$, and $(ee+\mu\mu)qq$ channels. Recent running at $\sqrt{s}=183$ GeV is sensitive to Higgs-boson masses up to about 82 GeV/c^2. Next year's running at 192 GeV should allow a search up to $M_H \approx 96$ GeV/c^2.

Clues about M_H

Precision electroweak measurements are sensitive to the Higgs-boson mass through radiative corrections. The constraints that arise on M_H depend on the selection and weighting of the data set and on assumptions made about the light-quark contribution to the vacuum polarization for $\alpha(M_Z)$. I quote three recent analyses by Erler and Langacker [19] to illustrate the range of possibilities.

Including all the precision electroweak data at face value and using a selection of measured cross sections for $e^+e^- \to$ light hadrons to determine $\alpha(M_Z)$, their best fit for the Higgs-boson mass is $M_H = 69^{+85}_{-43}$ GeV/c^2, which corresponds to the bounds

$$M_H < \begin{Bmatrix} 236 \\ 287 \\ 413 \end{Bmatrix} \text{ GeV}/c^2 \text{ at } \begin{Bmatrix} 90\% \\ 95\% \\ 99\% \end{Bmatrix} \text{ CL.}$$

The central value lies in the range already excluded by direct searches for the standard-model Higgs boson. Using instead perturbative QCD to compute $\delta\alpha^{(5)}_{\text{had}}$, they find a best fit of $M_H = 97^{+79}_{-48}$ GeV/c^2, which implies the bounds

$$M_H < \begin{Bmatrix} 229 \\ 273 \\ 377 \end{Bmatrix} \text{ GeV}/c^2 \text{ at } \begin{Bmatrix} 90\% \\ 95\% \\ 99\% \end{Bmatrix} \text{ CL.}$$

In spite of the shift of the central value, the upper bounds are relatively stable against the change in $\alpha(M_Z)$.

However, we may notice that the implications of individual precision measurements are not entirely consistent. For example, SLD's measurement of A_{LR} favors very low—unphysically low—values of M_H. Having no basis to exclude any measurements, one can follow the Particle Data Group prescription and rescale the weights of all the inconsistent measurements. Using measured cross sections for $e^+e^- \to$ light hadrons to determine $\alpha(M_Z)$, Erler and Langacker then find $M_H = 122^{+134}_{-77}$ GeV/c^2, which leads to the noticeably different bounds

$$M_H < \begin{Bmatrix} 329 \\ 408 \\ 613 \end{Bmatrix} \text{ GeV}/c^2 \text{ at } \begin{Bmatrix} 90\% \\ 95\% \\ 99\% \end{Bmatrix} \text{ CL.}$$

I have reviewed this work at some length to show the fragility of our current estimates of the Higgs-boson mass. I will nevertheless focus on the case of a light Higgs boson, because only a light Higgs boson will be accessible at the Tevatron.

The branching fractions of a light Higgs boson are shown in Figure 1. The most promising channel for searches at the Tevatron will be the $b\bar{b}$ mode, for which the branching fraction exceeds about 50% throughout the region preferred by supersymmetry and the precision electroweak data.

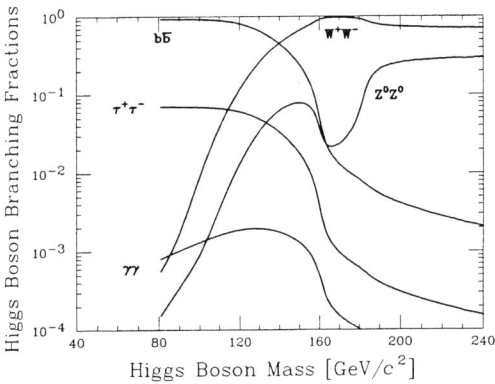

FIGURE 1. Branching fractions of a light Higgs boson.

Tevatron Search Strategies

At the Tevatron, the direct production of a light Higgs boson in gluon-gluon fusion $gg \to H \to b\bar{b}$ is swamped by the ordinary QCD production of $b\bar{b}$ pairs. Even with an integrated luminosity of 30 fb^{-1}, the experiments anticipate only $< 1\text{-}\sigma$ excess, with plausible invariant-mass resolution. It will be possible to calibrate the $b\bar{b}$ mass resolution over the region of the Higgs search in Run II: the electroweak production of $Z^0 \to b\bar{b}$ should stand well above background and be observable in Run II.

The high background in the $b\bar{b}$ channel means that special topologies must be employed to improve the ratio of signal to background and the significance of an observation. The high luminosities that can be contemplated for a future run argue that the associated-production reactions

$$\bar{p}p \to HW + \text{anything}$$
$$\phantom{\bar{p}p \to HW}\hookrightarrow \ell\nu$$
$$\phantom{\bar{p}p \to H}\hookrightarrow b\bar{b}$$

and

$$\bar{p}p \to HZ + \text{anything}$$
$$\phantom{\bar{p}p \to HZ}\hookrightarrow \ell^+\ell^- + \nu\bar{\nu}$$
$$\phantom{\bar{p}p \to H}\hookrightarrow b\bar{b}$$

are plausible candidates for a Higgs discovery at the Tevatron [20]. The Feynman diagrams for these processes are shown in Figure 2.

The prospects for exploiting these topologies were explored in detail in connection with the TeV2000 and TeV33 study groups at Fermilab [21]. Taking into account what is known, and what might conservatively be expected,

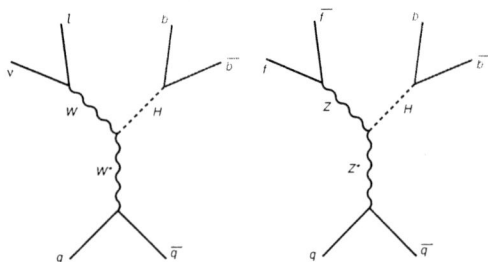

FIGURE 2. Feynman diagrams for the associated production of a Higgs boson and an electroweak gauge boson.

about sensitivity, mass resolutions, and background rejection, these investigations show that it is unlikely that a standard-model Higgs boson could be observed in Tevatron Run II. (Note, however, that the ability to use $W \to q\bar{q}$ decays would markedly increase the sensitivity.) The expected number of signal and background events in Run II are collected in Table 2. The prospects are much brighter for Run III. Indeed, the sensitivity to a light Higgs boson is what motivates the integrated luminosity of 30 fb^{-1} specified for Run III. The number of events projected for Run III, collected in Table 3, show that a Tevatron experiment could explore the range of Higgs-boson masses up to about 125 GeV/c^2, covering the entire range favored by light-scale supersymmetry.

We can make this result a little more transparent by plotting, in Figure 3, the luminosity needed for a three- or five-standard-deviation observation of the Higgs boson at the Tevatron. We see that, in the WH modes discussed, an integrated luminosity of 2 fb^{-1} is insufficient to detect the standard-model Higgs boson at an interesting mass. About 10 fb^{-1} would permit the observation of a Higgs boson discovered at LEP2, while 30 fb^{-1} would make it possible to explore masses up to about 125 GeV/c^2. With about 10 fb^{-1}, one could expect a 3-σ indication for the Higgs boson throughout the low-mass

TABLE 2. Number of signal and background events in Run II (2 fb^{-1}) for WH and ZH processes, and signal significance [22].

M_H[GeV/c^2]	60	80	90	100	110	120
WH signal S	45	28		15		8
Background B	139	84		53		30
$S/\sqrt{B}$	3.8	3.1		2.1		1.4
ZH signal S			7	6	5	3
Background B			36	33	31	25
$S/\sqrt{B}$			1.2	1.1	1.0	0.7

TABLE 3. Number of signal and background events in Run III (30 fb^{-1}) for WH and ZH processes, and signal significance [22].

M_H[GeV/c^2]	60	80	90	100	110	120
WH signal S	681	420		228		117
Background B	2085	1260		789		456
S/B	0.33	0.33		0.29		0.26
$S/\sqrt{B}$	14.9	11.8		8.1		5.5
ZH signal S			108	92	82	51
Background B			533	495	462	378
S/B			0.20	0.19	0.18	0.13
$S/\sqrt{B}$			4.7	4.1	3.8	2.6

régime.

A slightly different cut on the same information is provided in Figure 4. There I show the significance of observations in the WH and ZH channels for runs of 2 and 30 fb^{-1}. While the ZH channel probably would not suffice for an independent discovery, it could provide good supporting evidence—and complementary measurements—to an observation in WH.

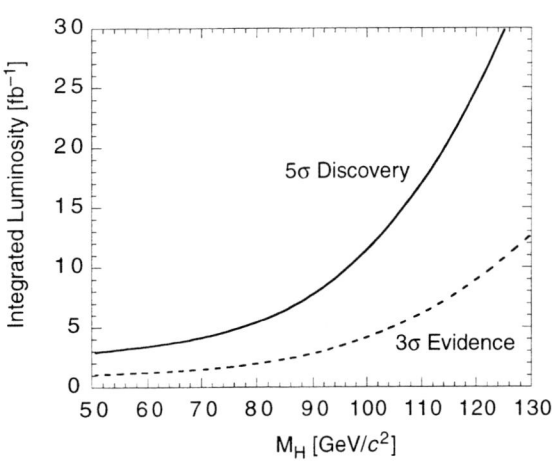

FIGURE 3. Luminosity required for the observation of a Higgs boson in WH associated production at the Tevatron.

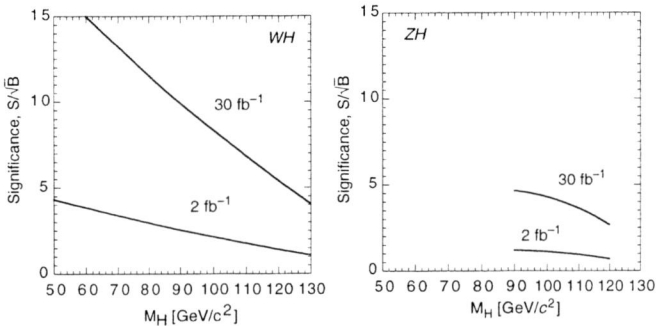

FIGURE 4. Significance of Higgs observation in Tevatron Run II & Run III.

Higgs at the Tevatron: Summary

If the Higgs boson is discovered at LEP2, then it should be observed at the Tevatron in WH with $\int \mathcal{L}dt \lesssim 10$ fb^{-1}. If the Higgs boson lies beyond the reach of LEP2, $M_H \gtrsim (95\text{-}100)$ GeV/c^2, then a 5-σ discovery will be possible in the WH channel in a future Run III of the Tevatron (30 fb^{-1}) for masses up to $M_H \approx 125$ GeV/c^2. This prospect is the most powerful incentive that we have for Run III. To support this discovery, a 3-σ observation will be possible in ZH in Run III for masses up to $M_H \approx 110$ GeV/c^2. In combination, the two observations at the Tevatron would imply a $\pm 15\%$ measurement of the ratio of couplings g_{WWH}^2/g_{ZZH}^2. If the coupling strength g_{ZZH} and the branching fraction $B(H \to b\bar{b})$ are known from experiments at LEP2, the observations at the Tevatron would make it possible to determine g_{WWH} to $\pm 10\%$. Over the range of masses accessible at the Tevatron, it should be possible to determine the mass of the Higgs boson to $\pm(1\text{-}3)$ GeV/c^2.

Higgs at the LHC: Summary

The capabilities of the LHC experiments to search for, and study, the Higgs boson are thoroughly documented in the Technical Proposals [23]. I will confine myself here to a few summary comments.

A 5-σ discovery is possible up to $M_H \approx 800$ GeV/c^2 in a combination of the channels

$$H \to Z\,Z$$
$$\hookrightarrow \ell^+\ell^-$$
$$\hookrightarrow \ell^+\ell^-,$$

$$H \quad W$$
$$\quad | \quad \hookrightarrow \ell\nu$$
$$\hookrightarrow b\bar{b}$$

and

$$H \to \gamma\gamma \text{ or perhaps } \tau^+\tau^-.$$

The reach of LHC experiments can be extended by making use of the channels

$$H \to Z\,Z$$
$$\quad | \quad \hookrightarrow \ell^+\ell^- \text{ or } \nu\bar{\nu}$$
$$\hookrightarrow \text{jet jet},$$

and

$$H \to W\,W$$
$$\quad | \quad \hookrightarrow \ell\nu$$
$$\hookrightarrow \text{jet jet}.$$

For Higgs-boson masses below about 300 GeV/c^2, it should be possible to determine the Higgs mass to 100-300 MeV/c^2 [24].

SUMMARY REMARKS

The Tevatron exists, and will produce important results on the top quark and Higgs boson through the next decade. We can expect considerable improvements in the determinations of m_t and M_W, as well as increasingly telling searches for nonstandard production and decay in Run II (2 fb^{-1}). In the realm of what might be possible thereafter, what we have called Run III (30 fb^{-1}) holds great promise for refining our knowledge of top properties, including the measurement of $|V_{tb}|$ in single-top production. Run III would also extend the search for a light Higgs boson throughout the low-mass region favored by supersymmetry. On a related note, if low-scale supersymmetry exists, there is every reason to expect that it should be found at the Tevatron.

During the week of this workshop, the United States sealed its commitment to participate in the construction of the Large Hadron Collider at CERN. The LHC will be a fountain of tops: ~ 8 million pairs will be produced per year at a luminosity of $\mathcal{L} = 10^{33}$ cm^{-2} s^{-1}; hundreds to thousands of interesting events will be detected each day. The LHC will extend the search for the agent of electroweak symmetry breaking toward 1 TeV. It will have good sensitivity to the standard-model Higgs boson throughout the interesting range. The LHC will explore the spectrum of superpartners up to ~ 1 TeV/c^2 and make possible detailed measurements of supersymmetric parameters. Opening a new energy frontier, the LHC will also offer many other possibilities for exploration.

ACKNOWLEDGEMENTS

It is a pleasure to thank Dave Cline and his staff for the stimulating and pleasant atmosphere of the workshop. I am grateful to Jens Erler, Steve Kuhlmann, Scott Willenbrock, and John Womersley for advice and assistance in the preparation of this talk.

REFERENCES

1. For general background to the present discussion, see C. Quigg, "Hadron Colliders, the Top Quark, and the Higgs Sector," FERMILAB–CONF–97/157–T (hep-ph/9707508), to appear in the Proceedings of the Advanced School on Electroweak Theory (Maó, Menorca, Spain, June 17–21, 1996).
2. F. Abe, et al. (CDF Collaboration), Phys. Rev. Lett. **74**, 2626 (1995); S. Abachi, et al. (DØ Collaboration), Phys. Rev. Lett. **74**, 2632 (1995).
3. F. Abe, et al. (CDF Collaboration), Phys. Rev. Lett. **79**, 1992 (1997); "Measurement of the Top Quark Mass," FERMILAB–PUB–97/284–E (hep-ex/9801014); "Measurement of the Top Quark Mass and $t\bar{t}$ Production Cross Section from Dilepton Events at the Collider Detector at Fermilab," FERMILAB–PUB–97/304–E. S. Abachi, et al. (DØ Collaboration), Phys. Rev. Lett. **79**, 1197 (1997); B. Abbott, et al. (DØ Collaboration), "Measurement of the top quark mass using dilepton events," FERMILAB–PUB–97/172–E (hep-ex/9706014).
4. I. Bigi, Yu. L. Dokshitzer, V. Khoze, J. Kühn, and P. Zerwas, Phys. Lett. **B181**, 157 (1986). For the QCD corrections, see M. Jeżabek and J. Kühn, Nucl. Phys. **B314**, 1 (1989), Phys. Rev. **D48**, 1910 (1993).
5. E. Laenen, J. Smith, and W. van Neerven, Nucl. Phys. **B369**, 543 (1992), Phys. Lett. **B321**, 254 (1994); S. Catani, et al., Phys. Lett. **B378**, 329 (1996), Nucl. Phys. **B478**, 273 (1996); E. L. Berger and H. Contopanagos, Phys. Lett. **B361**, 115 (1995); Phys. Rev. **D57**, 253 (1998). For the NLO predictions without resummation, see R. K. Ellis, Phys. Lett. **B259**, 492 (1991).
6. F. Abe, et al. (CDF Collaboration), "Measurement of the Top Quark Mass and $t\bar{t}$ Production Cross Section from Dilepton Events at the Collider Detector at Fermilab," FERMILAB–PUB–97/304–E.
7. S. Abachi, et al. (DØ Collaboration), Phys. Rev. Lett. **79**, 1203 (1997).
8. R. Frey, et al., "Top Quark Physics: Future Measurements," in Snowmass '96.
9. G.F. Tartarelli for the CDF collaboration, "Direct Measurement of $|V_{tb}|$ at CDF," FERMILAB–CONF–97/401–E, to be published in the proceedings of HEP 97, Jerusalem, Israel, 19-26 Aug 1997.
10. S. Willenbrock, "Top Quark Physics for Beautiful and Charming Physicists" (hep-ph/9709355).
11. Standard-model (and beyond) predictions for the $t \to gc, Z^0 c$, and γc decay rates are given by G. Eilam, J. L. Hewett, and A. Soni, Phys. Rev. **D44**, 1473 (1991).

12. For a recent treatment of the effects of anomalous $Zt\bar{c}$ couplings, see T. Han, R. D. Peccei, and X. Zhang, *Nucl. Phys.* **B454**, 527 (1995).
13. F. Abe, *et al.* (CDF Collaboration), *Phys. Rev. Lett.* **79**, 357 (1997); M. Strovink, "Top Quark Results from DØ," to appear in the proceedings of HEP 97, Jerusalem, Israel, 19-26 Aug 1997.
14. E. Eichten and K. Lane, *Phys. Lett.* **B327**, 129 (1994).
15. C. T. Hill and S. J. Parke, *Phys. Rev.* **D49**, 4454 (1994). The topcolor model is developed in C. T. Hill, *Phys. Lett.* **B345**, 483 (1995).
16. B. W. Lee, C. Quigg, and H. B. Thacker, *Phys. Rev.* **D16**, 1519 (1977).
17. L. Maiani, G. Parisi, and R. Petronzio, *Nucl. Phys.* **B136**, 115 (1978).
18. Lower bounds on the Higgs mass date from the work of A. D. Linde, *Zh. Eksp. Teor. Fiz. Pis'ma Red.* **23**, 73 (1976) [*JETP Lett.* **23**, 64 (1976)]; S. Weinberg, *Phys. Rev. Lett.* **36**, 294 (1976). For useful updates in light of the large mass of the top quark ($m_t \approx 175$ GeV/c^2), see G. Altarelli and G. Isidori, *Phys. Lett.* **B337**, 141 (1994); J. Espinosa and M. Quirós, *Phys. Lett.* **B353**, 257 (1995).
19. J. Erler and P. Langacker, October 1997 update of "Electroweak model and constraints on new physics," for the *Review of Particle Physics*; "Bounds on the Standard Higgs Boson," UPR–791–T (hep-ph/9801422).
20. A. Stange, W. Marciano and S. Willenbrook, *Phys. Rev.* **D49**, 1354 (1994); *Phys. Rev.* **D50**, 4491 (1994).
21. "Future Electroweak Physics at the Fermilab Tevatron: Report of the TeV2000 Group," edited by D. Amidei and R. Brock, FERMILAB–PUB–96/082. For light Higgs, see http://www-theory.fnal.gov/TeV2000/chapter5_higgs.ps. The report of the TeV33 Working Group is available at http://www-theory.fnal.gov/tev33.ps; for the TeV33 Convenor Reports, see http://www-theory.fnal.gov/tev33/convenerreports.html. See also S. Kim, S. Kuhlmann, and W.-M. Yao, "Improvement of signal significance in $WH \to \ell + \nu + b + \bar{b}$ search at TeV33," in *Snowmass '96*.
22. Based on J. Womersley, "Discovering the Higgs at TeV33: a Status Report," DØ Note 3227, April 1997 (unpublished).
23. The ATLAS Technical Proposal can be retrieved from ftp://www.cern.ch/pub/Atlas/TP/tp.html. The CMS Technical Proposal is available at http://cmsinfo.cern.ch/cmsinfo/TP/TP.html.
24. J. F. Gunion, L. Poggioli, R. Van Kooten, C. Kao, P. Rowson, *et al.*, "Higgs Boson Discovery and Properties," *Snowmass '96*.

Top Quark Physics at a Polarized Muon Collider

Stephen Parke

Theoretical Physics Department
Fermi National Accelerator Laboratory
Batavia, IL 60510
USA
e-mail: parke@fnal.gov

Abstract. Top quark pair production is presented at a polarized Muon Collider above the threshold region. The off-diagonal spin basis is the natural basis for this discussion as the top quark pairs are produced in an essentially unique spin configuration for 100% polarization. Modest polarization, say 30%, can lead to 90% of all top quark pair events being in one spin configuration. This will lead to sensitive tests on anomalous top quark couplings.

Recently Parke and Shadmi [1] have shown that at a 100% polarized lepton collider that the top and anti-top quark pairs are produced in essentially a unique spin configuration. This spin basis has been called the "off-diagonal" basis and it interpolates between the beam direction at threshold and the top quark direction, i.e. helicity, far above threshold. The differential cross section using this basis is given by

$$\frac{d\sigma}{d\cos\theta^*}(\mu_L^- \mu_R^+ \to t_\uparrow \bar{t}_\uparrow \text{ or } t_\downarrow \bar{t}_\downarrow) = 0,$$

$$\frac{d\sigma}{d\cos\theta^*}(\mu_L^- \mu_R^+ \to t_\uparrow \bar{t}_\downarrow \text{ or } t_\downarrow \bar{t}_\uparrow) = \left(\frac{3\pi\alpha^2}{8s}\beta\right)$$

$$\left[f_{LL}(1+\beta\cos\theta^*) + f_{LR}(1-\beta\cos\theta^*) \right.$$

$$\left. \pm \sqrt{(f_{LL}(1+\beta\cos\theta^*) - f_{LR}(1-\beta\cos\theta^*))^2 + 4f_{LL}f_{LR}(1-\beta^2)} \right]^2 \quad (1)$$

where the f_{IJ}'s are the sum of the photon and Z-boson couplings corrected for the difference in the propogators. Details of this basis can be found in reference [1]. Figure 1 is the spin components for top quark pair production in the off-diagonal basis for both LR and RL incoming lepton helicities for a $\sqrt{s} = 400\ GeV$ collider. The sub-leading terms have been amplified by a factor of 100 so it is clear that the dominant configuration makes up more than 99% of the total cross section.

In the helicity basis the differential cross section is [2]

$$\frac{d\sigma}{d\cos\theta^*}(\mu_L^- \mu_R^+ \to t_L \bar{t}_L \text{ or } t_R \bar{t}_R) = \left(\frac{3\pi\alpha^2}{8s}\beta\right)(1-\beta^2)\sin^2\theta^* |f_{LL} + f_{LR}|^2,$$

$$\frac{d\sigma}{d\cos\theta^*}(e_L^- e_R^+ \to t_R \bar{t}_L \text{ or } t_L \bar{t}_R) = \left(\frac{3\pi\alpha^2}{8s}\beta\right)(1\mp\cos\theta^*)^2$$
$$\times |f_{LL}(1\mp\beta) + f_{LR}(1\pm\beta)|^2. \quad (2)$$

Figure 2 is the corresponding plot for the helicity basis. Here the dominant spin configuration is less than 60% of the total.

In figure 3 I have plotted the dominant spin component's fraction of the total as function of the polarization of the beams,

$$\frac{(1+P^-)(1-P^+)\sigma_{LR\to UD} + (1-P^-)(1+P^+)\sigma_{RL\to UD}}{(1+P^-)(1-P^+)\sigma_{LR}^{tot} + (1-P^-)(1+P^+)\sigma_{RL}^{tot}} \quad (3)$$

for the off-diagonal basis. Here $P \equiv (N_L - N_R)/(N_L + N_R)$ for the μ^+ and μ^- beams. Two different machines are included, the Muon Collider and the NLC. The Muon Collider is assumed to have equal but opposite polarization for the μ^+ and μ^- beams whereas the NLC is an electron-positron collider with only the electron beam polarized. From these curves a modest amount

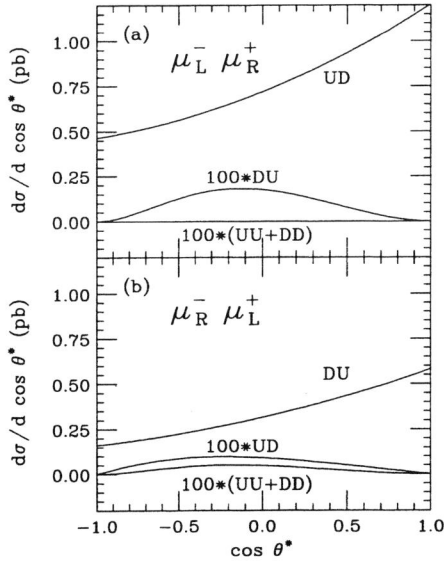

FIGURE 1. The spin configurations using the off-diagonal basis for both $\mu_L^+ \mu_R^-$ and $\mu_R^+ \mu_L^-$ for $\sqrt{s} = 400$ GeV. Note that the sub-leading configurations have been amplified by a factor of 100 in these figures.

of polarization, say 30%, at a Muon Collider can make the dominant spin configuration close to 90% of the total. Whereas at an electron-positron machine one requires 55% polarization to achieve the same goal.

Figure 4 is the corresponding plot for the helicity configuration. If one uses this spin basis the dominant spin configuration ranges from 41% to 52% of the total. Clearly polarization is not as important here without further cuts.

Since the top-quark pairs are produced in a nearly unique spin configuration, and the electroweak decay products of polarized top-quarks are strongly correlated to the spin axis, the top-quark events at $\mu^+\mu^-$ collider have a very distinctive topology. Deviations from this topology would signal anomalous couplings. In the Standard Model, the predominant decay mode of the top-quark is $t \to bW^+$, with the W^+ decaying either hadronically or leptonically. For definiteness we consider here the decay $t \to bW^+ \to be^+\nu$. The differential decay width of a polarized top-quark depends non-trivially on three angles. The first is the angle, χ_w^t, between the top-quark spin and the direction of motion of the W-boson in the top-quark rest-frame. Next is the angle between the direction of motion of the b-quark and the positron in the W-boson rest-frame. We call this angle $\pi - \chi_e^w$. Finally, in the top-quark rest-frame, we have the azimuthal angle, Φ, between the positron direction of motion and the top-quark spin around the direction of motion of the W-boson.

The differential polarized top-quark decay distribution in terms of these three angles is given by

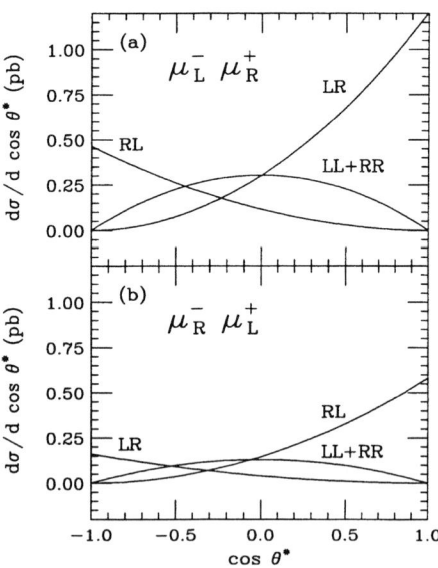

FIGURE 2. The spin configurations using the helicity basis for a $\sqrt{s} = 400\ GeV$ collider.

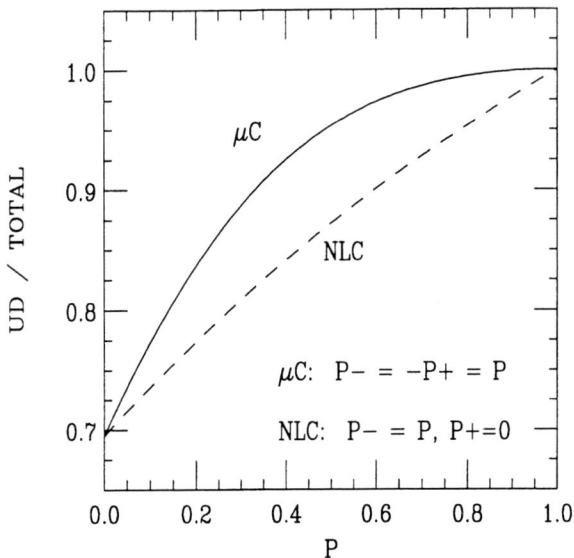

FIGURE 3. Fraction of the total cross section in the off-diagonal basis' Up-Down spin configuration as a function of the polarization. Both beams are assumed to be polarized for the Muon Collider (μC) but only one beam for the NLC.

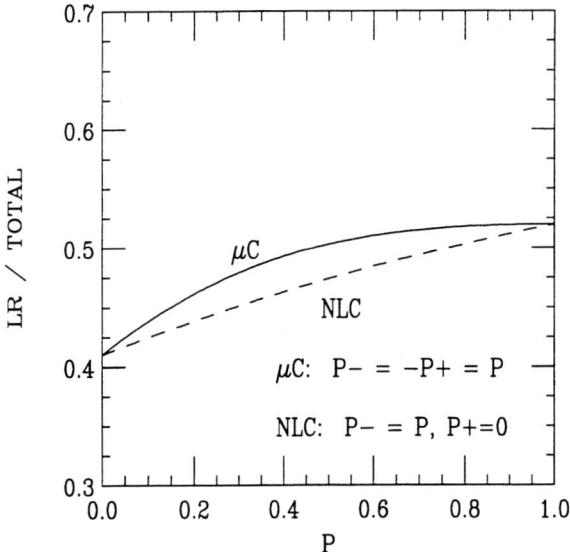

FIGURE 4. Same as Fig. 3 but the helicity basis is used.

$$\frac{1}{\Gamma_T} \frac{d^3\Gamma}{d\cos\chi_w^t \, d\cos\chi_e^w \, d\Phi} = \frac{3}{8\,(m_t^2 + 2m_W^2)}$$

$$\left[m_t^2(1+\cos\chi_w^t)\sin^2\chi_e^w + m_W^2(1-\cos\chi_w^t)(1-\cos\chi_e^w)^2 \right.$$

$$\left. + 2m_t m_W(1-\cos\chi_e^w)\sin\chi_e^w \sin\chi_w^t \cos\Phi \right], \quad (4)$$

where m_t is the top-quark mass, m_W is the W mass, and Γ_T is the total decay width (we neglect the b-quark mass). The first and second terms in (4) give the contributions of longitudinal and transverse W-bosons respectively. The interference term, given by the third term in (4), does not contribute to the total width, but its effects on the angular distribution of the top-quark decay products are sizable. Fig. 5 shows contour plots of the differential angular decay distribution in the $\chi_e^w - \chi_w^t$ plane after integrating over the azimuthal angle Φ. The peak at the center of the right hand side of this figure is due to the longitudinal W-bosons whereas the peak at the bottom left hand corner is caused by the transverse W-bosons.

There are also significant correlations of the angle between the top-quark spin and the momentum of the i-th decay product, χ_i^t, measured in the top-quark rest-frame, see figure 6. The differential decay rate of the top-quark is given by

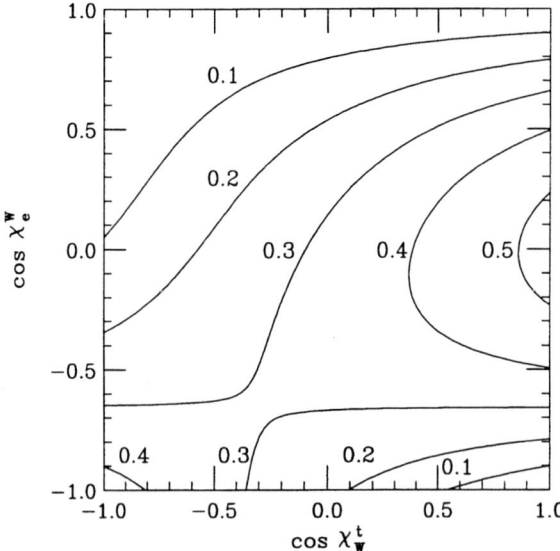

FIGURE 5. Correlations between the W-boson and the top spin direction in the top rest frame ($\cos\chi_W^t$) and the positron and the minus b-quark direction in the W-boson rest frame ($\cos\chi_e^W$).

$$\frac{1}{\Gamma_T}\frac{d\Gamma}{d\cos\chi_i^t} = \frac{1}{2}\left[1 + \alpha_i \cos\chi_i^t\right], \qquad (5)$$

where $\alpha_b = -\alpha_W = -0.41$, $\alpha_\nu = -0.31$ and $\alpha_{e^+} = 1$, for $m_t = 175$ GeV, see ref. [3]. The interference between the longitudinal and transverse W-bosons is very importnat in determining these correlations. Note, the positron is more highly correlated with the spin of the top quark than its parent the W-boson!

Given that we know the spin configuration of the top quark pairs and the correlations of the top quark decay products there are many correlations studies that can be performed in top quark pair product at a lepton collider looking for anomalous couplings of the top quarks. These studies have been performed for the helicity basis [4] but need to be redone using the superior off-diagonal spin basis.

QCD effects modify this picture in only a minor fashion. The reason being that soft gluons cannot flip the spin of the heavy top quarks. Detailed studies of the effects of one loop calculations show that the dominant spin configuration is still more than 99% of the total even when QCD corrections are included [5].

In conclusion top quark pairs above the threshold region at Muon Colliders are a great place to search for anomalous couplings of the top quark. For $\sqrt{s} < 1\,TeV$ the off-diagonal basis is superior to the helicity basis in describing the events in the simplest possible terms. Polarization of the incoming

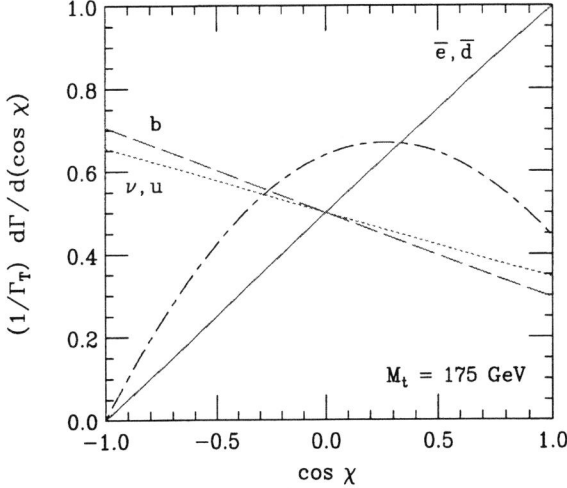

FIGURE 6. The straight lines are the correlations between the top quark decay products and the top quark spin direction in the rest frame of the top ($\cos\chi_i^t$). Whereas the curved line is the correlation between the b-quark and the positron or d-quark in the rest frame of the W-boson ($-\cos\chi_e^W$).

beams enhances this effect. Detailed studies of the one loop QCD corrections have been recently completed, showing no qualitative difference than the tree level analysis. An extensive analysis of the sensitivity to anomalous couplings of the top quark, using the off-diagonal basis, is now needed.

ACKNOWLEDGMENTS

Special thanks to the local organizers of this conference. Fermi National Accelerator Laboratory is operated by the Universities Research Association, Inc., under contract DE-AC02-76CHO3000 with the United States of America Department of Energy.

REFERENCES

1. S. Parke and Y. Shadmi, Phys. Lett. **B387**, 199, (1996); hep-ph/9606419.
2. M.E. Peskin and C.R. Schmidt, in *Physics and Experiments at Linear Colliders*, R. Orava, P. Eeorla and M. Nordberg, eds. (World Scientific, 1992).
3. M. Jeżabek and J.H. Kühn, Phys. Lett. **B329**, 317, (1994).
4. T.L. Barklow and C.R. Schmidt, in *DPF '94: The Albuquerque Meeting*, S. Seidel, ed. (World Scientific, Singapore, 1995);
   C.R. Schmidt Phys. Rev. **D54**, 3250, (1996).
5. J. Kodaira, T. Nasuno and S. Parke manuscript in preparation.
   M. Hori, Y. Kiyo, J. Kodaira, T. Nasuno and S. Parke hep-ph/9801370.

Threshold Cross Section Measurements

M. S. Berger

Indiana University
Bloomington Indiana 47405

Abstract. Accurate measurements of particles masses, couplings and widths are possible by measuring production cross sections near threshold. We discuss the prospects for performing such measurements at a high luminosity muon collider.

INTRODUCTION

A muon collider is particularly well suited to the threshold measurement because the spread in energy of the beam is very small [1]. Pair production of W-bosons, $t\bar{t}$ production and the Bjorken process $\mu^+\mu^- \to ZH$ have been considered as possible places to study thresholds at a muon collider [2–4]. Threshold production of chargino pairs at a muon collider offers a possible way of accurately measuring the chargino mass [5].

We assume here that the muon collider has a relatively modest beam energy spread of $R = 0.1\%$, where R is the rms spread of the energy of a muon beam. We assume that 100 fb^{-1} integrated luminosity is available and that this amount of luminosity could be accumulated at the relevant energies for the measuring the threshold cross sections; high luminosity is essential if the threshold measurements are to prove interesting.

M_W MEASUREMENT AT THE $\mu^+\mu^- \to W^+W^-$ THRESHOLD

The threshold cross section is most sensitive to M_W just above $\sqrt{s} = 2M_W$, but a tradeoff exists between maximizing the signal rate and the sensitivity of the cross section to M_W. Detailed analysis [6] shows that if the background level is small and systematic uncertainties in efficiencies are not important, then the optimal measurement of M_W is obtained by collecting data at a single energy

$$\sqrt{s} \sim 2M_W + 0.5 \text{ GeV} \sim 161 \text{ GeV},$$

where the threshold cross section is sharply rising.

At a muon collider with high luminosity, systematic errors arising from uncertainties in the background level and the detection/triggering efficiencies will be dominant unless some of the luminosity is devoted to measuring the level of the background (which automatically includes somewhat similar efficiencies) at an energy below the W^+W^- threshold. Then, assuming that efficiencies for the background and W^+W^- signal are sufficiently well understood that systematic uncertainties effectively cancel in the ratio of the above-threshold to the below-threshold rates, a very accurate M_W determination becomes possible.

We analyzed [2] the possible precision obtainable for the W mass via just two measurements: one at center of mass energy $\sqrt{s} = 161$ GeV, just above threshold, and one at $\sqrt{s} = 150$ GeV. The optimal M_W measurement is obtained by expending about two-thirds of the luminosity at $\sqrt{s} = 161$ GeV and one-third at $\sqrt{s} = 150$ GeV. Combining the three modes, an overall precision of $\Delta M_W = 6$ MeV should be achievable with 100 fb^{-1} integrated luminosity.

HIGGS BOSON MEASUREMENT AT THE $\mu^+\mu^- \to Zh$ THRESHOLD

The SM Higgs boson is easily discovered in the Bjorken Higgs-strahlung process [7] $\ell^+\ell^- \to Zh$ running the machine well above threshold, e.g. at $\sqrt{s} = 500$ GeV. For $m_h \lesssim 2M_W$ the dominant Higgs boson decay is to $b\bar{b}$ and most backgrounds can be eliminated by b-tagging. A very accurate determination of m_h could then obtained by measuring the threshold cross section of Zh production, which rises rapidly as shown in Fig. 1(a) since the threshold behavior is S-wave.

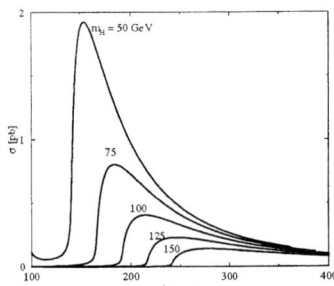 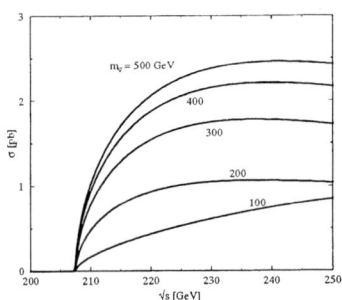

FIGURE 1. The cross section vs. $\sqrt{s}$ for (a) the process $\mu^+\mu^- \to Z^\star h \to f\bar{f}h$ for a range of Higgs masses, and for (b) $\mu^+\mu^- \to \tilde{\chi}^+\tilde{\chi}^-$ for various sneutrino masses and $m_{\tilde{\chi}^\pm} = 103.7$ GeV.

The sensitivity to the SM Higgs boson mass is maximized by a single measurement of the cross section at $\sqrt{s} = M_Z + m_h + 0.5$ GeV, just above the real particle threshold provided that the normalization of the measured Zh cross section as a function of $\sqrt{s}$ can be precisely predicted, including efficiencies and systematic effects. We employed b-tagging and cuts in order to reduce the background to a very low level. These cuts and other systematic uncertainties are discussed in more detail in Ref. [3]. The background is very much smaller than the signal unless m_h is close to M_Z. The electroweak radiative corrections to the cross section are estimated to be less than 1% for $m_H \sim 100$ GeV [8], and the measurement of the cross section described here is at the 2% level. We found a precision of the SM Higgs mass determination to within 45 MeV for $m_h = 100$ GeV may be achievable at a muon collider. More generally the precision ranges from 20-100 MeV for $m_h < 150$ GeV.

Beyond the Standard Model the cross section generally depends on the ZZH coupling (g_{ZZh}) and the total Higgs width (Γ_H) in addition to m_h. In order to simultaneously determine these three quantities, measurements could be made at the three c.m. energies $\sqrt{s} = m_h + M_Z + 20$ GeV, $\sqrt{s} = m_h + M_Z + 0.5$ GeV, and $\sqrt{s} = m_h + M_Z - 2$ GeV. With a three-parameter fit to m_h, $g_{ZZh}^2 B(h \to b\bar{b})$ and Γ_H, the attainable error in m_h is about 110 MeV at the 1σ level for a 100 GeV Higgs. Measurements that would simultaneously determine m_h, $\sigma(Zh)B(h \to b\bar{b})$ and Γ_H could be done at a level of accuracy that could distinguish a Standard Model Higgs boson from its many possible (e.g. supersymmetric) extensions [3].

TOP-QUARK MASS MEASUREMENT AT THE $\mu^+\mu^- \to t\bar{t}$ THRESHOLD

The top-quark threshold cross section is calculable since the large top-quark mass puts one in the perturbative regime of QCD [9]. One can perform scan of the threshold curve by devoting to 10 fb^{-1} integrated luminosity to measuring the cross section at each of ten energies in 1 GeV intervals. Then the top-quark mass can be determined to within $\Delta m_t \sim 70$ MeV, provided systematics and theoretical uncertainties are under control. Considerable progress has been made recently in the theoretical calculations of the some NNLO corrections to the threshold cross section [10]. The remaining theoretical uncertainties [11] in the threshold cross section are still fairly large and make it difficult to fully exploit the large luminosity for determining say the strong coupling α_s or a light Higgs boson mass (and the top quark Yukawa coupling) from the size of the cross section. Furthermore there is theoretical ambiguity in the mass definition of the top quark. The theoretical ambiguity in relating quark pole mass to other definitions of the top quark mass (that might be relevant as input to radiative correction calculations) is of order Λ_{QCD}, i.e., or a few hundred MeV [12]. So it is not clear that an extraction of the top-quark mass better than this is useful, at least at the present time.

CHARGINO SIGNAL AND BACKGROUND

The mass of the lighter chargino in the minimal supersymmetric standard model (MSSM) can be determined accurately by measuring the cross section[1] for

$$\mu^+\mu^- \to \chi^+\chi^- \tag{1}$$

near the threshold [5]. The precision that can be obtained in the chargino mass depends substantially on the mass of the chargino mass itself: the heavier the chargino the smaller the production cross section. The cross section also depends on the mass of the sneutrino which appears in the t-channel since this contribution interferes destructively with the s-channel graphs. The cross section is displayed in Fig. 1(b) for several values of the sneutrino mass. If the lightest chargino is gaugino-dominated, then changing the parameters of the chargino mass matrix essentially changes the mass but not the chargino couplings significantly. The width of the lightest chargino is usually less than a few MeV, and often substantially less when two-body decays are kinematically impossible. Therefore one can envision a measurement of the cross section that depends on just two parameters: the chargino mass $m_{\tilde{\chi}^\pm}$ and the sneutrino mass $m_{\tilde{\nu}}$.[2]

As in the other threshold measurements, the statistical precision on the chargino mass is maximized just above $2m_{\tilde{\chi}^\pm}$. A simultaneous measurement of the chargino and sneutrino masses requires a sampling of the cross section at at least two points. It turns out to be advantageous for the chargino mass measurement to choose this higher energy measurement at a point where the chargino cross section is not flat.

The chargino decay mode is $\tilde{\chi}^\pm \to \tilde{\chi}^0 f\overline{f}'$ provided the chargino is lighter than the muon sneutrino. The cross section is reduced near threshold, so the cuts to reduce backgrounds need to be reoptimized. The backgrounds to chargino pair-production have been investigated in Refs. [16,17] where the signal efficiencies have been obtained for the various final states when the center-of-mass energy is $\sqrt{s}$ = 500 GeV. The primary background is W pair production which is very large, but can be effectively eliminated because the W's are produced in the very-forward direction. However, if the energy is reduced so that the collider is operating in the chargino threshold region, then the effectiveness of these cuts might be reduced (the signal events might be expected to be more spherical as well). Therefore the efficiencies were reinvestigated for the threshold measurement.

A further advantage of the threshold measurement is that the chargino mass measurement is somewhat isolated from its subsequent decays. Distributions in the final state observables, say e.g. E_{jj} from the decay $\tilde{\chi}^\pm \to \tilde{\chi}^0 jj$ [16], depend on

---

[1] The measurement of the chargino mass via the threshold cross section has been considered previously for electron-positron machines in Ref. [13,14]. We consider the measurement at a muon collider with high luminosity, carefully taking into account the beam effects and reoptimizing cuts to eliminate the background in the threshold region.

[2] The overall normalization of the cross section could also depend on radiative corrections which could be substantial in some cases [15].

TABLE 1. Precison of mass measurements assuming 100 fb^{-1} luminosity. The ranges considered for the Higgs and chargino masses are also shown.

Particle	Mass Measurement (MeV)	Mass Range (GeV)
W	6	–
t	70	–
h	20-150	50-200
$\chi^\pm$	30-200	100-200[a]

[a] $m_{\tilde{\nu}} > 300$ GeV

the neutralino mass. The cross section for chargino pair production, on the other hand, is independent of the final state particles, and only the branching fractions and detector efficiencies for the various final states impact this measurement (as indicated above, if $m_{\tilde{\chi}^\pm} - m_{\tilde{\chi}^0} > M_W$ the branching fractions of chargino decay is given essentially in terms of the W branching fractions).

The chargino production cross section decreases with increasing chargino mass. Therefore the precision with which the mass can be measured is better at smaller values of the mass with precisions of as small as 30 MeV possible for $m_{\tilde{\chi}^\pm} = 100$ GeV. For $m_{\tilde{\chi}^\pm} = 200$ GeV the chargino mass can be determined to 100 (200) Mev for $m_{\tilde{\nu}} = 500$ (300) GeV. The sneutrino mass can be measured to about 6 GeV accuracy for $m_{\tilde{\nu}} = 300$ GeV and to about 20 GeV accuracy for $m_{\tilde{\nu}} = 500$ GeV. This provides an indirect method of measuring the sneutrino mass (the sneutrino might be too heavy to produce directly).

CONCLUSION

A muon collider would provide an opportunity for precision mass measurements in the respective threshold regions[3]. The precisions that can be obtained for particle masses is shown in Table 1 assuming an integrated luminosity of 100 fb^{-1}. The precisions for the Higgs and chargino measurements are correlated with the (as of yet unknown) mass, so the ranges we considered are shown as well. To utilize the highest precision measurements achievable at the statistical level, theoretical uncertainties and other systematics need to be under control in all cases. The muon sneutrino mass can also be simultaneously measured to a few GeV if it is less than 500 GeV in the process $\mu^+\mu^- \to \chi^+\chi^-$.

---

[3] The most recent TESLA design envisions a beam energy spread of $R = 0.2\%$ [18] while the NLC design expects a beam energy spread of $R = 1.0\%$. A high energy e^+e^- collider in the large VLHC tunnel would have a beam spread of $\sigma_E = 0.26$ GeV [19] which should give numbers precisions comparable to those considered here.

ACKNOWLEDGMENTS

I thank V. Barger, J. F. Gunion and T. Han for a pleasant collaboration on the issues reported here. This work was supported in part by the U.S. Department of Energy under Grant No. DE-FG02-91ER40661.

REFERENCES

1. $\mu^+\mu^-$ *Collider: A Feasibility Study*, Snowmass, Colorado, July, 1996.
2. V. Barger, M.S. Berger, J.F. Gunion and T. Han, Phys. Rev. **D56**, 1714 (1997).
3. V. Barger, M. S. Berger, J. F. Gunion and T. Han, Phys. Rev. Lett. **78**, 3991 (1997).
4. M.S. Berger, talk presented at the *Workshop on Particle Theory and Phenomenology: Physics of the Top Quark*, Iowa State University, May 25–26, 1995, hep-ph/9508209.
5. V. Barger, M.S. Berger and T. Han, hep-ph/9801410.
6. Z. Kunszt and W.J. Stirling *et al.*, hep-ph/9602352, in *Proceedings of the Workshop on Physics at LEP2*, eds. G. Alterelli, T. Sjostrand and F. Zwirner, CERN Yellow Report CERN-96-01 (1996), Vol. 1, p. 141; W.J. Stirling, Nucl. Phys. **B456**, 3 (1995).
7. J.D. Bjorken, *Proceedings of the Summer Institute on Particle Physics*, ed. M. Zipf (Stanford, 1976).
8. B.A. Kniehl, Z. Phys. **C55**, 605 (1992); R. Hempfling and B. Kniehl, Z. Phys. **C59**, 263 (1993).
9. V.S. Fadin and V.A. Khoze, JETP Lett. **46**, 525 (1987); Sov. J. Nucl. Phys. **48**, 309 (1988).
10. A. H. Hoang, Phys. Rev. **D56**, 5851 (1997); A. H. Hoang, Talk at the Workshop on Physics at the First Muon Collider and at the Front End of the Muon Collider, Batavia, IL, 6-9 Nov 1997, hep-ph/9801273; A. H. Hoang and T. Teubner, hep-ph/9801397.
11. M. Jezabek, et al., hep-ph/9801419.
12. M. C. Smith and S. Willenbrock, Phys. Rev. Lett. **79**, 3825 (1997).
13. A. Leike, Int. J. Mod. Phys. **A3**, 2895 (1988).
14. *Physics with e^+e^- Linear Colliders*, by ECFA/DESY LC Physics Working Group (E. Accomando et al.), DESY-97-100, May 1997, hep-ph/9705442.
15. P. Chankowski, Phys. Rev. **D41**, 2877 (1990); M. M. Nojiri, K. Fujii and T. Tsukamoto, Phys. Rev. **D54**, 6756 (1996); H.-C. Cheng, J. L. Feng and N. Polonsky, Phys. Rev. **D56**, 6875 (1997); Phys. Rev. **D57**, 152 (1998); M. A. Diaz, S. F. King and D. A. Ross, hep-ph/9711307.
16. T. Tsukamoto, K. Fujii, H. Murayama, M. Yamaguchi and Y. Okada, Phys. Rev. **D51**, 3153 (1995).
17. J.-F. Grivaz, preprint LAL 91-63, Talk at the Workshop on Physics and Experiments with Linear Colliders, Saariselka, Finland, 9-14 September 1991.
18. D. Miller, private communication.
19. J. Norem, private communication and http://www-ap.fnal.gov/VLHC/electrons/index.html.

Physics Potential of the CMS/LHC

S. Dasu
for The CMS Collaboration

Department of Physics, University of Wisconsin[1]
Madison, Wisconsin 53706

Abstract. Physics potential of the Compact Muon Solenoid, CMS, detector at the Large Hadron Collider, LHC, is discussed. Particular emphasis is placed on the searches for the Standard Model and the various Minimal Supersymmetric Standard Model Higgs bosons.

INTRODUCTION

The Large Hadron Collider at the CERN laboratory is the next large facility for the exploration of the high energy frontier. Although the Standard Model is very successful in describing all particle physics data now available, some fundamental aspects of it are not understood. Foremost amongst the unexplored portions of the Standard Model is the mechanism for providing masses for the known particles. The Higgs mechanism elegantly provides the needed masses for weakly mediating vector bosons, and through Yukawa couplings of the resultant bosonic Higgs field to the quarks and leptons. Unfortunately, the mass of the Higgs boson is not predicted in theory well. The top quark mass measurements from the Tevatron experiments can be combined with the precise data from LEP and SLD experiments in a global analysis of the radiative corrections in the Standard Model to determine the range of valid Higgs boson masses [1]. The negative evidence for the Higgs searches at the LEPII can push the lower limit of the Higgs mass to 80 GeV. The current knowledge still leaves room for a Higgs boson of mass between 80 and 1000 GeV. New physical phenomenon must occur at TeV scale if the Standard Model Higgs boson does not appear in this mass range.

Although the Standard Model is able to account for all the data in particle physics very well, it has a draw back of needing as many as nineteen parameters. It is appealing to consider theories at higher energies that approximate to the Standard Model at lower energies. Supersymmetry, i.e., a symmetry between gauge boson and quark-lepton fermion fields, is particularly interesting because it is the only

---

[1] This research is funded by the U.S. Department of Energy Grant DE-FG02-95ER40896.

known theoretical framework that can include gravity. If supersymmetry is related to the electroweak symmetry breaking, the masses of some of the super particles and its Higgs sector should be less than a TeV.

The Large Hadron Collider at CERN laboratory with its 14 TeV center of mass proton-proton collisions running at up to luminosities of 10^{34} cm^{-2} s^{-1} will definitely enable study of TeV quark-quark interactions. Any Higgs bosons and super particles of masses up to a TeV can be produced. Detailed studies of top quark physics and high E_t QCD physics are feasible because those events are copiously produced even at the initial low luminosity runs of the LHC. Unfortunately, both the rate of the Standard Model strong interaction physics at few hundred GeV scale and the hadron production per event are large. Extraction of signals, particularly those of the Higgs decays, from the profusely produced Standard Model background requires exploration of low branching fraction leptonic or photonic modes with good energy resolution. Therefore, a detector with good resolution and high degree of segmentation, that can stand a very high rate environment, is necessary. Further, a trigger and data acquisition system that can weed out the well understood background, while retaining the interesting high energy physics is required. The CMS detector at the LHC is designed to address this TeV physics. While the CMS is designed to address all high P_t physics at the LHC [3], here we concentrate on its ability to measure the Higgs sector.

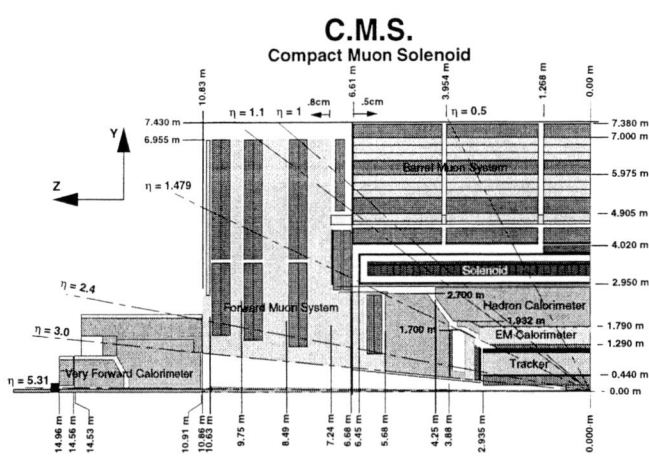

FIGURE 1. The longitudinal view of a quarter section of the CMS detector showing the detector subsystems.

CMS DETECTOR

The Compact Muon Solenoid (CMS) detector [2] is illustrated in the Figure 1. This detector is designed around a single high field (4 Tesla) superconducting solenoid. The solenoid encloses the copper-scintillator hadron calorimeter (HCAL), PbWO$_4$ crystal EM calorimeter (ECAL) and a tracker system. The return for the magnetic field is provided by the iron plate structure that also houses the muon detectors. Tracking of the high momentum muons in the high field within the tracker and in the reverse bend in the return yoke enables a good muon momentum measurement in this compact design.

The tracking system composed of micro-strip gas chambers, silicon strip and silicon pixel detectors, is highly segmented with a total of about 10^7 channels to enable track pattern recognition even in high multiplicity (1000 tracks) events that will occur in high luminosity runs. Track resolution of $\Delta P_t/P_t \approx 0.1 P_t$, where P_t is in TeV, with charge measurement for particles of momenta up to 2 TeV is feasible. This tracking system also provides separated vertex tagging with impact parameter resolution of 20 μm in transverse and 100 μm in longitudinal direction.

The CMS ECAL is made of about 100000 PbWO$_4$ crystals covering pseudorapidity range $|\eta| < 2.6$. It provides a resolution of about $2\%/\sqrt{E}$ stochastic, 0.7% constant and 200 MeV/E noise terms in the barrel region. In the forward region the stochastic term is degraded to about $5\%/\sqrt{E}$. The angular resolution of EM showers is 50 mrad/$\sqrt{E}$ in the barrel region. The ECAL is designed with these parameters so that the Higgs signal in its decays to two photons can be measured over the immense π^0 background.

The Cu-Scintillator HCAL provides hermetic coverage up to $|\eta| < 5$ with a granularity of $0.087\phi \times 0.087\eta$ with matching ECAL and muon chamber segmentation. The multi-jet and the missing transverse energy signals that are expected in SUSY events provide a strong motivation for a good coverage from HCAL although the energy resolutions themselves are not as critical.

The muon system has a coverage of up to $|\eta| < 2.4$ with at least sixteen interaction lengths of material for suppressing hadron leakage. The steel yoke is interspersed with layers of detectors, the drift tubes in the barrel and the cathode strip chambers in the forward region. Additional resistive plate chambers are used to further strengthen the trigger system. Together with the tracking system the muon system provides momentum resolution of 0.5% at 10 GeV increasing to few % at 100 GeV and 20% at few TeV. Correct charge assignment at 99% confidence level is feasible up to 7 TeV.

Triggers at hadron collider and in particular at the CMS/LHC are a challenge. Strong interaction physics with 30 GeV E_t particles occur at MHz rate in CMS. Whereas, the decays of Higgs bosons of mass 80 - 120 GeV also yield 30 GeV particles but at far lower rate. The data acquisition bandwidth is limited to 100 Hz due to costs involved in permanent data storage and analysis computing. Therefore, sophisticated triggering is required to save only the interesting physics while suppressing the profuse Standard Model background. CMS reduces the interaction rate from 40 MHz to 100 kHz using custom electronics at the first level trigger, and further down to 100 Hz with higher level triggers running on scalable and programmable computers. Efficiencies and rate capability of the first level triggers is particularly problematic and has been simulated extensively. Trigger energy cutoffs can be maintained at about 30 GeV for single electrons and photons, 20 GeV for single muons, 20 GeV each for double electrons and photons, 10 GeV each for double muons, 80 GeV for missing transverse energy and 100 GeV for single jets down to 30 GeV each for multiple jets, while limiting the background rate. High efficiencies are realized for several physics processes studied. Trigger cutoffs can be lowered for initial low luminosity running to study lower P_t physics.

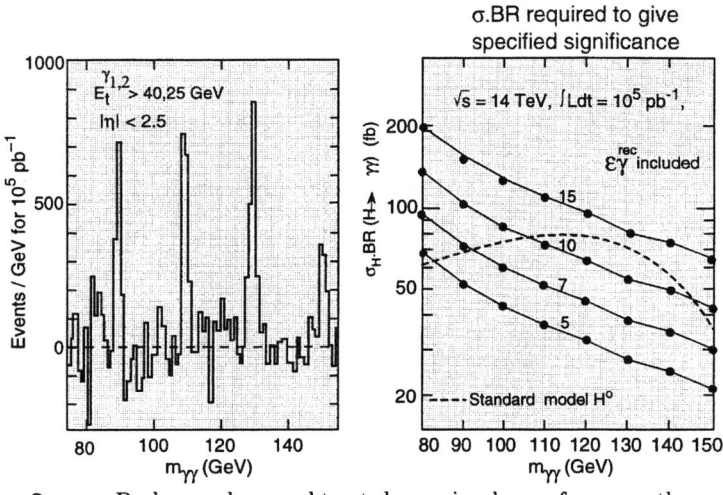

FIGURE 2. Background subtracted signals from the Standard Model Higgs ($M_H = 90, 110, 130, 150$ GeV) decay to two photons and expected cross section times the branching ratio to this channel are plotted versus the reconstructed diphoton invariant mass.

STANDARD MODEL HIGGS

The Standard Model Higgs boson production at LHC is dominated by gluon fusion for low masses of Higgs. Weak boson fusion and bremsstrahlung also contribute at larger Higgs masses. The production cross section varies from about 10 pb for 80 GeV Higgs to about 0.1 pb for TeV Higgs. Although tens of thousands of Higgs bosons will be produced every year at the LHC, the observable fraction of events is small. Higgs can only be extracted from those decay modes involving either photons or leptonic decays of weak bosons. The branching fractions for these decay modes are small and the cuts needed to suppress the background are rather stringent resulting in only a hand full of events.

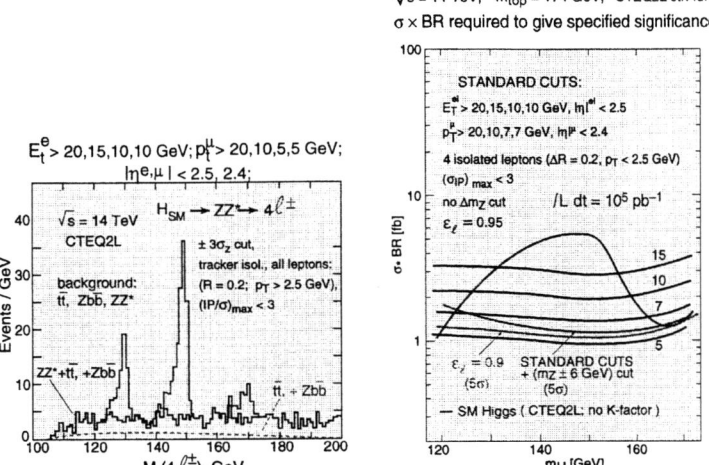

FIGURE 3. Signals from the Standard Model Higgs ($M_H = 130, 150, 170$ GeV) decay to four leptons, background, and expected cross section times the branching ratio to this channel are plotted versus the reconstructed quadlepton invariant mass.

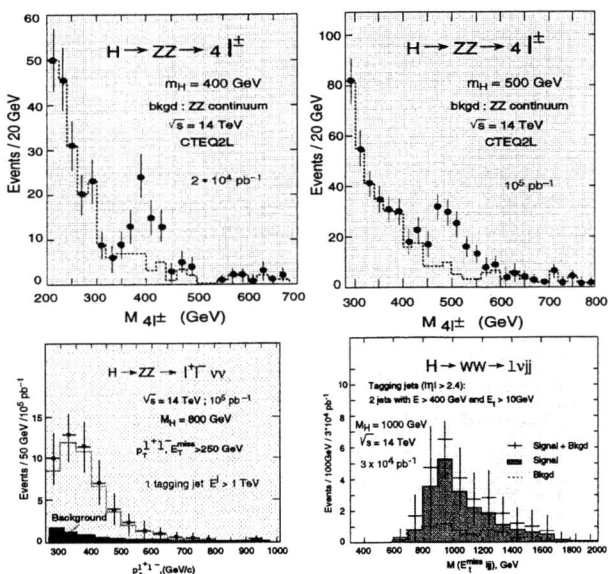

FIGURE 4. Four lepton invariant mass spectrum is shown for Higgs signals ($M_H = 400, 500$ GeV) and background. For $M_H = 800$ GeV for events where one of the Z bosons decayed invisibly, the P_t of the dileptons is plotted showing the signal above the expected background. For $M_H = 1000$ GeV, WW decay mode is considered and the invariant mass is reconstructed using the missing E_t.

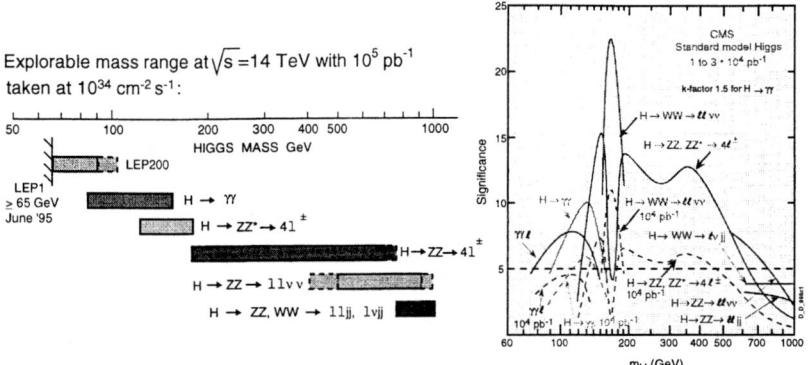

FIGURE 5. The Standard Model Higgs performance summary.

Higgs decay to two photons is the most promising discovery mode for low masses. The crystal ECAL resolutions dominate the signal measurement capability. Figure 2 shows the diphoton invariant mass spectrum, after statistical subtraction of the background, for one year of running at high luminosity, i.e., $\mathcal{L} = 10^5$ pb^{-1}. Better than 5σ significance discovery is feasible in the Higgs mass range 80-150 GeV even after including the detector resolutions and reconstruction efficiency of 64%.

For the intermediate mass range Higgs decays to a pair of Z bosons followed by their leptonic decay is the discovery mode. Z decays to either $e^+ - e^-$ or $\mu^+ - \mu^-$ can be reconstructed well. The mass range 120-180 GeV can be explored where one of the Z bosons is off mass shell. The reconstructed four lepton invariant mass is shown along with the background in Figure 3. Better than 5σ discovery is feasible in the range 120-180 GeV after one year of running at high luminosity.

Higgs bosons with masses greater than twice the mass of the Z boson are well reconstructed as shown in Figure 4. Beyond about 500 GeV the event rate is very low, and this channel can be supplemented with those events where one of the Z bosons decays to an invisible neutrino channel or those events where Higgs decays to two W bosons followed by one hadronic and one leptonic W decay. The explorable mass range is indicated in Figure 5 along with the significance of the signal in various Higgs decay modes.

MINIMAL SUPERSYMMETRIC STANDARD MODEL

The Minimal Supersymmetric Standard Model Higgs sector consists of three neutral bosons h, H and A, and two charged bosons H$^\pm$. Two independent parameters, e.g., M_A and the ratio of vacuum expectation values, $\tan\beta$, are sufficient to describe the Higgs sector. The lightest of the Higgs bosons, h, may still be massive enough that it will be discovered at LHC. Even if h is discovered at LEPII, LHC is still required to sort out the MSSM Higgs sector. The neutral Higgs can be studied in the same channels as in the Standard model. However, the diphoton

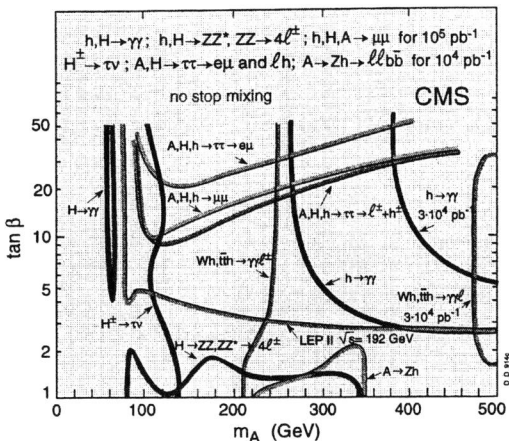

FIGURE 6. The summary of coverage of MSSM parameter space with various Higgs decay modes.

and quadlepton modes are not sufficient to fully span the MSSM parameter space. However, there are other modes, e.g., those with μ, τ and top final states that allow extend coverage of the MSSM parameter space. 5σ significance discovery contours for all these modes in the explorable MSSM parameter space are shown in Figure 6. Integrated luminosity of $\mathcal{L} = 10^5$ pb^{-1} can exclude the entire $M_A - \tan\beta$ plane at 95% confidence level. However, ensuring a 5σ discovery over the entire plane requires more luminosity.

SUMMARY

The CMS detector at LHC is being built to explore exciting TeV physics frontier in the next decade. The simulation studies indicate that the CMS detector has the potential to make definitive discoveries. Studies discussed here indicate that both the Standard Model and the Minimal Supersymmetric Model Higgs sector can be fully explored by CMS at LHC. Simulation of other new physics processes not discussed here also indicate high potential for discovery. Although, the detector and the trigger systems are optimized with the studies of the benchmark processes in the Standard Model and MSSM, they are designed to be flexible to enable discoveries in hitherto unexpected fronts.

REFERENCES

1. G. Altarelli et al., CERN-TH/97-290, Submitted to Int. Journal of Mod. Physics A.
2. The Compact Muon Solenoid Technical Proposal, CERN/LHCC 94-39, Dec. 1994.
3. I. Hinchliffe and J. Womersley, Proceedings of the 1996 DPF/DPB Summer Study on New Directions for High-Energy Physics, Snowmass, CO, Jul. 1996.

R-Parity Violation and Sneutrino Resonances at Muon Colliders [1]

Jonathan L. Feng

*Theoretical Physics Group, Lawrence Berkeley National Laboratory
and Department of Physics, University of California, Berkeley, CA 94720*

Abstract. In supersymmetric models with R-parity violation, sneutrinos may be produced as s-channel resonances at $\mu^+\mu^-$ colliders. We demonstrate that, for R-parity violating couplings as low as 10^{-4}, sneutrino resonances may be observed and may be exploited to yield high precision SUSY parameter measurements. The excellent beam energy resolution of muon colliders may also be used to resolve MeV level splittings between CP-even and CP-odd sneutrino mass eigenstates.

Low-energy supersymmetry (SUSY) is a leading candidate for physics beyond the standard model (SM). When exploring the physics opportunities at a muon collider, it is therefore important to consider its potential for discovering supersymmetric particles and determining SUSY parameters.

For many studies, the muon collider's potential parallels that of more extensively studied e^+e^- colliders, with obvious modifications for differences in luminosity and beam polarization. In fact, as studies of LEP II typically assume $\sqrt{s} \sim 190$ GeV and a total integrated luminosity of ~ 1 fb^{-1}, characteristics similar to those of the proposed First Muon Collider (FMC), many interesting results from these studies apply equally well to the FMC. For example, from chargino production at LEP II or the FMC, gaugino mass unification and the viability of the LSP as a dark matter candidate can be tested in a highly model-dependent manner [1].

There are, however, essential differences that warrant more careful study. Most obviously, if a muon collider reaches $\sqrt{s} \sim 4$ TeV, a great number of complicated sparticle signals may be present [2], as well as a number of backgrounds that have not yet been intensively studied. In addition, the excellent beam energy resolution of muon colliders is promising for precise mass measurements, whether through threshold scans [3] or kinematic endpoints [4].

In this study, we will consider what a muon collider may bring to the study of

---

[1] This work was supported in part by the Director, Office of Energy Research, Office of High Energy and Nuclear Physics, Division of High Energy Physics of the DOE under Contracts DE-AC03–76SF00098 and by the NSF under grant PHY-95-14797.

R-parity violating ($\not{R}_P$) SUSY theories. When R-parity is violated, the distinction between neutral Higgs bosons and scalar neutrinos is blurred, and so scalar neutrinos are also produced as s-channel resonances at lepton colliders. As with Higgs resonances, such resonances may be highly suppressed at electron colliders. At muon colliders, however, we will see that even if $\not{R}_P$ couplings are comparable to their Yukawa coupling counterparts, sneutrino resonances may be exploited to yield high precision measurements of SUSY parameters. Further details may be found in Ref. [5]. (See also Ref. [6].)

R-parity is defined to be $R_P = +1$ and -1 for SM particles and their superpartners, respectively. If R-parity is conserved, all superpartners must be produced in pairs. However, renormalizable gauge-invariant interactions that explicitly violate R-parity and lepton number are also allowed by the superpotential

$$W = \lambda LLE^c + \lambda' LQD^c$$
$$= \lambda_{ijk}(N_i E_j E_k^c - E_i N_j E_k^c) + \lambda'_{lmn}(N_l D_m D_n^c - V^*_{pm} E_l U_p D_n^c) \,, \quad (1)$$

where the lepton and quark chiral superfields $L = (N, E)$, E^c, $Q = (U, D)$, and D^c contain the SM fermions f and their scalar partners $\tilde{f}$, V is the CKM matrix, $i < j$, and all other generational indices are arbitrary. With the couplings of Eq. (1), superpartners may be produced singly at colliders. In particular, sneutrinos $\tilde{\nu}$ may be produced as s-channel resonances at lepton colliders through the λ couplings [7,8]. Such resonance production is unique in that it probes supersymmetric masses up to $\sqrt{s}$. As sneutrinos are likely to be among the lighter superparticles, even a first stage muon collider with $\sqrt{s} = 80 - 250$ GeV will cover much of the typically expected mass range. We will explore the potential of a muon collider to study such resonances, assuming luminosity and beam resolution options $(\mathcal{L}, R) = (1 \text{ fb}^{-1}/\text{yr}, 0.1\%)$ and $(0.1 \text{ fb}^{-1}/\text{yr}, 0.003\%)$.

At muon colliders, sneutrinos $\tilde{\nu}_e$ and $\tilde{\nu}_\tau$ may be produced in the s-channel. They can then decay through λ (λ') couplings to charged lepton (down-type quark) pairs or through R_P-conserving decays, such as $\tilde{\nu} \to \nu \chi^0$. In the latter case, the lightest neutralino χ^0 subsequently decays to three SM fermions through $\not{R}_P$ interactions. The phenomenology of sneutrino resonances is thus rather complicated in full generality. However, in analogy with the Yukawa couplings, $\not{R}_P$ couplings involving higher generational indices are usually expected to be larger. We therefore focus on $\tilde{\nu}_\tau$ production through the coupling λ_{232}, and, in addition to the decay $\tilde{\nu}_\tau \to \mu^+\mu^-$, consider the possibility of $\tilde{\nu}_\tau \to b\bar{b}$ decays governed by λ'_{333}. For simplicity, we take these two $\not{R}_P$ couplings to be real and assume that all other $\not{R}_P$ parameters are negligible. The current bounds on these couplings arising from a variety of sources [5] are $\lambda_{232} \lesssim 0.06$, $\lambda'_{333} \lesssim 1$, and $\lambda_{232}\lambda'_{333} \lesssim 0.001$. We will also consider a scenario in which the R_P-conserving decay $\tilde{\nu}_\tau \to \nu_\tau \chi^0$ is important. Fig. 1 shows representative decay widths for the three modes.

The cross section for resonant $\tilde{\nu}_\tau$ production is

$$\sigma_{\tilde{\nu}_\tau}(\sqrt{s}) = \frac{8\pi \Gamma(\tilde{\nu}_\tau \to \mu^+\mu^-)\Gamma(\tilde{\nu}_\tau \to X)}{(s - m_{\tilde{\nu}_\tau}^2)^2 + m_{\tilde{\nu}_\tau}^2 \Gamma_{\tilde{\nu}_\tau}^2} \,, \quad (2)$$

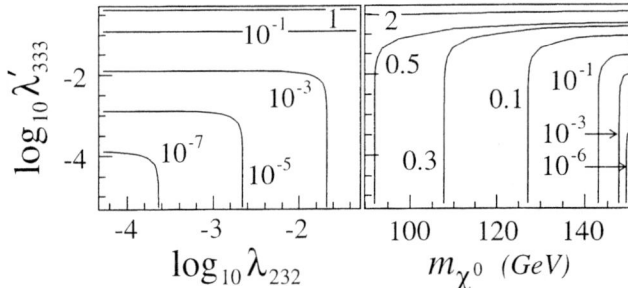

FIGURE 1. Contours of total decay width $\Gamma_{\tilde{\nu}_\tau}$ in GeV (i) for $m_{\tilde{\nu}_\tau} = 100$ GeV, assuming only $\not{R}_P$ decays $\tilde{\nu}_\tau \to \mu^+\mu^-, b\bar{b}$ are open, and (ii) for $m_{\tilde{\nu}_\tau} = 150$ GeV, assuming that $\tilde{\nu}_\tau \to \nu_\tau \chi^0$ decays are also allowed, with $\chi^0 = \tilde{B}$ and fixed $\lambda_{232} = 5 \times 10^{-5}$.

where a factor of 2 has been explicitly included to account for both $\tilde{\nu}_\tau$ and $\tilde{\nu}_\tau^*$ exchange, X denotes a generic final state from $\tilde{\nu}_\tau$ decay, and $\Gamma_{\tilde{\nu}_\tau}$ is the total sneutrino decay width. The effective cross section $\bar{\sigma}_{\tilde{\nu}_\tau}$ is obtained by convoluting $\sigma_{\tilde{\nu}_\tau}(\sqrt{s})$ with the collider's $\sqrt{s}$ distribution. Neglecting (for purposes of discussion) bremsstrahlung and beamstrahlung, this distribution is well-approximated by a Gaussian distribution with rms width $\sigma_{\sqrt{s}} = 7$ MeV $[R/0.01\%][\sqrt{s}/100$ GeV$]$, where R is the beam energy resolution factor. In two extreme limits, $\bar{\sigma}_{\tilde{\nu}_\tau}$ can be expressed in terms of branching fractions B as

$$\Gamma_{\tilde{\nu}_\tau} \ll \sigma_{\sqrt{s}}: \quad \bar{\sigma}_{\tilde{\nu}_\tau}(m_{\tilde{\nu}_\tau}) \simeq \frac{\sqrt{8\pi^3}}{m_{\tilde{\nu}_\tau}^2} \frac{\Gamma_{\tilde{\nu}_\tau}}{\sigma_{\sqrt{s}}} B(\mu^+\mu^-) B(X),$$

$$\Gamma_{\tilde{\nu}_\tau} \gg \sigma_{\sqrt{s}}: \quad \bar{\sigma}_{\tilde{\nu}_\tau}(m_{\tilde{\nu}_\tau}) \simeq \frac{8\pi}{m_{\tilde{\nu}_\tau}^2} B(\mu^+\mu^-) B(X). \quad (3)$$

If only highly suppressed $\not{R}_P$ decays are present, $\bar{\sigma}_{\tilde{\nu}_\tau} \propto \Gamma_{\tilde{\nu}_\tau}/\sigma_{\sqrt{s}}$. The small values of $\sigma_{\sqrt{s}}$ possible at a muon collider thus provide an important advantage for probing small $\not{R}_P$ couplings. At a muon collider, the effects of bremsstrahlung are small (but are included in our numerical results); beamstrahlung is negligible.

The signals for $\tilde{\nu}_\tau$ production depend on the $\tilde{\nu}_\tau$ decay patterns. We consider two well-motivated scenarios. In the first, $m_{\tilde{\nu}_\tau} < m_{\chi^0}$, and $\tilde{\nu}_\tau$ decays only through $\not{R}_P$ operators. Neglecting $\not{R}_P$ couplings other than λ_{232} and λ'_{333}, the signal is $\mu^+\mu^-$ or $b\bar{b}$ pairs in the final state. For concreteness, we consider $m_{\tilde{\nu}_\tau} = 100$ GeV.

The dominant backgrounds are Bhabha scattering and $\mu^+\mu^- \to \gamma^*, Z^* \to \mu^+\mu^-, b\bar{b}$. To reduce these, we apply the following cuts: for the $\mu^+\mu^-$ ($b\bar{b}$) channel, we require $60° < \theta < 120°$ ($10° < \theta < 170°$) for each muon (b quark). The stronger θ cuts in the $\mu^+\mu^-$ channel are needed to remove the forward-peaked Bhabha scattering. We also require $|m_{f\bar{f}} - m_{\tilde{\nu}_\tau}| < 7.5$ GeV in both channels to reduce background from radiative returns to the Z. After the cuts above and including beam energy spread and bremsstrahlung, the background cross sections at $\sqrt{s} = 100$ GeV are $\sigma(\mu^+\mu^-) = 3.5 \times 10^4$ fb and $\sigma(b\bar{b}) = 2.0 \times 10^5$ fb.

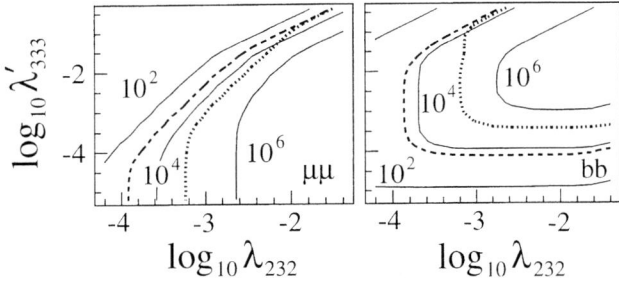

FIGURE 2. Contours for (i) $\sigma(\mu^+\mu^- \to \tilde{\nu}_\tau \to \mu^+\mu^-)$ and (ii) $\sigma(\mu^+\mu^- \to \tilde{\nu}_\tau \to b\bar{b})$ (solid) in fb after cuts for the $m_{\tilde{\nu}_\tau} < m_{\chi^0}$ scenario, with $\sqrt{s} = m_{\tilde{\nu}_\tau} = 100$ GeV and $R = 0.003\%$. The dashed and dotted contours give the optimistic and pessimistic/scan 3σ discovery boundaries, respectively, for total integrated luminosity $L = 0.1$ fb^{-1}. (See discussion in text.)

In this scenario, $\Gamma_{\tilde{\nu}_\tau}$ is unknown *a priori*, but a very small $\Gamma_{\tilde{\nu}_\tau}$ is possible. We choose the $(\mathcal{L}, R) = (0.1 \text{ fb}^{-1}/\text{yr}, 0.003\%)$ option, which maximizes $S/\sqrt{B}$ if $\Gamma_{\tilde{\nu}_\tau}$ is indeed small. With this choice, signal cross sections after cuts are given by the solid contours in Fig. 2. We see that the cross sections may be extremely large (> 1 nb) in some regions of the allowed parameter space.

In Fig. 2 we also give sneutrino resonance discovery contours for two extreme possibilities. In the most optimistic case, the sneutrino mass is exactly known and the total luminosity is applied at the sneutrino resonance peak. The corresponding "optimistic" 3σ discovery contours are given by dashed lines. (In calculating $S/\sqrt{B}$ for the $b\bar{b}$ mode here and below, we include a 75% efficiency for tagging at least one b quark.) More realistically, the sneutrino mass will be known only approximately from other colliders with some uncertainty $\pm\frac{1}{2}\Delta m_{\tilde{\nu}_\tau}$; we assume $\Delta m_{\tilde{\nu}_\tau} = 100$ MeV using the fully reconstructable $\not{R}_P$ decays. The dotted contours of Fig. 2 represent the 3σ "pessimistic/scan" $\tilde{\nu}_\tau$ discovery boundaries, where the effects of having to scan over the allowed sneutrino mass interval are included. (See Ref. [5] for details.) The actual discovery limit will lie between the dashed and dotted contours. We see that $\tilde{\nu}_\tau$ resonance observation is possible for $\not{R}_P$ couplings as low as $10^{-3} - 10^{-4}$.

We now consider a second scenario in which $m_{\tilde{\nu}_\tau} > m_{\chi^0}$. In addition to $\not{R}_P$ decays, decays $\tilde{\nu}_\tau \to \nu_\tau \chi^0$ are now also allowed and typically dominate, with χ^0 then decaying to $\nu_\tau \mu\mu$ or $\nu_\mu \mu\tau$ through the λ_{232} coupling, or $\nu_\tau b\bar{b}$ through the λ'_{333} coupling. The final signals are then $\mu^+\mu^- + \not{E}_T$, $\mu^\pm \tau^\mp + \not{E}_T$, and $b\bar{b} + \not{E}_T$. For this scenario, we consider masses $m_{\tilde{\nu}_\tau} = 150$ GeV and $m_{\chi^0} = 100$ GeV.

The leading backgrounds to the $\nu\chi^0$ channels are from $WW^{(*)}$ and $ZZ^{(*)}$. To reduce these, we require $\not{E}_T > 25$ GeV, that the visible final state fermions have $p_T > 25$ GeV and $60° < \theta < 120°$ for the lepton modes ($40° < \theta < 140°$ for the $b\bar{b} \not{E}_T$ mode), and that the invariant mass of the two visible fermions be > 50 GeV. With these cuts, the total combined background in the $\nu\chi^0$ channels is ~ 1 fb.

The signal cross sections for the $\nu\chi^0$ channel (without cuts) and the direct $\not{R}_P$ $b\bar{b}$ channel (after cuts as in Fig. 2) are plotted in Fig. 3. We also give 3σ discovery contours for both optimistic and pessimistic/scan cases as before, where we choose

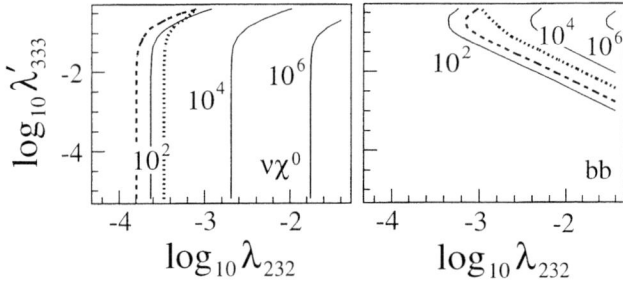

FIGURE 3. Contours for (i) $\sigma(\mu^+\mu^- \to \tilde{\nu}_\tau \to \nu\chi^0)$ (no cuts) and (ii) $\sigma(\mu^+\mu^- \to \tilde{\nu}_\tau \to b\bar{b})$ (after cuts) in fb assuming $m_{\tilde{\nu}_\tau} = 150$ GeV, $m_{\chi^0} = 100$ GeV, and $\chi^0 = \tilde{B}$. The optimistic (dashed) and pessimistic/scan (dotted) discovery contours assume $L = 1$ fb^{-1} and $R = 0.1\%$.

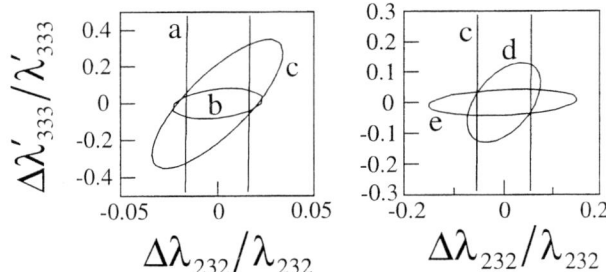

FIGURE 4. $\chi^2 = 1$ contours in the $(\Delta\lambda_{232}/\lambda_{232}, \Delta\lambda'_{333}/\lambda'_{333})$ plane for (i) the $m_{\tilde{\nu}_\tau} = 100$ GeV $< m_{\chi^0}$ scenario, assuming $L = 0.3$ fb^{-1}, $R = 0.003\%$ and (ii) the $m_{\tilde{\nu}_\tau} = 150$ GeV $> m_{\chi^0} = 100$ GeV scenario, assuming $L = 3$ fb^{-1}, $R = 0.1\%$. Contours are for $\lambda_{232} = 5 \times 10^{-4}$ and $\lambda'_{333} =$: (a) 10^{-5}; (b) 5×10^{-4}; (c) 10^{-2}; (d) 10^{-1}; (e) 0.3.

the $(\mathcal{L}, R) = (1$ fb^{-1}/yr, 0.1%) option to maximize $\mathcal{L}$. For the "pessimistic/scan" discovery contours, we assume $\Delta m_{\tilde{\nu}_\tau} \sim 2$ GeV from kinematic endpoints. We see that the nearly background-free $\nu\chi^0$ mode makes possible a dramatic improvement in discovery reach compared to the $m_{\tilde{\nu}_\tau} < m_{\chi^0}$ scenario. The $\tilde{\nu}_\tau$ resonance may be discovered for $\lambda_{232} \gtrsim 10^{-4}$, irrespective of the value of λ'_{333}.

Once we have found the sneutrino resonance via the scan described, the crucial goal will be to precisely measure the relevant $\not{R}_P$ couplings. In the $m_{\tilde{\nu}_\tau} < m_{\chi^0}$ scenario, the discovery scan gives a precise determination of $m_{\tilde{\nu}_\tau}$ (and, if $\Gamma_{\tilde{\nu}_\tau} > 2\sigma_{\sqrt{s}}$, a rough determination of $\Gamma_{\tilde{\nu}_\tau}$). We then envision accumulating $L = 0.1$ fb^{-1} ($R = 0.003\%$) at each of the three points $\sqrt{s} = m_{\tilde{\nu}_\tau}$, $m_{\tilde{\nu}_\tau} \pm \Delta\sqrt{s}/2$, where $\Delta\sqrt{s} = \max[2\sigma_{\sqrt{s}}, \Gamma_{\tilde{\nu}_\tau}]$. The off-resonance points ensure good sensitivity to $\Gamma_{\tilde{\nu}_\tau}$. This is especially crucial when $\Gamma_{\tilde{\nu}_\tau} > \sigma_{\sqrt{s}}$, as in this case a single measurement of $\bar{\sigma}_{\tilde{\nu}_\tau}$ at $\sqrt{s} = m_{\tilde{\nu}_\tau}$ determines $B(\tilde{\nu}_\tau \to \mu^+\mu^-)$ but not $\Gamma(\tilde{\nu}_\tau \to \mu^+\mu^-)$; see Eq. (3). In the $m_{\tilde{\nu}_\tau} > m_{\chi^0}$ scenario, we noted that $\Gamma_{\tilde{\nu}_\tau}$ can be computed with good precision from observations at other colliders; we assume a $\pm 5\%$ error for $\Gamma_{\tilde{\nu}_\tau}$. We would then run only at $\sqrt{s} \simeq m_{\tilde{\nu}_\tau}$ and accumulate $L = 3$ fb^{-1} ($R = 0.1\%$). In Fig. 4, the resulting $\chi^2 = 1$ error contours are plotted for each of the two scenarios. We find that 1σ

fractional errors at the few percent level can be achieved, even for a small value of $\lambda_{232} = 5 \times 10^{-4}$, which is not very far inside the discovery regions.

As a final remark, we note that $\not{R}_P$ interactions can split the complex scalar $\tilde{\nu}_\tau$ into a real CP-even and a real CP-odd mass eigenstate. This splitting is generated both at tree-level (from sneutrino-Higgs mixing) and radiatively, and both contributions depend on many SUSY parameters. However, such $\not{R}_P$ terms also generate neutrino masses, and it is generally true that the sneutrino splittings generated are $\mathcal{O}(m_\nu)$ [9]. Given the current bound $m_{\nu_\tau} < 18.2$ MeV [10], we see that τ sneutrino splittings may be as large as $\mathcal{O}(10 \text{ MeV})$. A muon collider with $R = 0.003\%$ is uniquely capable of resolving resonance peak splittings at or below the MeV level.

In summary, we have demonstrated that a muon collider is an excellent tool for discovering sneutrino resonances and measuring their R-parity violating couplings. Note that for small $\not{R}_P$ couplings, absolute measurements through other processes and at other colliders are extremely difficult, as they typically require that $\not{R}_P$ effects be competitive with a calculable R_P-conserving process. For example, $\not{R}_P$ neutralino branching ratios constrain only ratios of $\not{R}_P$ couplings. In addition, a muon collider is unique in its ability to resolve the splitting between the CP-even and CP-odd sneutrino components when this splitting is as small as expected given the current bounds on neutrino masses.

I thank J. Gunion and T. Han for the collaboration upon which this talk was based and the U.C. Davis theory group for hospitality during the course of this work.

REFERENCES

1. J. L. Feng and M. J. Strassler, Phys. Rev. D **51**, 4661 (1995); *ibid.*, **55**, 1326 (1997).
2. J. Kelly, talk presented at the Workshop on Physics at the First Muon Collider, Fermilab, November 6–9, 1997; J. F. Gunion, these proceedings.
3. M. S. Berger, talks presented at the Workshop on Physics at the First Muon Collider, Fermilab, November 6–9, 1997, hep-ph/9712486 and hep-ph/9712474.
4. J. Lykken, these proceedings.
5. J. L. Feng, J. F. Gunion, and T. Han, hep-ph/9711414.
6. See also S. Raychaurhuri, talk presented at the Workshop on Physics at the First Muon Collider, Fermilab, November 6–9, 1997.
7. S. Dimopoulos and L. J. Hall, Phys. Lett. B **207**, 210 (1988); S. Dimopoulos, R. Esmailzadeh, L. J. Hall, J.-P. Merlo, and G. D. Starkman, Phys. Rev. D **41**, 2099 (1990); H. Dreiner and S. Lola, in *Workshop on e^+e^- Collisions at 500 GeV*, 1991; V. Barger, G. F. Giudice, and T. Han, Phys. Rev. D **40**, 2987 (1989); J. Kalinowski, R. Rückl, H. Spiesberger, and P. M. Zerwas, Phys. Lett. B **406**, 314 (1997).
8. J. Erler, J. L. Feng, and N. Polonsky, Phys. Rev. Lett. **78**, 3063 (1997).
9. See, *e.g.*, N. Polonsky, hep-ph/9708325; Y. Grossman, hep-ph/9710276.
10. M. Girone, ALEPH Collaboration, talk #1003 presented at the International Europhysics Conference on High Energy Physics, 19–26 August 1997, Jerusalem, Israel.

New Particles and Interactions at High Energy Muon Colliders

Stephen Godfrey

Ottawa-Carleton Institute for Physics[1],
Department of Physics, Carleton University, Ottawa Canada K1S 5B6

Abstract. I give an overview of the ability of a high energy $\mu^+\mu^-$ collider to discover new particles and interactions. I start with heavy fermions which will be the most straightforward to produce and observe. I then discuss single leptoquark production which is produced via the quark content of the photon and the discovery potential for extra gauge bosons which will manifest themselves via deviations of observables from their standard model values. Finally, contact interactions are studied as the generalization of looking for new interactions via deviations from the standard model.

INTRODUCTION

Although the Standard Model (SM) of particle physics is in complete agreement with present experimental data, it is believed to leave many questions unanswered. This belief has resulted in numerous models that approximate the SM at presently accessible energies but which have a much richer particle spectrum above 100 GeV. Some models extend the SM gauge group by either embedding the extra gauge groups in a Grand Unified Group (GUT) or not embedding them. GUT theories also come in supersymmetric varieties which leads to further phenomenological consequences, in particular all the supersymmetric partners of the "conventional" particles and gauge bosons [1]. Another broad class of models are the various composite models where the gauge bosons are composite, the fermions are composite, or the Goldstone bosons that become the longitudinal components of the massive gauge bosons are composite (eg. technicolour models).

These models lead to many types of new particles such as; extra gauge bosons (Z''s and W''s); new fermions which come in many forms such as 4th generation fermions, mirror fermions, vector fermions, and singlets like massive neutrinos; leptoquarks, bileptons and diquarks; extended Higgs sector; excited fermions which would signify substructure; and other truly weird particles that we have yet to imagine.

---

[1] This Research is supported by the Natural Sciences and Engineering Research Council of Canada

To reveal what lies beyond the SM we need to elucidate and complete the TeV particle spectrum. In the remainder of this contribution I will survey the capability of high energy $\mu^+\mu^-$ colliders to discover new particles and interactions. Because this is such a broad topic the survey is necessarily incomplete. A good source of recent results is the contributions of the New Phenomena working group at the 1996 Snowmass Study on High Energy Physics [2].

NEW FERMIONS

New fermions [3] are generally classified by the quantum numbers of their chiral components. Fourth generation fermions are massive duplicates of SM fermions. In contrast the left and right handed components of vector fermions are in $SU(2)_L$ and $SU(2)_R$ doublets respectively and mirror fermions have their left handed components in $SU(2)_L$ singlets and their right handed components in $SU(2)_R$ doublets. Except for singlet neutrinos new fermions couple to the photon and/or weak bosons with full strength allowing for pair production with unambiguous cross section. Fermion-antifermion pairs are produced via $\mu^+\mu^- \to F\bar{F}$ through s-channel γ or Z^0 so the cross section goes approximately like the QED point cross section. Fermions can be pair produced in sufficient numbers for discovery up to close to the kinematic limit, $\sqrt{s}/2$.

New fermions with conventional quantum numbers can mix with their SM partners. The mixing is severely constrained by the non-observation of FCNC. Nevertheless if the mixing is not too small new fermions can be produced singly in association with their light partners. This results in a significantly higher search limit, almost $\sqrt{s}$ of the collider.

LEPTOQUARKS

Leptoquarks are colour triplets or anti-triplets carrying both baryon and lepton quantum numbers and can have spin 0 or spin 1. They appear in a wide variety of models such as GUT's, technicolour, and composite models [4]. Leptoquarks reveal themselves with a dramatic signal of a high p_T lepton balanced by a jet.

In addition to being pair produced like the fermions of the previous sections [4] leptoquarks can also be produced singly via the quark content of a Weissacker-Williams photon radiated off an incoming muon [5]. The cross-section for the process is found by convoluting the quark distribution inside the photon with the $q + \mu \to LQ$ cross section:

$$\sigma(s) = \int f_{q/\gamma}(z, M_s^2)\hat{\sigma}(\hat{s})dz = f_{q/\gamma}(M_s^2/s, M_s^2)\frac{2\pi^2 \kappa \alpha_{em}}{s} \quad (1)$$

where the leptoquark couplings are replaced by a generic Yukawa coupling g which is scaled to electromagnetic strength $g^2/4\pi = \kappa \alpha_{em}$. The resulting cross-section is then convoluted with the photon distribution to obtain the total cross section:

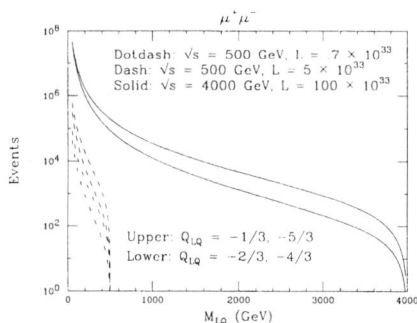

FIGURE 1. Event rates for leptoquark production at high energy muon colliders. The results were obtained using the GRV distribution functions for the quark content of the photon [6].

$$\sigma(\mu^+\mu^- \to XS)$$
$$= \frac{2\pi^2 \alpha_{em} \kappa}{s} \int_{M_s^2/s}^{1} \frac{dx}{x} f_{\gamma/\mu}(x, \sqrt{s}/2) f_{q/\gamma}(M_s^2/(xs), M_s^2) \qquad (2)$$

The number of expected events ($L \times \sigma$) for various muon collider parameters are shown in Fig. 1 where we have taken $\kappa = 1$. Because this is a muon collider we are considering 2nd generation LQ's so that we use the s and c-quark content of the photon as appropriate. Basing discovery on the production of 100 LQ's leads to the search limits quoted in Table I. The OPAL [7] and DELPHI [8] collaborations have used this process to obtain limits on LQ's at LEP200.

TABLE 1. LQ discovery limits at $\mu^+\mu^-$ colliders for the given $\sqrt{s}$ and integrated luminosity. The Scalar and Vector refers to the LQ spin and the $-1/3 - 5/3$ etc. refers to its charge.

$\sqrt{s}$ (TeV)	L fb^{-1}	Scalar		Vector	
		-1/3, -5/3	-4/3, -2/3	-1/3, -5/3	-4/3, -2/3
0.5	7	250	170	310	220
0.5	50	400	310	440	360
4.0	1000	3600	3000	3700	3400

NEW GAUGE BOSONS

New gauge bosons are a generic prediction of models with extensions of the SM gauge group [9]. They contribute to $\mu^+\mu^-$ cross-sections in the s-channel [10]. The cross-sections for various Z''s are shown in Fig. 2. It is clear from this figure that if production of real Z''s is kinematically accessible it will be produced in a sufficiently large quantity so that its properties can be investigated in detail. For the highest energy muon colliders being contemplated this translates into production

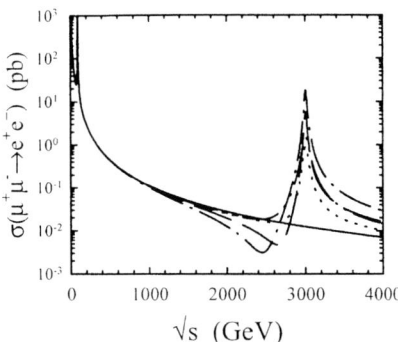

FIGURE 2. $\mu^+\mu^-$ cross-section as a function for $\sqrt{s}$ for the SM (solid), Z_χ (dashed), Z_{LR} (dotted), Z_{ALR} (dot-dashed), and Z_{SSM} (dot-dot-dashed).

of Z''s of $M_{Z'} = 4$ TeV (or 5 TeV depending on the actual $\sqrt{s}$ of the machine). By comparison, the LHC can achieve a discovery reach of 4-5 TeV, depending on the specific Z', based on roughly 10 dilepton pairs clustering at the same invariant mass. Thus, the main advantage of the muon collider is that it could produce enough Z''s to study them in detail.

Searches for Z''s can be extended to masses much higher than $\sqrt{s}$ by looking for deviations from SM observables. This is illustrated in the $\sigma(\mu^+\mu^- \to e^+e^-)$ plotted in Fig. 2 where significant deviations from the SM occur below the Z' pole due to interference of the Z' propagator with the γ and Z^0 propagators.

To represent a meaningful signal of new physics, deviations should be observed in as many observables as possible. Observables are constructed from cross sections to specific final state fermions. A set of such observables are; σ^f, the cross sections, A_{FB}^f, the forward-backward asymmetries, and A_{LR}^f, the left-right polarization asymmetries, where $f = \mu, \tau, c, b,$ and had =sum over hadrons. To obtain discovery limits for new physics we look for statistically significant deviations from standard model expectations. In Fig. 3 a number of observables are shown with their standard model values and for various Z''s as a function of the Z' mass. The $1 - \sigma$ error bars shown are based on the statistics expected in the standard model. What is important to note is that the different observables have different sensitivities to different models. For example, of the models shown, $\sigma(\mu^+\mu^- \to e^+e^-)$ is most sensitive to Z_{ALR} while R^{had} is most sensitive to Z_χ. Therefore to have the highest possible reach for the largest number of possible models it is important to include all possible observables. We quantify the sensitivity to an extra gauge boson by comparing the predictions for various observables assuming the presence of a Z' to the predictions of the standard model and constructing the χ^2 figure of merit. The "discovery" limits were obtained by including the ten observables: $\sigma^\mu, \sigma^\tau, \sigma^c$, $\sigma^b, R^{had}, A_{FB}^\mu, A_{FB}^\tau, A_{FB}^c, A_{FB}^b$, and P_τ. In calculating the χ^2 we assumed 35% c-tagging efficiency and 60% b-tagging efficiency. The 99% C.L. discovery limits are

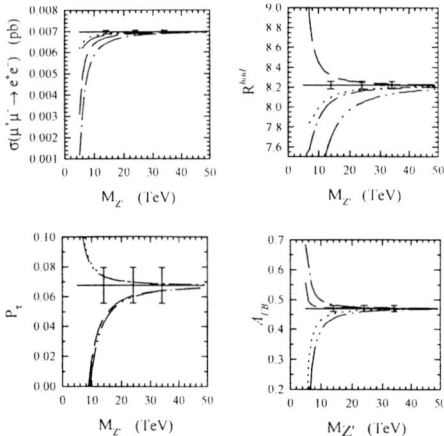

FIGURE 3. Some $\mu^+\mu^-$ observables as a function of $M_{Z'}$ at $\sqrt{s} = 500$ GeV for the SM (solid), Z_χ (dashed), Z_η (dotted), Z_{LR} (dot-dashed) and Z_{ALR} (dot-dot-dashed). The error bars are based on the statistical error assuming an integrated luminosity of 50 fb^{-1}.

shown in Fig. 4 [10]. Only statistical errors are considered in obtaining the limits shown. We did not consider observables involving polarization of the initial state leptons for the muon colliders (although they were included for the e^+e^- collider results). A very exciting development discussed at this meeting was the possibility of very high muon polarization without too large a decrease in the luminosity. Polarization asymmetries are in many cases the most sensitive observables so that polarization is potentially very important for searches for Z''s.

CONTACT INTERACTIONS

In the previous section we described how the existence of Z''s might reveal themselves through deviations from the SM. For very massive Z's the Z' propagator can be described by a 4-Fermi interaction [11]:

$$\frac{g_{Z'}^2}{s - M_{Z'}^2} \xrightarrow{M_{Z'} \gg \sqrt{s}} \frac{g_{Z'}^2}{M_{Z'}^2}. \tag{3}$$

Likewise, leptoquark exchange in the t-channel in processes like $\mu^+\mu^- \to q\bar{q}$ can also be described this way

$$\frac{\kappa\alpha_{em}}{t + M_{LQ}^2} \xrightarrow{M_{LQ} \gg \sqrt{s}} \frac{\kappa\alpha_{em}}{M_{LQ}^2}. \tag{4}$$

Form factors or residual effective interactions associated with fermion substructure is also often parametrized by contact terms in the low-energy Lagrangian. Thus

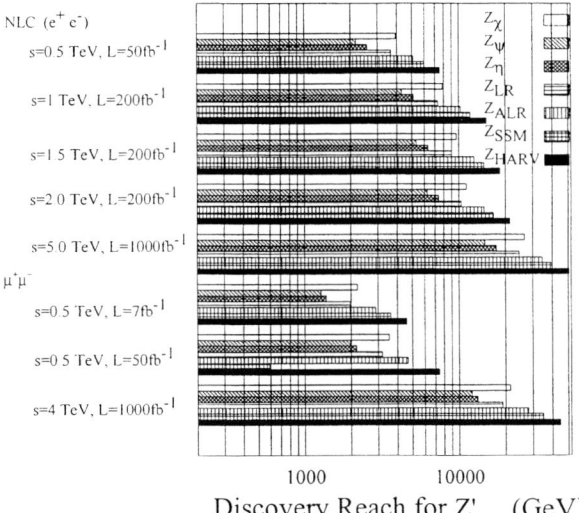

FIGURE 4. Search limits for extra neutral gauge bosons at high energy lepton colliders. The criteria for obtaining these limits are described in the text.

four fermion contact interactions represents a useful parametrization of many types of new physics originating at a high energy scale.

These contact interactions are described by non-renormalizable operators in the effective low-energy lagrangian. The lowest order four-fermion contact terms are dimension-6 and hence have dimensionful coupling constants proportional to g_{eff}^2/Λ^2. They are often written in the form [11]:

$$\mathcal{L} = \frac{4\pi}{2\Lambda^2}[\eta_{LL}(\bar{e}_L\gamma_\mu e_L)(\bar{f}_L\gamma^\mu f_L) \\ + \eta_{LR}(\bar{e}_L\gamma_\mu e_L)(\bar{f}_R\gamma^\mu f_R) + \eta_{RL}(\bar{e}_R\gamma_\mu e_R)(\bar{f}_L\gamma^\mu f_L) \\ + \eta_{RR}(\bar{e}_R\gamma_\mu e_R)(\bar{f}_R\gamma^\mu f_R)]. \quad (5)$$

Interference between the contact terms and the usual gauge interactions can lead to observable deviations from SM predictions at energies lower than Λ. The effects of a contact interaction are illustrated in Fig. 5 where the differential cross-section for $\mu^+\mu^- \to b\bar{b}$ is plotted for various values of Λ.

To gauge the sensitivity to the compositeness scale we assume that the SM is correct and perform a χ^2 analysis of the $\cos\theta$ angular distribution. To perform this we choose the detector acceptance to be $|\cos\theta| < 0.94$ (corresponding to $\theta = 20°$) [12]. We note that angular acceptance of a typical muon collider detector is expected to be reduced due to additional shielding required to minimize the radiation backgrounds from the muon beams. We assume canonical LEP values, $\epsilon_b = 25\%, \epsilon_c = 5\%$ but warn the reader that these numbers are quite arbitrary and

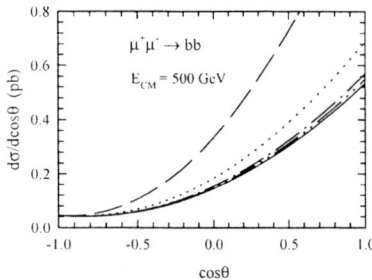

FIGURE 5. The $\cos\theta$ distribution for $\mu^+\mu^- \to b\bar{b}$ at $E_{CM} = 0.5$ TeV with $\eta_{LL} = +1$ for the SM (solid), $\Lambda = 5$ TeV (dashed), $\Lambda = 10$ TeV (dotted), $\Lambda = 20$ TeV (dot-dashed), $\Lambda = 30$ TeV (dot-dot-dashed).

are only used for illustrative purposes. We divide the angular distribution into 10 equal bins. The χ^2 distribution is evaluated by the usual expression.

The 95% C.L. bounds on Λ are shown graphically in Fig. 6. Quite generally, high luminosity $\mu^+\mu^-$ colliders are quite sensitive to contact interactions with discovery limits ranging from 5 to 50 times the center of mass energy. As in the discussion of Z's, polarization will be important, especially in unravelling the chirality of deviations if they are observed.

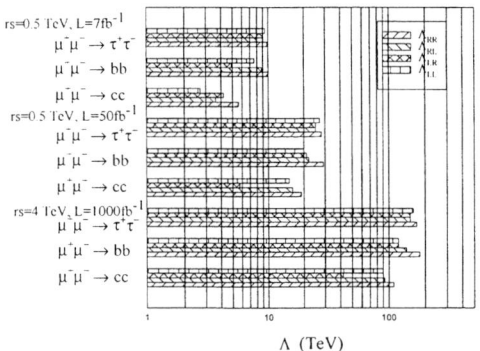

FIGURE 6. Sensitivity to the new physics scale, Λ, at high energy muon colliders. The criteria for obtaining these limits are described in the text.

FINAL COMMENTS

The main attractions of a muon collider are its high energy reach in a relatively clean environment. For certain types of physics a high energy muon collider could play a unique role. For example, if the LHC discovered a Z' with mass of 4 TeV a muon collider of sufficiently high energy would be able to do detailed studies of its properties. Another example is the existence of heavy leptons. These are notoriously difficult and maybe impossible to discover at a hadron collider. Yet for a high energy muon collider this would be straightforward.

The workshop discussed the likelihood of producing highly polarized beams. These would play an important role in identifying the nature of a new particle or interaction, whether it be a leptoquark or Z'. Identification studies using polarization would be a useful excerise.

Finally, we should keep our minds open to the possibility of genuine surprises which we have not yet imagined.

REFERENCES

1. See for example H. Baer, these proceedings; J. Gunion, these proceedings; J. Feng, thse proceedings; and references therein.
2. S. Godfrey, J. Hewett, and L. Price, *Proceedings of the 1996 DPF/DPB Summer Study on New Directions for High Energy Physics* - Snowmass96, Snowmass, CO, 25 June - 12 July, 1996, p. 846, [hep-ph/9704291] and references therein. See also Ref. [3] and Ref. [9]
3. For a recent review see A. Djouadi, J. Ng, and T.G. Rizzo, *Electro-Weak Symmetry Breaking and Beyond the Standard Model*, eds. T. Barklow, S. Dawson, H. Haber and J. Seigrist (World Scientific, 1996) p. 416.
4. T.G. Rizzo, *Proceedings of the 2nd International Workshop on e^-e^- Interactions at TeV Energies*, Santa Cruz CA, September 22-24 1997 [hep-ph/9710350].
5. M.A. Doncheski and S. Godfrey, *Proceedings of the 1996 DPF/DPB Summer Study*, [hep-ph/9612385]. See also M. A. Doncheski and S. Godfrey, Phys. Rev. **D49**, 6220 (1994) [hep-ph/9608368]; **D51**, 1040 (1995) and references therein.
6. M. Glück, E. Reya and A. Vogt, Phys. Lett. **B222**, 149 (1989); Phys. Rev. **D45**, 3986 (1992); Phys. Rev. **D46**, 1973 (1992).
7. OPAL Collaboration, *The 11th International Workshop on Photon-Photon Collisions*, Egmond aan Zee Netherlands, 10-15 May 1997, [hep-ex/9706003].
8. DELPHI Collaboration, Submitted to the *Lepton Photon'97 Conference*, Hamburg Germany, July 28- Aug 1, 1997, DELPHI 97-112 CONF 94.
9. For a recent review see M. Cvetic and S. Godfrey, *Electro-Weak Symmetry Breaking and Beyond the Standard Model*, eds. T. Barklow, S. Dawson, H. Haber and J. Seigrist (World Scientific, 1996) p. 383 [hep-ph/9504216].
10. S. Godfrey, *Proceedings of the 1996 DPF/DPB Summer Study*, p. 883, [hep-ph/9612384]. See also S. Godfrey, Phys. Rev. **D51**, 1402 (1995); S. Capstick and S. Godfrey, Phys. Rev. **D37**, 2466 (1988); G. Bélanger and S. Godfrey, Phys. Rev. **D34**, 1309-1315 (1986); T.G. Rizzo, *Proceedings of the 2nd International Workshop on e^-e^- Interactions at TeV Energies*, Santa Cruz CA, September 22-24 1997 [hep-ph/9710229] and references therein.
11. E. Eichten, K. Lane, and M. Peskin, Phys. Rev. Lett. **50**, 811 (1983).
12. K. Cheung, S. Godfrey, and J. Hewett, *Proceedings of the 1996 DPF/DPB Summer Study*, p. 989, [hep-ph/9612257].

Sparticle masses from kinematic fitting at a muon collider

Joseph D. Lykken*

*Theoretical Physics Department[1]
Fermi National Accelerator Laboratory
P.O. Box 500
Batavia, IL 60510

Abstract. Three case studies are presented of slepton pair production followed by two-body or quasi-two-body decays at a muon collider. Precision mass measurements are possible using a variety of kinematic fitting methods. Standard Model and supersymmetric backgrounds are easily controlled by kinematic cuts. In all three cases it appears that detector resolutions, not backgrounds or statistics, will dominate the final error bars.

Introduction

A muon collider is in principle an excellent machine for precision studies of weak scale supersymmetry. Depending on $\sqrt{s}$ and the SUSY mass spectrum, it may be possible to observe pair production of a half-dozen or more distinct sparticles. For R parity preserving SUSY, sparticle pair production is kinematically underconstrained, due to the pair of unmeasured LSP's. However in many cases each sparticle in the pair has a significant branching fraction for what is essentially a two-body decay:

$$\text{sparticle} \to \text{LSP} + \text{particle} \quad , \tag{1}$$

where "particle" refers to a fully reconstructible Standard Model particle (e, μ, W, Z, and possibly h_0, t), while the LSP is assumed to be the lightest neutralino $\tilde{\chi}_1^0$.

In these cases there are a variety of kinematic fitting methods for extracting sparticle masses. In this talk I will report on two such methods applied to smuon, selectron, and sneutrino production at a muon collider. Chargino production is not discussed, since for light fermionic sparticles the best method for a precision mass

---

[1] Research supported by the Fermi National Accelerator Laboratory, which is operated by Universities Research Association, Inc., under contract no. DOE-AC02-76CHO3000.

TABLE 1. Sparticle and Higgs spectrum for LHC Point 5, which corresponds to minimal sugra parameters m_0=100 GeV, $m_{1/2}$=300 GeV, A_0=0, $\tan\beta$=2.1, and sgn(μ)=1.

Particle	Mass (GeV)	Particle	Mass (GeV)
$\tilde{\chi}_1^0$	119	$\tilde{\chi}_2^0$	228
$\tilde{\chi}_1^\pm$	228	$\tilde{\chi}_2^\pm$	565
$\tilde{e}_R$	157	$\tilde{e}_L$	241
$\tilde{\mu}_R$	157	$\tilde{\mu}_L$	241
$\tilde{\nu}_L$	232	$\tilde{g}$	754
$\tilde{t}_1$	448	$\tilde{b}_1$	604
h_0	94	H_A	657

measurement is a threshold scan [1]. An interesting challenge for future investigation is the production of staus, stops, and the heavier chargino and neutralinos.

Sparticle production at a muon collider is similar in many respects to sparticle production at an e^+e^- machine. For the present analysis the most important differences are that the muon collider has (i) much higher energy reach, (ii) significantly lower advertised luminosity at comparable energies, (iii) little or no polarization available without taking a significant hit in luminosity, and (iv) large detector backgrounds from muon decays. These detector backgrounds are generally soft, but large fluctuations could cause problems for precision SUSY measurements. They will also impact on isolation cuts, determinations of missing E_T, and detector resolutions generally. These problems will be left to future study.

At a muon collider smuon pairs arise from both s and t channel production; the s channel production is through a virtual photon or Z, while the t channel diagram involves the exchange of a neutralino. The s and t channel contributions interfere destructively, but this effect will not be important for the examples considered here, where the t channel production is dominant. Selectron production proceeds only through the s channel, and is thus suppressed in the examples. Muon sneutrino production proceeds only through the t channel, and is thus competitive with smuons.

In both supergravity (sugra) and gauge mediated models, the $\tilde{l}_R$'s are lighter than the $\tilde{l}_L$'s. Independent of the SUSY model, $\tilde{l}_R$'s decay almost 100% via a single two-body mode: $\tilde{\mu}_R \to \tilde{\chi}_1^0 \mu$. The branching fractions of the $\tilde{l}_L$'s and $\tilde{\nu}_L$'s are model dependent. The important decay modes are: $\tilde{\mu}_L \to \tilde{\chi}_1^0 \mu$, $\tilde{\chi}_2^0 \mu$, and $\tilde{\chi}_1^\pm \nu_\mu$.

Kinematics

The basic kinematics can be understood by considering pair production of $\tilde{\mu}_R$:

$$\mu^+\mu^- \to \tilde{\mu}_R(p_1)\tilde{\mu}_R(p_2)$$
$$\tilde{\mu}_R(p_1) \to \tilde{\chi}_1^0(p_3)\mu(p_4), \quad \tilde{\mu}_R(p_2) \to \tilde{\chi}_1^0(p_5)\mu(p_6) \quad . \tag{2}$$

TABLE 2. Sparticle and Higgs spectrum for the heavy sugra point, which corresponds to minimal sugra parameters m_0=500 GeV, $m_{1/2}$=350 GeV, A_0=0, $\tan\beta$=2, and sgn(μ)=-1.

Particle	Mass (GeV)	Particle	Mass (GeV)
$\tilde{\chi}_1^0$	145	$\tilde{\chi}_2^0$	290
$\tilde{\chi}_1^\pm$	290	$\tilde{\chi}_2^\pm$	809
$\tilde{e}_R$	519	$\tilde{e}_L$	561
$\tilde{\mu}_R$	519	$\tilde{\mu}_L$	561
$\tilde{\nu}_L$	558	$\tilde{g}$	886
$\tilde{t}_1$	597	$\tilde{b}_1$	763
h_0	84	H_A	1083

Each event consists of an acoplanar dimuon pair plus missing E_T. Six measurements are made, i.e., the 3-momenta of the two muons. The event is characterized by 13 kinematic variables: the four 3-momenta of the final state plus the common LSP mass. There are 5 kinematic constraints: one from the assumption that the two smuons have the same mass, and the rest from the known initial state 4-momentum. This leaves 2 undetermined variables in the event, which we may take as $M_{\tilde{\mu}}$, $M_{\rm LSP}$.

The kinematic endpoint method arises from the expression for the energy of each muon as measured in the rest frame of its parent smuon. The maximum and minimum boosts from this frame to the lab frame then provides us with two kinematic endpoints $E_\mu^{\rm max}$, $E_\mu^{\rm min}$, in the muon energy spectrum. A precision measurement of both endpoints allows us to extract both $M_{\tilde{\mu}}$ and $M_{\rm LSP}$. Note that this method requires good statistics to be useful, and does not take advantage of all the kinematic information in the event.

Another kinematic method, developed by Feng and Finnell for e^+e^- studies, extracts $M_{\tilde{\mu}}$ assuming a precision value of $M_{\rm LSP}$ is already known from other sources and thus can be used as an input. This method starts with the relation

$$M_{\tilde{\mu}}^2 = \frac{1}{4}s - |\vec{p}_3|^2 - |\vec{p}_4|^2 - 2|\vec{p}_3||\vec{p}_4|\cos\theta_{34} \quad . \tag{3}$$

For given input value of $M_{\rm LSP}$, the only unknown on the right hand side is θ_{34}, the angle between the 3-vectors $\vec{p}_3$ and $\vec{p}_4$. This angle is then estimated, event by event, by a certain function of measured variables. This function has the property that the error of the estimate goes to zero in the limit that the two LSP's are back-to-back in the lab frame. For $\sqrt{s}/2 \gg M_{\tilde{\mu}} \gg M_{\rm LSP}$, the mass estimates peak strongly around the true value, and precise results are possible even for rather sparse data.

A third kinematic method, which is currently under investigation, involves adapting the likelihood methods developed for extracting the top quark mass from the dilepton channel. This method is also well-suited to sparser data sets.

Simulations

Simulations were performed using PYTHIA v6.1 [3] coupled to the ATLFAST v1.25 [4] fast detector simulator. Note that the small differences in the sparticle spectra produced by PYTHIA and ISAJET [5] make a difference for the analysis done here. The ATLFAST defaults were used for lepton isolation and jet reconstruction. Smearing was not included, and no attempt was made to include detector backgrounds. Thus "precision" here refers only to statistics and to SUSY signal versus Standard Model (SM) backgrounds and SUSY backgrounds.

Sleptons at LHC point 5

This first study overlaps with the analysis presented by Frank Paige at the Fermilab workshop [6]. LHC point 5 is a mimimal supergravity reference point described in Table 1. For dimuon and dielectron production at $\sqrt{s}=600$ GeV, cuts were imposed similar to those of [6]: exactly two isolated e or μ leptons and no jets, $E > 10$ GeV and $|\eta| < 1.3$ for each lepton, $\Delta\phi_{1,2} < 0.9\pi$, $|\vec{p}_{T,1} + \vec{p}_{T,2}| > 10$ GeV, and missing $E_T > 20$ GeV.

The signal acceptance with these cuts is approximately 40%. The cuts are very efficient at eliminating backgrounds. The simulations included the six most impor-

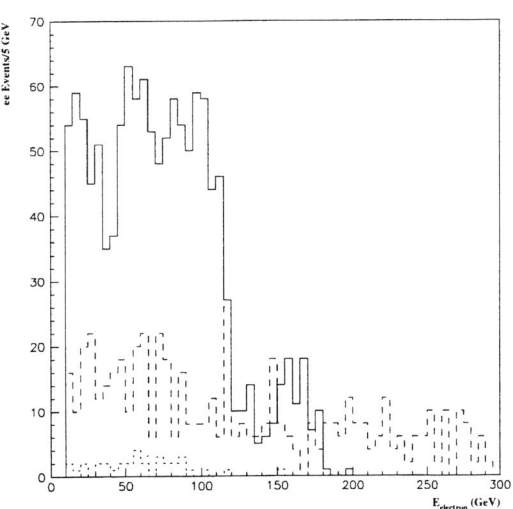

FIGURE 1. Dielectron production after cuts, 20 fb^{-1} at $\sqrt{s}=600$ GeV for LHC point 5. The solid line is the total selectron signal. The dashed line is the sum of the Standard Model backgrounds; the dotted line is the background from chargino pairs.

tant SM backgrounds. The main SUSY background is from chargino pair production, with both charginos decaying leptonically.

Figure 1 shows the dielectron event rate plotted versus electron energy, with 5 GeV bins. The SM backgrounds after cuts are rather flat and encouraging small. The SUSY background is negligible. For 20fb^{-1} of integrated luminosity, Figure 1 also reflects the rather poor statistics of selectron production. This is not surprising given that the total cross section is only 64 fb. The situation is noticeably better for smuon production, where the cross section is 400 fb.

Figure 2 shows the $\mu\mu$–ee flavor subtracted slepton signal, after cuts, broken down into the its three components: RR, RL+LR, and LL. The integrated luminosity is 100 fb^{-1} to enhance the statistics. As discussed in [6], this figure shows a rather complicated structure, reflecting the fact that there are eight distinct kinematic endpoints affecting the distribution. These are: $\tilde{l}_R\tilde{l}_R$: 118 GeV, 9 GeV, $\tilde{l}_R\tilde{l}_L$: 105 GeV, 11 GeV, $\tilde{l}_L\tilde{l}_R$: 208 GeV, 40 GeV, and $\tilde{l}_L\tilde{l}_L$: 181 GeV, 46 GeV. Comparing with Figures 1, it appears that with 20 fb^{-1} and a perfect detector, one can determine the endpoints at 118, 208, and 181 GeV to an accuracy of one bin or better. The other endpoints look very challenging.

The situation improves if we include the Feng-Finnell estimate for the smuon mass. This is shown in Figure 3, plotted with 1 GeV bins. The SM background shown is completely negligible. Because of the strong peaking, which actually

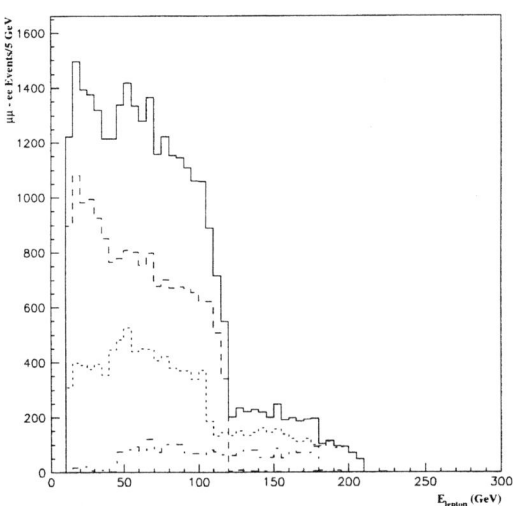

FIGURE 2. Flavor subtracted slepton signal, 100 fb^{-1} at $\sqrt{s}$=600 GeV for LHC point 5. The solid line is the total smuon + selectron signal. The dashed, dotted, and dot-dashed lines are from $\tilde{l}_R\tilde{l}_R$, $\tilde{l}_R\tilde{l}_L+\tilde{l}_L\tilde{l}_R$, and $\tilde{l}_L\tilde{l}_L$, respectively.

resembles a sharp edge, it is trivial to extract the $\tilde{\mu}_R$ mass with an accuracy of one bin or better. This assumes that the $\tilde{\chi}_1^0$ mass is already known to within 1 GeV. Similar results are obtained for the $\tilde{e}_R$, with somewhat worse statistics.

Heavy sleptons

The second study is for the heavy sugra point described in Table 2. The results are for dimuon and dielectron production at $\sqrt{s} = 1400$ GeV, using the same cuts as in the previous example.

Figure 4 shows the flavor subtracted slepton signal corresponding to 1000 fb^{-1} of integrated luminosity. Comparing with Figure 2, one notes several differences. In the present case the signal is completely dominated by RR production. This is because the branching fraction for $\tilde{\mu}_L$ or $\tilde{e}_L$ decay to muon or electron plus $\tilde{\chi}_1^0$ is only 16%. At this heavy sugra point, the $\tilde{\mu}_L$ decays predominantly to either $\tilde{\chi}_1^\pm \nu_\mu$ or $\tilde{\chi}_2^0 \mu$. Subsequent decays in these modes are unlikely to pass the cuts.

Since RR production now dominates, there are effectively only two kinematic endpoints: 539 GeV and 106 GeV. Note that the lower endpoint is now sufficiently large not to be distorted or hidden by the cuts. Both edges are very sharp in Figure 4. The SM backgrounds after cuts are negligible. Thus with a perfect detector one could extract the masses of both $\tilde{\mu}_R$ and $\tilde{\chi}_1^0$ with an accuracy better than 5 GeV.

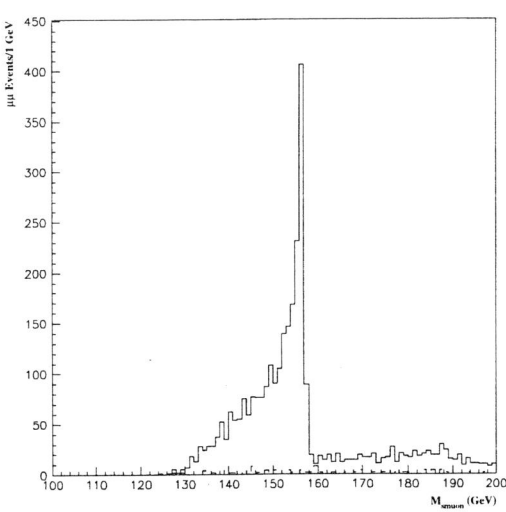

FIGURE 3. Dimuon production after cuts, 20 fb^{-1} at $\sqrt{s}$=600 GeV for LHC point 5. The solid line is the total smuon signal, plotted versus the Feng-Finnell estimate for the smuon mass. The dashed line is the sum of the Standard Model backgrounds.

Figure 5 shows the Feng-Finnell plot for the heavy sugra point. The SM backgrounds shown are negligible. Again we see strong edgelike peaking around the actual $\tilde{\mu}_R$ mass of 519 GeV. It is clearly possible to extract the mass with an accuracy of one bin or better. This assumes that the $\tilde{\chi}_1^0$ mass is already known to within 1 GeV. Similar results are obtained for the $\tilde{e}_R$, but with poor statistics.

Sneutrino pair production

Muon sneutrino pair production fits our kinematic scenario, provided that the sneutrino has a substantial branching fraction to $\tilde{\chi}_1^\pm \mu$. The chargino will decay predominantly to $\tilde{\chi}_1^0$ plus jets. Thus the signature is an acoplanar dimuon pair plus missing E_T plus jets. Note that, in the presence of the R parity violating coupling LLE, s-channel resonant production of single sneutrinos may also be possible at a muon collider [7].

Here we have studied the sugra point described in Table 3, for production at $\sqrt{s} = 800$ GeV. The branching fraction for the muon $\tilde{\nu}_L$ into $\tilde{\chi}_1^\pm \mu$ is 56%, while the branching fraction for $\tilde{\chi}_1^\pm$ into $\tilde{\chi}_1^0$ plus jets is 65%. We will employ the same cuts as previously, except that we now require two or more reconstructed jets (cone radius $R = 0.4$).

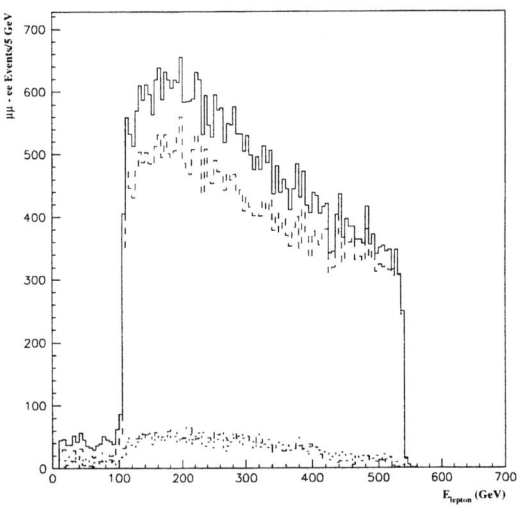

FIGURE 4. Flavor subtracted slepton signal, 1000 fb^{-1} at $\sqrt{s}$=1400 GeV for the heavy sugra point. The solid line is the total smuon + selectron signal. The dashed, dotted, and dot-dashed lines are from $\tilde{l}_R\tilde{l}_R$, $\tilde{l}_R\tilde{l}_L + \tilde{l}_L\tilde{l}_R$, and $\tilde{l}_L\tilde{l}_L$, respectively.

Figure 6 shows the Feng-Finnell mass estimate after cuts plotted in 1 GeV bins. Shown is the total signal from all SUSY production mechanisms. The SM background after cuts is negligible. The signal acceptance for muon sneutrino pairs after cuts is about 4%. Thus, despite a rather large cross section (over 500 fb) the plot has rather poor statistics. Nevertheless we again see strong edgelike peaking at the true $\tilde{\nu}_L$ mass of 262 GeV.

It is interesting to note that a previous study of sneutrino pair production at e^+e^- colliders [8] relied on the trilepton plus missing E_T plus jets channel to kill SM backgrounds. For our study point this does not appear to be necessary. Furthermore, our complementary dilepton channel has five times the rate, before cuts, as the trilepton channel.

Conclusions

A variety of precision sparticle mass measurements are possible at a muon collider using kinematic methods such as those discussed here. Polarized beams are not necessary to control SM backgrounds. However, without polarization it may be difficult in some cases to disentangle $\tilde{l}_R$ from $\tilde{l}_L$ signals.

It seems likely that in most cases detector resolutions, not backgrounds or statistics, will dominate the final error bars. Thus it will be crucial to perform simulations

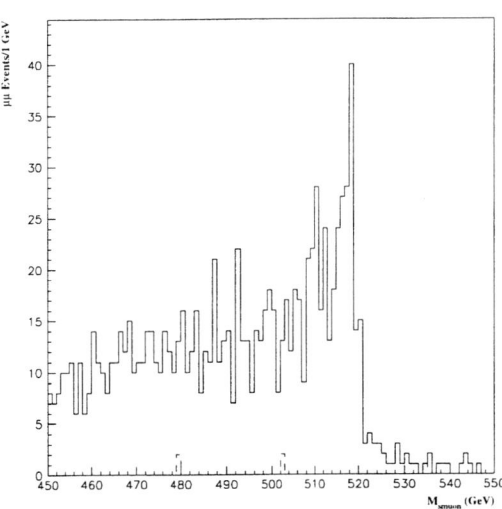

FIGURE 5. Dimuon production after cuts, 100 fb^{-1} at $\sqrt{s}$=1400 GeV for the heavy sugra point. The solid line is the total smuon signal, plotted versus the Feng-Finnell estimate for the smuon mass. The dashed line is the sum of the Standard Model backgrounds.

TABLE 3. Sparticle and Higgs spectrum for the third sugra point, which corresponds to minimal sugra parameters m_0=225 GeV, $m_{1/2}$=200 GeV, A_0=0, $\tan\beta$=2, and sgn(μ)=1.

Particle	Mass (GeV)	Particle	Mass (GeV)
$\tilde{\chi}_1^0$	77	$\tilde{\chi}_2^0$	146
$\tilde{\chi}_1^\pm$	144	$\tilde{\chi}_2^\pm$	449
$\tilde{e}_R$	240	$\tilde{e}_L$	270
$\tilde{\mu}_R$	240	$\tilde{\mu}_L$	270
$\tilde{\nu}_L$	262	$\tilde{g}$	536
$\tilde{t}_1$	310	$\tilde{b}_1$	450
h_0	88	H_A	560

with a realistic mock-up of a muon collider detector.

Adequate statistics for the type of analysis presented here correspond to integrated luminosities of at least 20 fb^{-1} for $\sqrt{s}$ in the range 500 to 800 GeV. For heavy sparticles and $\sqrt{s} \gtrsim 1$ TeV, the minimum useful integrated luminosity is about 100 fb^{-1}.

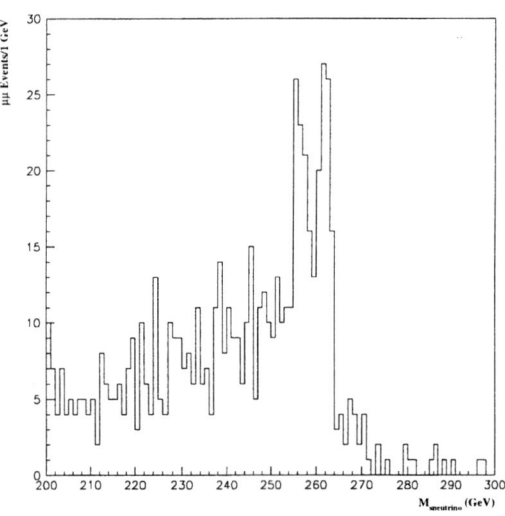

FIGURE 6. Dimuons plus two or more jets, after cuts, 20 fb^{-1} at $\sqrt{s}$=800 GeV for the second sugra point. Shown is the total SUSY signal, plotted versus the Feng-Finnell estimate for the sneutrino mass.

REFERENCES

1. Berger, M.S., these proceedings; hep-ph/9802213.
2. Feng, J., and Finnell, D., *Physical Review* **D49** 2369 (1994).
3. Sjostrand, T., *Computer Phys. Commun.* **82** 74 (1994).
4. Richter-Was, E., Froidevaux, D., and Poggioli, L., ATLAS internal note PHYS-NO-079, 1996.
5. Baer, H., Paige, F., Protopopescu, S., and Tata, X., in *Physics at Current Accelerators and Supercolliders*, ed. Hewett, J., White, A., and Zeppenfeld, D., Argonne National Lab, 1993.
6. Paige, F., proceedings of *Workshop on Physics at the First Muon Collider and at the Front End of a Muon Collider*, Fermilab, 6-9 Nov. 1997; hep-ph/9801396.
7. Feng, J., these proceedings; hep-ph/9801248.
8. Baer, H., Munroe, R., and Tata, X., *Physical Review* **D54** 6735 (1996).

The Physics of Like-sign Muon Collisions at High Energy

Clemens A. Heusch

Institute for Particle Physics, University of California, Santa Cruz, CA 95064

INTRODUCTION

As we discuss the proper place that the muon collider is liable to occupy in the future of our field, it behooves us to take note of the facts of life that will govern its progress in the years prior to any potential realization of this project. They can be summarized very succinctly in terms of the machines that we expect to give us experimental data beyond our presently assured knowledge in the next decade or so.

LEP 200 and the Tevatron have years of operation ahead of them. TeV33 is likely to turn on in 2003. The LHC has been formally approved and will become operational in 2006. It is highly probable that some version of a TeV electron collider, which we will generically call NLC in this text, will come online in about the year 2008. If we accept this premise, what do we need the muon collider for?

The standard list of advantages this machine is expected to offer in comparison with what has been realistically projected for electron colliders includes, above all, its potential energy reach. Next, the greater rest mass of the muon does away with most of the vexing radiative tail that floods detectors and smears out the center-of-mass energy definition of electron-electron collisions. If planned cooling schemes can be successfully implemented, high luminosities should be attainable. That may not easily remain true if a good polarization definition is demanded. But in principle, at least, the origin of the accelerated muons from weak hadron decay suggests the possibility of high degrees of polarization as well as the availability of either positive or negative beams with identical characteristics.

In contrast to the much higher energies foreseen at hadron colliders, muon machines present the advantage of clean, point-like initial states where there is no need for a reconstruction of (parton) subenergies. The absence of partonic fellow travelers leads to a cleaner and simpler disentangling of final states as well, making it easier to look for novel phenomena that will not have to be culled from massive amounts of detector hits and reconstruction ambiguities. This implies more promising chances of looking for the validity of various conservation laws that are telling for novel phenomena. It also means that exotic quantum numbers can lead to key discoveries; and exotic quantum numbers are most tellingly provided by like-sign muon (or electron) collisions.

It is in this spirit that we will review a few physics processes that should, by their potential promise not equally well attainable in lepton–antilepton collisions, argue in favor of a serious study of the inclusion of a like-sign option in the muon collider design.

POTENTIALLY IMPORTANT FEATURES OF THE LIKE-SIGN OPTION

As we try to understand the mechanism of electroweak symmetry breaking, the added energy reach of the muon collider well beyond the 1.5 TeV presently seen as a reasonable high-energy version of an NLC, will be important. It is well understood that, should there be no sign of an elementary Higgs boson up to 1 TeV, we expect the emergence of a new strong interaction of longitudinal W bosons [1] acting as Goldstone bosons: in this case, the like-sign collider version is uniquely poised to probe the I = 2 channel and its potential resonance structure in the energy regime that may elude the electron collider. An extended Higgs sector will similarly have to be investigated in a doubly charged mode [2], where either WW fusion or a hard-to-determine direct coupling $H^{++} \to \mu^+\mu^+$ will have to be invoked. While we do not venture to prejudge the strength of this coupling, it clearly makes sense to presume it may well be considerably stronger for muon pairs than for electrons, so that the mass-or generation-dependent coupling adds a second reason to look forward to this investigation in the context of the muon collider – again in its like-sign (++ or, equivalently, −−) version. We discussed these issues in Ref. [3], where we also covered compositeness studies, anomalous gauge couplings, and generic new contact interactions. In the present study, I will mention those aspects of possible new physics where recent discussion has added novel aspects.

If high luminosities and clean experimentation can be assured, the muon collider in its like-sign version is assured to open new avenues. Many features of like-sign charged-lepton collisions have recently been investigated in the e^-e^- case; it is important to note that for electron–electron colliders, the features of easy availability of high electron currents in well-defined and easily inverted polarization states permits a decisive look at a number of the physics questions we are discussing here – but the limitation to energies (with present technology) below 2 TeV, to the first lepton generation only, and to highly polarized beams in one charge state only imposes constraints. By the same token, *mutatis mutandis*, the potential availability of second-generation charged leptons in two charge states at higher energies, with the requisite higher luminosities, and with equal (if uncertain) polarization possibilities will decisively add to the physics reach of lepton colliders at the cutting edge of particle experimentation. Add to this the distinctive detection tool provided by high-transverse-momentum muons in the final state, and an appropriately designed detector will clearly help us broaden the promise offered by a well-defined initial state with the quantum numbers shown in Table 1, to a remarkable degree.

TABLE 1. Additive quantum numbers of like-sign muon scattering initial states. Cleanly helicity-selected (i.e., highly polarized) beams (rows 1-3, 5-7) permit the experimenter to choose chiral couplings and select (or enhance) potentially rare or exotic final states. Mixed helicity beam collisions carry less information (rows 4,8).

$\|in\rangle$	Q_{el}	S_z	L	L_e	$L\mu$	I_3^W	Y^W
$\mu_R^+\mu_R^+$	+2	1	-2	0	-2	1	2
$\mu_R^+\mu_L^+$	+2	0	-2	0	-2	1/2	3
$\mu_L^+\mu_L^+$	+2	-1	-2	0	-2	0	4
$\mu^+\mu^+$	+2	1,0,-1	-2	0	-2	1,1/2,0	2,3,4
$\mu_L^-\mu_L^-$	-2	-1	2	0	2	-1	-2
$\mu_R^-\mu_L^-$	-2	0	2	0	2	$-1/2$	-3
$\mu_R^-\mu_R^-$	-2	1	2	0	2	0	-4
$\mu^-\mu^-$	-2	1,0,-1	-2	0	2	0,$-1/2$,-1	-2,-3,-4

This expectation is slightly damped by the fact that, to date, no gold-plated method has been found to procure, at the same time, good luminosities and high degrees of polarization [4]. Table 1 illustrates to what extent it is imperative that we overcome this limitation: to cull novel, Beyond-the-Standard-Model features from the massive data densities expected, it will be very helpful if we can choose the relevant chiral couplings of the incident muons freely by means of selecting their helicities. Similarly, the conservation of weak I_3 and hypercharge can add significant constraints, as we will show below.

As column 2 of Table 1 implies, the selective couplings of left-handed (right-handed) negative (positive) muons to W_L^- (W_R^+) alone will permit us to eliminate many Standard-Model backgrounds if we come in with a polarized beam *and* can switch the polarization: left-handed incoming μ^-, like electrons, will produce high-transverse-momentum interactions by means of W^- exchange, right-handed ones will not. This permits, in the electron case, a quick check on whether some indication of a new effect is actually due to such a background, by means of a polarization reversal. For muons, there is no way to invert polarization quickly. It is therefore all the more important to investigate the chances of polarizing muon beams by a clever way of momentum selection prior to phase-space cooling. Columns 6,7 of the same table illustrate that all phenomena that can be traced by the conservation of weak isospin and hypercharge, particularly the identification of exotic representations, will profit from the choice of these eigenvalues as they are available from the choice of helicities in the incoming beams.

THE HIGGS SECTOR

The discovery of either the basic SM Higgs scalar or the MSSM h^0 will likely not await the appearance of the muon collider; but it turns out that this machine,

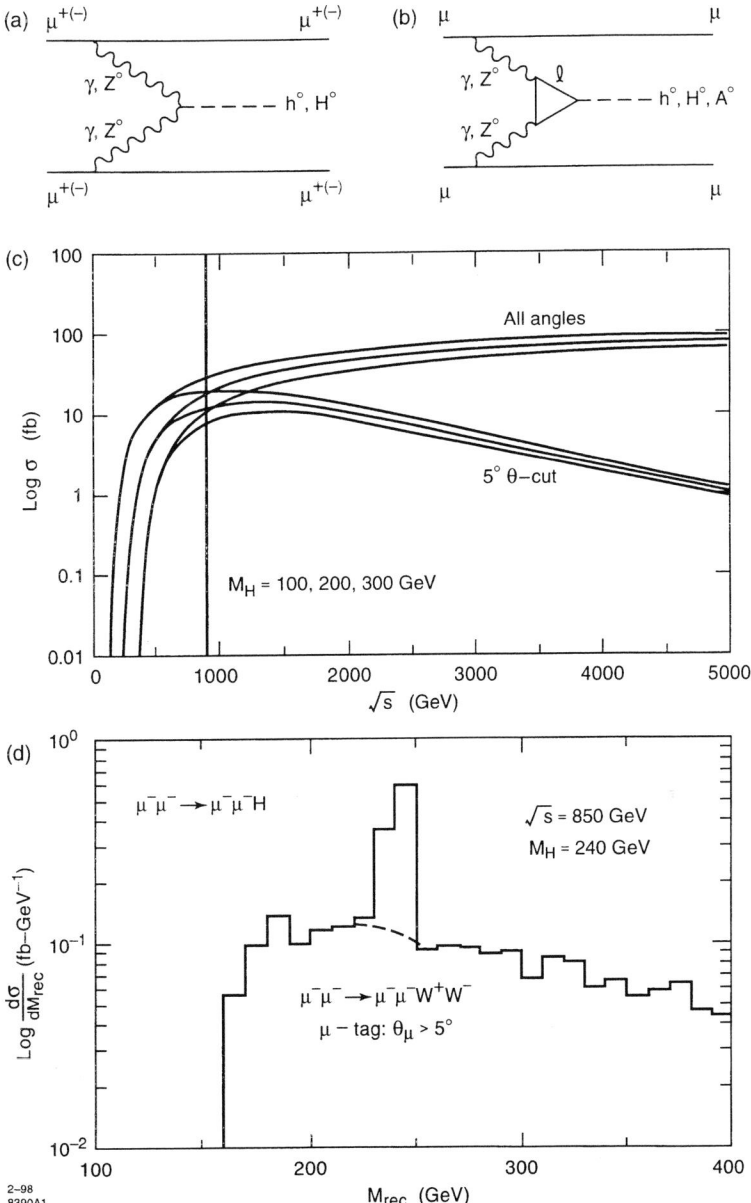

FIGURE 1. a) Central production of neutral bosons by gamma-gamma or ZZ fusion. b) same as a), but the lepton loop also permits the production of CP-odd bosons. c) Production cross section for the kinematical parameters given: the saturation at higher energies is obvious; a reasonable detector-originated cut of 5 degrees is entirely acceptable. d) Simulated signal (on a log scale). for the parameters given, irrespective of the boson decay modes. It can almost act as a "Higgs trigger".

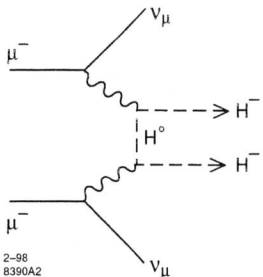
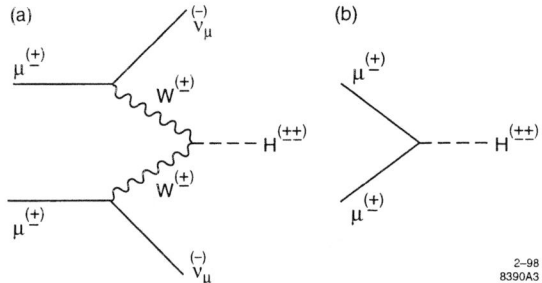

FIGURE 2. Pair production graph for charged Higgs bosons, by WW fusion.

FIGURE 3. a) Production graph for doubly charged Higgs boson from appropriately handed muons, by WW fusion; b) Direct coupling, of unknown strength, of incident muon pair to doubly charged Higgs boson.

even in its high-energy version, will be most useful for detailed studies of the Higgs sector: central neutral Higgs production, studied by means of momentum analysis of both recoiling muons, is a powerful tool for resolving even "invisibly" decaying Higgs bosons. The cross section saturates at $\sqrt{s} \gtrsim 2.5 m_H$ [5]. Figure 1 illustrates the point well: the basic graph (1a) shows ZZ fusion, with known couplings; the resulting cross section, including an angular cut due to detector acceptance, is shown in (1c), and available signal-to-background ratios can be gleaned from the simulated examples given in Fig. (1d). This beats any Higgs detection possibility at a hadron collider, and is more stable over a broad parameter range than the standard NLC modes from $e^+e^- \to ZH^0$. For an inclusion of the CP-odd A^0, the graph (1b) has to be added. Also note that the Higgs factory version of the $\mu^+\mu^-$ colliders [6] has much narrower focus.

In the MSSM, as in other two-Higgs-doublet scenarios, there is an attractive way of looking for charged Higgs bosons by way of H^-H^- (or H^+H^+) production via WW fusion in the muon helicity combinations that assure the desired couplings, as in Fig. 2 [7]. Similarly, more extended Higgs sectors may well contain triplets with tell-tale doubly charged scalar bosons that are liable to be produced at a good rate by the WW fusion graph (Fig. 3a) with standard couplings and sizeable, calculable cross sections [3,8]. Note that a direct $H^{++}W^+W^+$ ($H^{--}W^-W^-$) coupling may well profit from the mass dependence of Higgs couplings; here, the like-sign muon collider has a clear advantage over the electron collider, irrespective of its energy edge. Here we also are helped decisively in our search by the conservation of weak hypercharge: the H^{--} (H^{++}) tends to be present in $Y = 2$ (-2) representations, available in only one $\mu^-\mu^-$ ($\mu^+\mu^+$) helicity combination of the $|in\rangle$ state.

Clearly, the muon collider in its like-sign incarnation can play a pivotal part in the unraveling of extended Higgs scenarios, which the hadron colliders will have a hard time cracking at all.

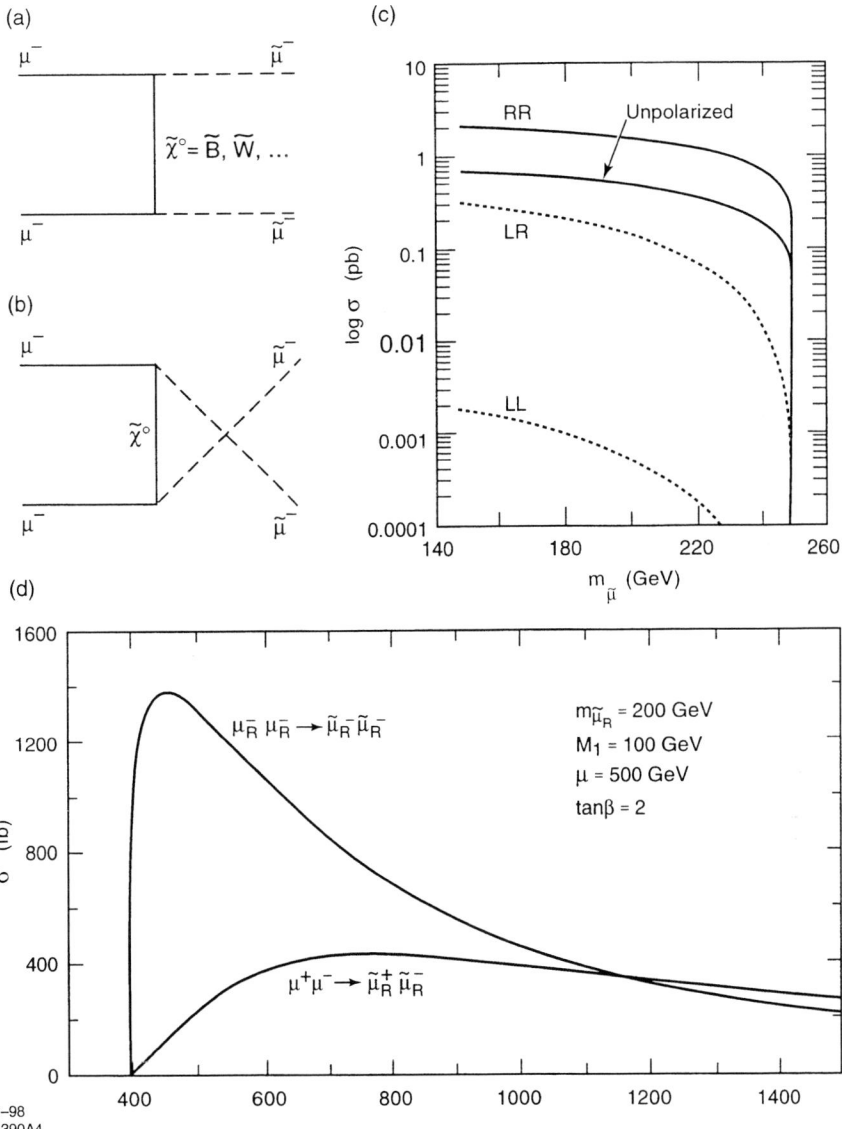

FIGURE 4. Smuon pair production from like-sign muon collisions: a), b) basic lowest-order graphs, by neutralino exchange; c) illustration of the importance of beam polarization for the cross sections due to the above graphs: it takes right-handed negative muons to reach large cross sections (the figure, from the second paper in ref. [9], applies to incident electrons at $E_{CM} = 0.5$ TeV, as a function of selectron mass. It is equally applicable to the muon collider; d) comparison of threshold cross sections in smuon pair production in s-wave $\mu^-\mu^-$ collisions vs. p-wave $\mu^+\mu^-$ interactions. The steep rise above threshold for the scalar pair permits very precise mass determinations for sleptons. From refs. [10,11].

SUPERSYMMETRY

Should Supersymmetry be Nature's choice for the breaking of Electroweak Symmetry, precision data on the slepton spectrum will likely be more accessible from like-sign lepton collisions than from any other collider: unlike in e^+e^- ($\mu^+\mu^-$) interactions, where the emerging slepton pairs materialize in p waves, they produce s-wave final-state configurations. Figs. 4a and b give the basic production graphs of smuon pairs by neutralino exchange, which generate strongly polarization-dependent cross sections [9] (Fig. 4c shows that for the e^-e^- case, fully analogous). The great advantage of this production method is the steep s-wave threshold rise of the cross section (Fig. 4d) that permits an optimal mass determination of the smuons; similarly, the subsequent decay $\tilde{\mu} \to \mu+$ neutralino then yields the chance of an equally impressive neutralino mass measurement [10].

Among other SUSY measurements that are set to take advantage of the specific features available at the same-sign muon collider are investigations of the sparticle couplings: the gauge boson and gaugino couplings in Figs. 5a,b are related by SUSY; in exact SUSY, $g_1 = h_1$. Recent work [10,11] has been probing precise measurements of this coupling relation not only as a general test of Supersymmetry, but also as a probe for heavier sparticles beyond the kinematical reach of the collision. Comparing such measurements in the reactions $e^-e^- \to \tilde{e}^-\tilde{e}^-$ and $\mu^-\mu^- \to \tilde{\mu}^-\tilde{\mu}^-$ may be very instructive as a probe of SUSY flavor structure. In fact, such gaugino couplings need not necessarily be flavor-diagonal: the disappearance of a high-transverse-momentum muon in the reaction

$$\mu^{-(+)} \mu^{-(+)} \to e^{-(+)} \mu^{-(+)} \tilde{B}\tilde{B} \tag{1}$$

will clearly establish lepton flavor violation, and thereby eliminate pure SUGRA and gauge-mediated SUSY as the choices of Nature.

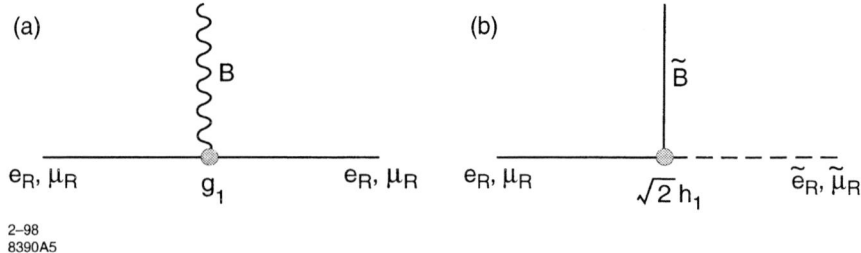

FIGURE 5. Couplings of leptons to gauge bosons a) and of lepton/slepton to gauginos b), are very sensitive to details of SUSY phenomenology. From ref. [11].

As a further point of interest, we mention that these studies will complement parallel efforts at an electron collider; backgrounds are smaller, and much smaller indeed than in e^+e^- or $\mu^+\mu^-$ collisions. But in this particular context, we do not need high degrees of polarization. What we do need for a successful background estimate is a clean beam definition and luminosity measurement.

EXOTICA: DILEPTONS

Among the various exotica that might spring from new kinematic and quantum number regimes to be explored, new gauge bosons with lepton number 2 and sharp mass definition [12] belong to the more attractive novelties, because they might well explain some of the underlying mysteries of the Standard Model, like the number of elementary fermion generations Nature realizes. Such gauge bosons may well have generation-dependent couplings to leptons, and their distinctive $I_3 = 1/2$ and weak hypercharge values make them a natural target for independent investigations in the electron and muon colliders. Since we might find s-channel structure in the like-sign muon collider from resonant doubly charged Higgs boson production, it will not only be the narrow shape of the dilepton that will tell the two effects apart: recall that $I_3 = 1/2$ is accessible only to incoming leptons of opposite helicities. The choice of incoming polarizations will then help us tell the difference IF we see an effect of promising structure in the total cross section as a function of dilepton mass.

MASSIVE MAJORANA NEUTRINOS

Lastly, I will mention a possibility that we have much belabored in recent discussions of like-sign electron collisions [13,14] – that there be TeV-level Majorana neutrinos beyond the reach of present detection in high-energy experiment or nuclear neutrinoless double beta decay searches. The relevant production cross section for the electron-initiated process mediated by their exchange,

$$e_L^- e_L^- \to W^- W^-, \qquad (2)$$

was shown to be [13]

$$\sigma \sim \frac{s^2}{M^4} |U_{\ell N}|^2, \qquad (3)$$

where M is the reduced mass of the two lightest heavy neutrinos N.

Figures 6a,b show the relevant diagrams. It is important to notice that this cross section becomes experimentally accessible only for left-handed incoming leptons (and longitudinal final-state W's). The couplings in Fig. 6a are defined by the neutrino mixing matrix, $U_{\ell N}$, and are quite probably generation dependent. There are experimental constraints on these from rare decay experiments.

Given that eq. (3) contains a strong energy dependence, there is plenty of motivation to run this search at the muon collider IF there is a like-sign version, and IF good polarization characteristics can be established. The mechanism at the root of this calculation assumes the existence of at least two N_M neutrinos of large mass: it is thinkable that the relevant mixing matrix element is larger for the heavier N_M and the muon, so that, beyond the reach of the NLC, an independent search for the Majorana particle that makes the process

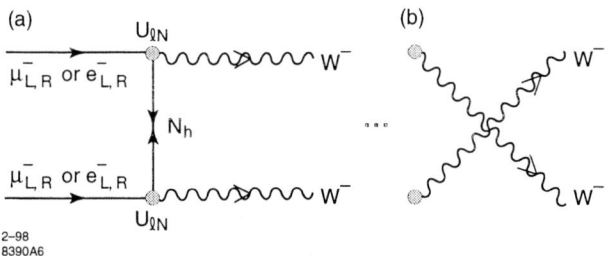

FIGURE 6. W pair production from equal-sign muon (or electron). Beams can proceed by means of heavy Majorana neutrino exchange (see text).

$$\mu_L^- \mu_L^- \to W^- W^- \qquad (4)$$

possible, is all the more indicated. We have stressed before that an investigation of this scenario in the electron collider case must be a major motivation for the realization of its like-sign option: it may well be the vital link in solving the Standard Model riddle of (approximately) massless "light" neutrinos and of the appropriate field-theoretical definition of the neutrino – Dirac vs. Majorana. The muon collider can significantly add to the promise of this investigation.

CONCLUSION

The scenario of physics motivations for the like-sign version(s) of the muon collider has a number of persuasive features, as we showed. Additional cogent arguments can be made on several grounds: We mentioned a potential strongly interacting W_L sector, compositeness studies in μ–μ Møller scattering, and should add leptoquark searches where, again, certain exotic features and generation-dependent couplings set the present channel apart. Trilinear and quartic gauge couplings can be searched for non-standard features beyond the regime accessible to the electron collider.

In summary, the muon collider project would strongly suffer if it did not include the field opened up by the like-sign option: if technical solutions can be found for the optimization of luminosities up to high energies while not abandoning the ambition to come in with high degrees of polarization, the only added technical feature we need for the like-sign version is a second final accelerator ring and an interaction region design that is compatible for all incoming charge states.

The broadening of the physics appeal certainly makes a cogent argument in favor of including these features in a realistic design for the muon collider, as was first argued in ref. [15].

ACKNOWLEDGMENT: The author wishes to acknowledge David Cline's continuing active enthusiasm for the muon collider and its potential success. He also thanks Joan George for her excellent work coordinating the December, 1997 meeting, where this presentation originated.

REFERENCES

1. M. Chanowitz, M.K. Gaillard, *Nucl. Phys.* **B261**, 379 (1985); T. Han, *Int. J. Mod. Phys.* **A11**, 1541 (1996).
2. J.F. Gunion, J. Grifols et al., *Phys. Rev.* **D40**, 1546 (1989). J.F, Gunion, *Int. J. Mod. Phys.* **A11**, 1551 (1996).
3. e^-e^- *1995, Proceedings of the Electron-Electron Linear Collider Workshop*, C.A. Heusch ed., *Int. J. Mod. Phys.* **A11**, 1523-1697 (1996).
4. B. Norum, R. Rossmanith, *Nucl. Phys. B (Proc. Suppl.)* **51A**, 191 (1996).
5. P. Minkowski, in *New Directions in High-Energy Physics. DPF Snowmass Study 1996*, D.G. Cassel et al, ed. (1997), p. 614.
6. D. Cline, these proceedings.
7. T. Rizzo, *Int. J. Mod. Phys.* **A11**, 1563 (1996).
8. J.F. Gunion, these proceedings.
9. W. Keung, L. Littenberg, Phys. Rev. D28, 1067 (1983); F. Cuypers, G.J. van Oldenborgh, R. Rueckl, *Nucl. Phys.* **B409**, 123 (1993).
10. J. Feng, in e^-e^- *1997, Proceedings of the Electron-Electron Linear Collider Workshop*, *Int. J. Mod. Phys. A*, to be published (1998).
11. H.C. Cheng, ibidem.
12. P.H. Frampton, *Phys. Rev. Lett.* **69**, 28899 (1992).; e^-e^- *1997, Proceedings of the Electron-Electron Linear Collider Workshop*, *Int. J. Mod. Phys. A*, to be published (1998).
13. C.A. Heusch, P. Minkowski, *Nucl. Phys.* **B416**, 3 (1994); *Phys. Lett.* **B374**, 116 (1996).
14. C.A. Heusch in *Neutrino 94: Proceedings of the 16th International Conference on Neutrino Physics and Astrophysics*, Eilat, Israel, 29 May–3 June 1994, *Nucl. Phys. (Proc. Suppl.)* **B38**, 313 (1995).
15. C.A. Heusch and F. Cuypers, *Proceedings of the 2nd Workshop on Physics Potential and Development of $\mu^+\mu^-$ Colliders, 17-19 Nov 1994, Sausalito, CA*, D.B. Cline, editor (AIP Conference Proceedings 352, 1996) p. 219.

Doubly Charged Particles at a $\mu^{\pm}\mu^{\pm}$ Collider

Subhash Rajpoot[1]

*Department of Physics & Astronomy, California State University
Long Beach, California 90840, USA*

Abstract. Doubly charged particles occur in many extensions of the standard model. These particles could either be spin-0 scalars, spin-1/2 fermions or spin-1 gauge bosons. A $\mu^{\pm}\mu^{\pm}$ collider is ideally suited for the study of such particles at energies well below their direct production at a $\mu^{+}\mu^{-}$ collider. The production of these particles are studied in the reaction $\mu^{-}\mu^{-} \to e^{-}e^{-}$. With high luminosities and beam energies to be attained at the first generation and second generation $\mu^{\pm}\mu^{\pm}$ colliders, it is found that the effects of doubly charged scalars and vector particles can lead to a sizable event rate while effects due to doubly charged fermions are suppressed either due to loop effects or by constraints imposed on unorthodox physics by conventional physics. Since muon beam polarisations are expected to be around 20%, studies of various asymmetries can pinpoint the nature of the doubly charged particles.

INTRODUCTION

It goes without saying that the work to be described here can also be studied at an $e^{\pm}e^{\pm}$ collider in the reaction $e^{-}e^{-} \to \mu^{-}\mu^{-}$.

The relative merits of constructing a $\mu\mu$ collider are by now well established. Highly collimated, intense muon beams, each of energy equal to 0.25 TeV to begin with and reaching energies of up to two TeV subsequently, are envisaged with luminosities around $10^{33} \text{cm}^{-2}\text{s}^{-1}$ in the initial stages and going upto $10^{35} \text{cm}^{-2}\text{s}^{-1}$ subsequently. Unlike the *ee* collider, these machines have the added advantage of low bremmstrahlung. Preliminary studies indicate that beam polarisations of up to 20% are feasible, making particle identification an easy task.

In a $\mu^{+}\mu^{-}$ collider, charge neutrality in the final states restricts particles to pair production. In a $\mu^{\pm}\mu^{\pm}$ collider, on the other hand, final state particles carry ± 2 units of electric charge. One implication of this is that the effects of new physics can be revealed at energies much lower than that at a $\mu^{+}\mu^{-}$ collider. Thus in many respects, the physics that can be studied at a $\mu^{\pm}\mu^{\pm}$ collider complements that at a $\mu^{+}\mu^{-}$ collider. This is particularly true in the case of doubly charged particles

---

[1] E-mail:rajpoot@csulb.edu

(DCPs), if they exist in nature. Their resonant production in a $\mu^+\mu^-$ collider requires center of mass energies equal to $2M_{--}$ (M_{--} = mass of a DCP) where as in a $\mu^\pm\mu^\pm$ collider, effects of DCP can be studied at energies well below $2M_{--}$.

The standard model does not require doubly charged particles. However, inspite of its many successes, issues remain that compels one to consider extensions beyond the standard model. These issues pertain to fermion masses in general and neutrino masses in particular, unification of particle interactions, baryon instability, CP violation and questions relating to "naturalness" of parameters in the scalar sector of the theory. An inevitable consequence of addressing these issues, or at least some of them, leads to the introduction of DCPs. These particles could be either spin-0 scalars, or spin-1/2 fermions or spin-1 gauge bosons. In the following, some models with doubly charged particles are considered. The prospects of their discovery at a $\mu^-\mu^-$ collider in the reaction $\mu^-\mu^- \to e^-e^-$ are then studied.

DOUBLY CHARGED SCALARS

The first model is motivated by the desire to have massive neutrinos. Today, massive neutrinos are required to explain a whole horde of experimental observations. These start at the observed deficit in the flux of electron antineutrinos from the sun and run through the depletition of the flux of atmospheric muon neutrinos; the controversial LSND result which reports an excess of $\bar{\nu}_e$ in a $\bar{\nu}_\mu$ beam from pion decays at rest, and end in cosmology and astrophysics where they serve as candidates for providing the necessary critical mass density for the closure of the universe and as candidates for providing the right admixture of hot and cold dark matter that can serve as the initial seeding for the formation of large scale structure. If neutrinos are treated on par with with the other fermions, then their masses must originate in spontaneous symmetry breaking. This requires the introduction of a triplet of scalars **T** carrying two units of weak hypercharge **Y**. The relevant term is

$$L = \sum_{ff'=e,\mu,\tau} \frac{Y_{f,f'}}{\sqrt{2}} \begin{pmatrix} f^T C & \nu_f^T C \end{pmatrix} \begin{pmatrix} \mathbf{T^0} & \mathbf{T^+} \\ \mathbf{T^+} & \mathbf{T^{++}} \end{pmatrix} \begin{pmatrix} f' \\ \nu_{f'} \end{pmatrix} + h.c, \qquad (1)$$

where T denotes transpose and C is the charge conjugation operator. The triplet scalar bosons all carry two units of lepton number. For simplicity and without loss of generality we will take all the non-diagonal yukawa couplings to be vanishingly small in what follows. This is justified since flavor changing lepton couplings are highly suppressed phenomenologically. When **T** develops a vacuum expectation, neutrinos develop masses $m_{\nu_e} \approx Y_{ee} <\mathbf{T^0}>/\sqrt{2}$, $m_{\nu_\mu} \approx Y_{\mu\mu} <\mathbf{T^0}>/\sqrt{2}$, $m_{\nu_\tau} \approx Y_{\tau\tau} <\mathbf{T^0}>/\sqrt{2}$. Constraints from the ρ parameter restricts $<\mathbf{T^0}> \leq 10$ GeV. With current limits on neutrino masses[1] from phenomenology ($m_{\nu_e} \leq 15$eV, $m_{\nu_\mu} \leq 0.17$ MeV, $m_{\nu_\tau} \leq 24$ MeV) we get $Y_{ee} \leq 1.5 \times 10^{-9}$, $Y_{\mu\mu} \leq 1.7 \times 10^{-5}$, $Y_{\tau\tau} \leq 2.4 \times 10^{-3}$. The DCSs couple to the charged leptons with couplings $Y_{ff'} = m_{\nu_f}\delta_{ff'}/<\mathbf{T^0}>$. The cross section for the reaction $\mu^-\mu^- \to T^{--} \to e^-e^-$ is given by

$$\sigma(\mu^-\mu^- \to e^-e^-) = \frac{Y_{ee}^2 Y_{\mu\mu}^2}{2\pi} \frac{s}{((s-M_{--}^2)^2 + M_{--}^2\Gamma^2)}. \quad (2)$$

Since the DCS decays into a pair of leptons the width for each mode is given by

$$\Gamma_f = \frac{Y_{ff}^2}{8\pi} M_{--}. \quad (3)$$

In the absence of any information on the mass scale of the DCSs, the total width Γ is a free parameter. However, since the DCSs couple to leptons only weakly and not many decay channels are open to the DCSs by construction, the width turns out to be narrow. With $\sqrt{s} = 0.5$ TeV, luminousity $= 10^{33}$ cm^{-2} s^{-1} and a time period of one year, the number of events expected are insignificantly small. For this model an improvement of two orders of magnitude in the luminosity will not improve matters since the yukawa couplings, suppressed by neutrino masses, remain tiny.

The second model with DCSs that we construct is motivated by the desire to have the parameters in the right range so that the discovery potential of DCSs at a $\mu^-\mu^-$ collider is enhanced. The standard model is supplemented with a DCS with four units of hypercharge. It couples only to right handed charged leptons. The relevant interaction term is given by

$$L = \sum_{ff'} \frac{Y_{ff'}}{\sqrt{2}} f_R^T C f_R' T^{++} + h.c, \quad (4)$$

where $f, f' = e, \mu, \tau$. This time the yukawa couplings are not suppressed by constraints from neutrino masses. We take $Y_{ee} = 10^{-1}$, $Y_{\mu\mu} = 10^{-1}$ and $M_{++} = M_{--} = 250$ GeV. This choice of the parameters ensures that muonium-antimuonium oscillation rate stays below the limits set by present day experiments. The total width of the DCS is around 0.25 GeV. With a luminosity of 10^{33} cm^{-2}s^{-1}, DCS mass $M_{--} = 0.25$ TeV, $\sqrt{s} = 0.5$ TeV, we get roughly 340 events per year. The signal is expected to be clean as there are no lepton number violating processes in the standard model that can serve as background to dilute the signal.

DOUBLY CHARGED FERMIONS

One way to motivate the introduction of doubly charged fermions (DCFs) is the observation that the standard model contains fermions only with weak hypercharge values **Y** equal to (−1,−2) for leptons and (1/3,4/3,−2/3) for quarks. There is no principle forbidding the existence of fermions representations with higher weak hypercharge quantum numbers. Explicit representations with doubly charged leptons (DCLs) are the singlet representation carrying =−4, the doublet representation carrying =−3 and the triplet representation carrying =−2:

$$\begin{array}{cccc} \mathbf{Y} = -4 & ; & \mathbf{Y} = -3 & ; & \mathbf{Y} = -2 \\ F^{--} & ; & \begin{pmatrix} F^- \\ F^{--} \end{pmatrix} & ; & \begin{pmatrix} F^0 & F^- \\ F^- & F^{--} \end{pmatrix} \end{array} \quad (5)$$

The models with these exotic representations can be rendered free from ABJ anomalies by requiring the representations to be vectorlike. Realistic models require additional particles. The general expectations of the above three models are similar and only differ in detail. In what follows, we consider the simplest models with DCL F^{--} ($\mathbf{Y} = -4$). Model one is a standard model extension with one DCL. In this model, the DCL has unorthodox particle interactions. The interaction term is

$$L = \sum_{f,f'=e,\mu,\tau} \frac{Y_{ff'}}{\sqrt{M}} f_R^T C f_R' F^{--} + h.c, \quad (6)$$

where M represents a mass scale so that the overall Lagrange density is of dimension four. It violates Lorentz invariance. The reaction $\mu^-\mu^- \to F^{--} \to e^-e^-$ proceeds at the tree level. Present limits on violation of Lorentz invariance impose the constraint $\frac{Y_{ff'}}{\sqrt{M}} \leq 10^{-12}$. In this model the event rate is highly suppressed. Model two is a standard model extension in which the DCL has conventional particle interactions. The additional particles are, one singlet DCF F^{--}, one singlet right handed neutral heavy lepton N_R^0 and one singlet scalar S^+. In this model the reaction $\mu^-\mu^- \to e^-e^-$ proceeds at the one loop level via the "box" diagram. If no fine tuning of parameters is assumed, the box contribution is suppressed leading to a small event rate. The box contribution can be enhanced by pushing the scalars and fermions masses in the TeV range. This is a general result in all models with DCLs.

DOUBLY CHARGED VECTOR BOSONS

Doubly charged vector bosons (DCVBs) or double charged vector particles (DCVPs) can be introduced into standard model extensions in two ways. Both extensions require the enlargement of the fundamental gauge symmetry G=SU(3)XSU(2)XU(1) either to a simple grand unifying symmetry[2] (GUT) or a semi-simple grand unifying symmetry. In simple GUTs, the gauge symmetry descends to G in one stage. The symmetry breaking stage is characterised by a mass scale M_G. Since quarks and leptons belong to a single representation, Baryon stability is a problem unless $M_G \geq 10^{15}$ GeV. Since the DCVBs belong to the coset space

$$\frac{G}{SU(3)XSU(2)XU(1)}, \quad (7)$$

they acquire masses of order M_G. Thus in GUT's in which G descends to $SU(3)XSU(2)XU(1)$ in either one stage or even two stages, DCVBs are out of the present reach of any collider leave alone the $\mu^-\mu^-$ collider. However, it is possible to construct standard model extensions with light DCVBs which can be accessed at the planned $\mu^-\mu^-$ collider. In one scheme, one unifies leptons and anti-leptons into one multiplet and is based on $SU(4)_l$ gauge symmetry[3] for the leptons. The lepton representation of the first family is given by $\Psi^T = (\nu_e, e^-, -e^{-c}, -N^c)_L$ where N_L^c is a heavy right handed neutral lepton required to give "see-saw" masses to the light neutrinos. The $SU(4)_l$ gauge boson structure depecting where the DCVBs sit is given by

$$\begin{pmatrix} W_L^0 & W_L^+ & X_1^- & X^0 \\ W_L^- & -W_L^0 & X^{--} & X_2^- \\ X_1^+ & X^{++} & -W_R^0 & W_R^+ \\ \bar{X}^0 & X_2^+ & W_R^- & W_R^0 \end{pmatrix} + \begin{pmatrix} J^0 & 0 & 0 & 0 \\ 0 & J^0 & 0 & 0 \\ 0 & 0 & -J^0 & 0 \\ 0 & 0 & 0 & -J^0 \end{pmatrix}. \tag{8}$$

All X particles carry two units of lepton number. The underlying symmetry of leptonic electroweak interactions is $SU(2)_L X SU(2)_R X U(1)$. The J^0 is the gauge boson of the extra $U(1)$ that represents fermion minus antifermion number in the model. The J^0 mixes with the Z^0 of the standard model. Constraints from precision electroweak data implies $M_{Z^0} \geq 200$ GeV. In the quark sector one requires $SU(3)_c$ for color and $SU(4)_q$ for quark flavor, $\Psi^q = (u, d, -d^c, -u^c)$. The full symmetry could descend from a grand unifying simple group like $SU(16)$. In a multi-stage descend, $SU(16) \to SU(12)_q$ X $SU(4)_l \to SU(3)_c$ X $SU(4)_q$ X $SU(4)_l$ $\to SU(3)_c$ X $(SU(2)_L X SU(2)_R X U(1))_q$ X $SU(4)_l \to SU(3)_c$ X $(SU(2) X U(1))_q$ X $SU(4)_l \to SU(3)_c$ X $(SU(2) X U(1))_q$ X $(SU(2)_L X SU(2)_R X U(1))_l \to SU(3)_c$ X $(SU(2) X U(1))_q$ X $(SU(2) X U(1))_l \to SU(3)_c$ X $SU(2)$ X $U(1)_q$ X $U(1)_l \to SU(3)_c$ X $SU(2)$ X $U(1)$, the X particles can stay light if the breaking of $SU(4)_l$ is delayed till one reaches energies close to the electroweak scale. A noteworthy feature in the chain of descent is the emergence of the unique sub-symmetries of the type[4] "$G_q X G_l$". The X^- particle contributes to muon decay through $\mu^- \to e^- \nu_e \bar{\nu}_\mu$. If the contribution is constrained to be less than 1%, we get $M_{X^-} \geq 120$ GeV for $g_4 \approx e$ where g_4 is the $SU(4)_l$ gauge coupling and e is the electric charge. The DCVB X^{--} contributes to muonium-antimuonium oscillations. Inorder to ensures that the DCVB contribution to muonium-antimuonium oscillation rate stays below the limits set by present day experiments, we require $M^{--} \geq 200$ GeV for $g_4 \approx e$. In analysing our results we will use $M_{--} = 250$ GeV. The interaction Lagrangian of DCVB is

$$L = \frac{g_4}{\sqrt{2}} X_\mu^{++} \bar{\mu}^c_L \gamma^\mu \mu_L + h.c. \tag{9}$$

The cross section for the reaction $\mu^-\mu^- \to X^{--} \to e^-e^-$ is given by

$$\sigma(\mu^-\mu^- \to e^-e^-) = \frac{g_4^4}{24\pi} \frac{s}{((s-M_{--}^2)^2 + M_{--}^2 \Gamma^2)}. \tag{10}$$

The total width is given by

$$\Gamma = \frac{3g_4^2}{16\pi} M_{--}. \tag{11}$$

With $g_4 \approx e$ and $M_{--} = 250$ GeV, the width is about 1.5 GeV, which is still quite narrow. With luminosity of 10^{33} cm^{-2}s^{-1} and $\sqrt{s} = 0.5$ TeV, the event rate is over 2000 per year!

CONCLUSION

A $\mu^-\mu^-$ collider offers the exciting possibility of studying DCPs. These could be scalar, fermionic or gauge particles. We find that models with DCSs and DCVBs can be constructed with unsuppressed interaction parameters so that the event rates in the reaction $\mu^-\mu^- \longrightarrow e^-e^-$ are within reach of the machine energy. Similar expectations in models with DCFs are suppressed. This is either due to loop effects or unconventional physics with suppressed parameters. Since initial beam polarisations of about 20% are expected, the nature of the DCPs can easily be distangled. Two distinct tests that can easily differenciate between a scalar boson and a vector boson exchange are (i) study of the variation of the differential cross section with respect to cm scattering angle θ. It is flat in the case of scalar exchange while it varies as $(1 + cos^2\theta)$ in the case of a vector exchange; (ii) study of asymmetries formed from the cm differential cross section, first with the both muon beams polarised in the same direction initially and then with the polarisation of one of the muon beams reversed finally. The asymmetries can be studied either as variation in θ or at fixed θ. With the anticipated 20% polarisation of the muon beams, one gets a factor of $\frac{1}{6}$ in the case of scalar exchange, while one gets a small factor of ≈ 0.04 in the case of vector exchange. If a doubly charged particle is discovered at a $\mu^-\mu^-$ collider, it will strongly support the widely held conviction that the standard model is certainly not the complete theory.

ACKNOWLEDGEMENTS

This work was supported by a grant from California State University, Long Beach. The work is dedicated to the memory of Mark A. Samuel. We thank Deepak, Jyoti and Ravi for reading the manuscript.

REFERENCES

1. Particle Date Group, Phys. D54, 275 (1996).

2. Rajpoot,S., Phys. D22, 2244 (1980).

3. Rajpoot,S., Phys. Lett. 95B, 253 (1980).

4. Rajpoot,S., Mod. Phys. Lett. A1, 645 (1986).

Neutrino Physics at Muon Colliders

Bruce J. King

Brookhaven National Laboratory
email: bking@bnl.gov

Abstract. An overview is given of the neutrino physics potential of future muon storage rings that use muon collider technology to produce, accelerate and store large currents of muons.

INTRODUCTION

This paper gives an overview of the neutrino physics possibilities at a future muon storage ring, which can be either a muon collider ring or a ring dedicated to neutrino physics that uses muon collider technology to store large muon currents. It summarizes a previous more detailed description of these topics by this author [1].

After a general characterization of the neutrino beam and its interactions, some crude quantitative estimates are given for the physics performance of a muon ring neutrino experiment (MURINE) consisting of a high rate, high performance neutrino detector at a 250 GeV muon collider storage ring.

NEUTRINO PRODUCTION AND EVENT RATES

Neutrinos are emitted from the decay of muons in the collider ring:

$$\mu^- \to \nu_\mu + \overline{\nu_e} + e^-,$$
$$\mu^+ \to \overline{\nu_\mu} + \nu_e + e^+. \tag{1}$$

The thin pencil beams of neutrinos for experiments will be produced from long straight sections in either the collider ring or a ring dedicated to neutrino physics. From relativistic kinematics, the forward hemisphere in the muon rest frame will be boosted, in the lab frame, into a narrow cone with a characteristic opening half-angle, θ_ν, given in obvious notation by

---

[1] Presented at the Fourth International Conference on the Physics Potential and Development of Muon Colliders, San Francisco, December 10-12, 1997. This work was performed under the auspices of the U.S. Department of Energy under contract no. DE-AC02-76CH00016.

$$\theta_\nu \simeq \sin\theta_\nu = 1/\gamma = \frac{m_\mu}{E_\mu} \simeq \frac{10^{-4}}{E_\mu(\text{TeV})}. \quad (2)$$

The large muon currents and tight collimation of the neutrinos results in extremely intense beams – intense enough even to constitute a potential off-site radiation hazard [2].

For the example of 250 GeV muons, the neutrino beam will have an opening half-angle of approximately 0.4 mrad. The final focus regions around collider experiments are important exceptions to equation 2 since the muon beam itself will have an angular divergence in these regions that is large enough to significantly spread out the neutrino beam.

For TeV-scale neutrinos, the neutrino cross-section is approximately proportional to the neutrino energy, E_ν. The charged current (CC) and neutral current (NC) interaction cross sections for neutrinos and antineutrinos have numerical values of [3]:

$$\sigma_{\nu N} \text{ for } \begin{pmatrix} \nu\_CC \\ \nu\_NC \\ \overline{\nu}-CC \\ \overline{\nu}-NC \end{pmatrix} \simeq \begin{pmatrix} 0.72 \\ 0.23 \\ 0.38 \\ 0.13 \end{pmatrix} \times \frac{E_\nu}{1\text{ TeV}} \times 10^{-35} \text{ cm}^2. \quad (3)$$

These cross sections are easily converted into approximate experimental event rates for the example used in reference [1] of a 250+250 GeV collider with a 200 meter straight section. For a general purpose detector subtending the boosted forward hemisphere of the neutrino beam:

$$\text{Number of } \begin{pmatrix} \nu_\mu - CC \\ \nu_\mu - NC \\ \overline{\nu}_e - CC \\ \overline{\nu}_e - NC \end{pmatrix} \text{events/yr} \simeq \begin{pmatrix} 2.6 \\ 0.8 \\ 1.4 \\ 0.5 \end{pmatrix} \times 10^7 \times l[\text{g.cm}^{-2}], \quad (4)$$

where l is the detector length. For a long baseline detector in the center of the neutrino beam:

$$\text{Number of } \begin{pmatrix} \nu_\mu - CC \\ \nu_\mu - NC \\ \overline{\nu}_e - CC \\ \overline{\nu}_e - NC \end{pmatrix} \text{events/yr} \simeq \begin{pmatrix} 1.4 \\ 0.4 \\ 0.7 \\ 0.2 \end{pmatrix} \times 10^7 \times \frac{M[\text{kg}]}{(L[\text{km}])^2}, \quad (5)$$

where M is the detector mass and L the distance from the neutrino source.

These event rates are several orders of magnitude higher than in today's neutrino beams from accelerators.

A GENERAL PURPOSE NEUTRINO DETECTOR

Figure 1 is an example of the sort of high rate general purpose neutrino detector that would be well matched to the intense neutrino beams. Note the contrast

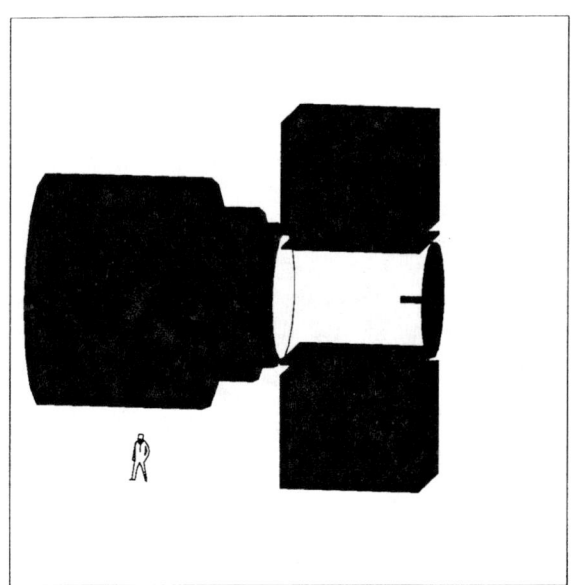

FIGURE 1. Example of a general purpose neutrino detector. A human figure in the lower left corner illustrates its size. The neutrino target is the small horizontal cylinder at mid-height on the right hand side of the detector. Its radial extent corresponds roughly to the radial spread of the neutrino pencil beam, which is incident from the right hand side. Further details are given in the text.

with the kilotonne-scale calorimetric targets used in today's high rate neutrino experiments.

The neutrino target is a 1 meter long stack of CCD tracking planes with a radius of 10 cm chosen to match the beam radius at approximately 200 meters from production for a 250 GeV muon beam. It contains 750 planes of 300 micron thick silicon CCD's, corresponding to a mass per unit area of approximately 50 g.cm^{-2}, about 2.5 radiation lengths and 0.5 interaction lengths. Equation 4 predicts a very healthy 2×10^9 CC interactions per year for this target.

Besides providing the mass for neutrino interactions, the tracking target allows precise reconstruction of the event topologies from charged tracks, including event-by-event vertex tagging of those events containing charm or beauty hadrons or tau leptons. Given the favorable vertexing geometry and the few-micron typical CCD hit resolutions, it is reasonable to expect almost 100 percent efficiency for b tagging, perhaps 70 to 90 percent efficiency for charm tagging and excellent discrimination between b and c decays.

The target in figure 1 is surrounded by a time projection chamber (TPC) tracker in a vertical dipole magnetic field. The characteristic dE/dx signatures from the tracks would identify each charged particle. Further particle ID is provided by the Cherenkov photons that are produced in the TPC gas then reflected by a spherical mirror at the downstream end of the tracker and focused onto a read-out plane at the upstream end of the target. The mirror is backed by electromagnetic and hadronic calorimeters and, lastly, by iron-core toroidal magnets for muon ID.

NEUTRINO INTERACTIONS AND THEIR EXPERIMENTAL INTERPRETATION

The dominant interaction of TeV-scale neutrinos is deep inelastic scattering (DIS) off nucleons (i.e. protons and neutrons). There are 2 types of DIS: neutral current (NC) and charged current (CC) scattering. In neutral current (NC) scattering, the neutrino is deflected by a nucleon (N) and loses energy with the production of several hadrons (X):

$$\nu + N \to \nu + X, \qquad (6)$$

This comprises about 25 percent of the total cross section and is interpreted as elastic scattering off one of the many quarks inside the nucleon through the exchange of a virtual neutral Z boson:

$$\nu + q \to \nu + q. \qquad (7)$$

Charged current (CC) scattering is similar to NC scattering except that the neutrino turns into its corresponding charged lepton:

$$\nu + N \to l^- + X,$$
$$\overline{\nu} + N \to l^+ + X, \qquad (8)$$

where l is an electron/muon for electron/muon neutrinos. At the more fundamental quark level a charged W boson is exchanged with a quark (q), which is turned into another quark species (q') whose charge differs by one unit.

$$\nu + q \to l^- + q',$$
$$\overline{\nu} + q' \to l^+ + q. \qquad (9)$$

The relativistically invariant quantities that are routinely extracted in DIS experiments are 1) Feynman x, the fraction of the nucleon momentum carried by the struck quark, 2) the inelasticity, $y = E_{\text{hadronic}}/E_\nu$, which is related to the scattering angle of the neutrino in the neutrino-quark CoM frame, and 3) the momentum-transfer-squared, $Q^2 = 2M_{proton}E_\nu xy$. MURINE's will have the further capability of reconstructing the hadronic 4-vector, resulting in a much better characterization of each interaction.

The final state quark always "hadronizes" at the nuclear distance scale, combining with quark-antiquark pairs to produce the several hadrons seen in the detector. Final state c and b quarks can be identified by vertex tagging of the decaying charm or beauty hadrons that contain them, and some statistically based flavor tagging will also be available for u, d or s final state quarks, using so-called "leading particle effect" [1].

PHYSICS OPPORTUNITIES

Neutrino interactions are interesting both in their own right and as probes of the quark content of nucleons, so a MURINE has wide-ranging potential to make advances in many areas of elementary particle physics. This section gives an overview for measurements involving the CKM quark mixing matrix, nucleon structure and QCD, electroweak measurements, neutrino oscillations and, finally, studies of charmed hadrons.

There is considerable theoretical interest in the mixture of final state quarks produced in CC interactions. The struck quark can be converted into any of the three final state quarks that differ by one unit of charge: a down (d), strange (s), or bottom(b) quark can be converted into an up (u), charmed (c), or top (t) quark and vice versa. In practice, production of the heavy top quark is kinematically forbidden at these energies and the production of other quark flavors is influenced by their mass. Beyond this, the Standard Model of elementary particle physics (SM) predicts the probability for the interaction to be proportional to the absolute square of the appropriate element in the so-called Cabbibo-Kobayashi-Maskawa (CKM) quark mixing matrix, a unitary matrix with 4 free parameters whose values are not predicted by the SM. Improved measurements involving the CKM matrix will test the SM hypothesis. It is of particular interest that one of the 4 parameters is a complex phase that is postulated as an explanation for CP violation – the intriguing experimental phenomenon that particles may have tiny deviations from the properties that mirror those of their antiparticles.

The experimentally determined values for the 9 mixing probabilities are given in table 1 [4], along with their current percentage uncertainties and speculative projections [1] for how the uncertainties could be reduced by a MURINE. From the large improvements in 4 of the 9 uncertainties it is clear that a MURINE has potential for tremendous improvements in measuring the quark mixing matrix, and more detailed studies are clearly desirable.

Another major motivation for MURINE's is the potential for greatly improved measurements of nucleon structure functions – the momentum distributions of quarks inside the nucleon. This provides [5] important tests of quantum chromodynamics (QCD) – the theory of the strong interaction that is widely accepted for its elegance and simplicity but which has not been experimentally verified at the level of the electroweak theory. A MURINE might well be the best single experiment of any sort for the examination of perturbative QCD [1].

Neutrino physics has also had an important historical role in measuring the electroweak mixing angle, which is simply related to the mass ratio of the W and Z intermediate vector bosons:

$$\sin^2 \theta_W \equiv 1 - \left(\frac{M_W}{M_Z}\right)^2. \tag{10}$$

Now that M_Z has been precisely measured at LEP, measurements of $\sin^2 \theta_W$ in neutrino physics can be directly converted to predictions for the W mass. The

TABLE 1. Quark mixing probabilities. Threshold suppression due to quark masses has been neglected. In practice, this will reduce the mixing probabilities to the heavier c and b quarks to below the values given in the table and will prevent any mixing to the top quark. The second row for each quark gives current percentage uncertainties in quark mixing probabilities and speculative projections of the uncertainties after analyses from a MURINE. The two uncertainties in brackets have not been measured directly from tree level processes. The uncertainties assume that no unitarity constraints have been used.

	d	s	b
u	0.95 ±0.1%	0.05 ±1.6%	0.00001 ±50% → 1-2%
c	0.05 ±15% → 0.2-0.5%	0.95 ±35% → ~1%	0.002 ±15% → 3-5%
t	0.0001 (±25%)	0.001 (±40%)	1.0 ±30%

comparison of this prediction with direct M_W measurements in collider experiments constitutes a precise prediction of the SM and a sensitive test for exotic physics modifications to the SM [5]. Reference [1] estimates that the predicted uncertainty in M_W from a MURINE analysis might be of order 10 MeV, which improves by more than an order of magnitude on today's neutrino experiments [5,6] and is comparable with the projected best direct measurements from future collider experiments.

A neutrino property that is currently drawing much interest is the question of whether neutrinos have a non-zero mass. If they do then it is possible for the 3 neutrino flavors to mix, perhaps producing neutrino oscillations that can be observed using a neutrino beam. The probability for an oscillation between two of the flavors is given by [7]:

$$\text{Oscillation Probability} = \sin^2\theta \times \sin^2\left(1.27\frac{\Delta m^2[\text{eV}^2].L[km]}{E_\nu[GeV]}\right), \tag{11}$$

where the first term gives the mixing strength and the second term gives the distance dependence.

Reference [1] obtains the following order-of-magnitude mass limit for an assumed long-baseline detector with reasonable parameters and with full mixing:

$$\Delta m^2|_{min} \sim O(10^{-4})\, eV^2, \tag{12}$$

independent of the distance to the detector. Similarly, a mixing probability sensitivity for 10^{10} events in a short-baseline detector is found to be as low as

$$\sin^2\theta|_{min} \sim O(10^{-7}), \tag{13}$$

for the most favorable value of Δm^2. Both of these estimates apply generically to all 3 possible mixings between 2 flavors: $\nu_e \leftrightarrow \nu_\mu$, $\nu_e \leftrightarrow \nu_\tau$ and $\nu_\mu \leftrightarrow \nu_\tau$. (See also reference [8] for another discussion of neutrino oscillations at a MURINE.)

The Δm^2 estimate is more than an order of magnitude better than any proposed accelerator or reactor experiments for $\nu_\mu \leftrightarrow \nu_\tau$ and $\nu_e \leftrightarrow \nu_\tau$, and is competitive with the best such proposed experiments for $\nu_e \leftrightarrow \nu_\mu$. The estimated value for $\sin^2\theta|_{min}$ is even more impressive – orders of magnitude better than in any other current or proposed experiment for each of the three possible oscillation.

As an interesting final topic, MURINE's should be rather impressive factories for the study of charm – with a clean, well reconstructed sample of several times 10^8 charmed hadrons produced in 10^{10} neutrino interactions. There are several interesting physics motivations for charm studies at a MURINE [9]. As an example, particle-antiparticle mixing has yet to be observed in the charm sector [10], and it is quite plausible [1] that a MURINE would provide the first observation of $D^0 - \overline{D^0}$ mixing.

SUMMARY

The intense neutrino beams at muon collider complexes should usher in an exciting new era of neutrino physics experiments, with great advances expected in both traditional and new areas of neutrino physics.

REFERENCES

1. B.J. King, *Neutrino Physics at a Muon Collider*, Proc. Workshop on Physics at the First Muon Collider and Front End of a Muon Collider, Fermilab, November 6-9, 1997.
2. B.J. King, *Assessment of the prospects for muon colliders*, paper submitted in partial fulfillment of requirements for Ph.D., Columbia University, New York (1994); B.J. King, *A Characterization of the Neutrino-Induced Radiation Hazard at TeV-Scale Muon Colliders*, BNL Center for Accelerator Physics internal report 162-MUON-97R, to be submitted for publication.
3. See, for example, Chris Quigg, *Neutrino Interaction Cross Sections*, FERMILAB-Conf-97/158-T.
4. Values extracted from Andrzej J. Buras, *CKM Matrix: Present and Future*, TUM-HEP-299/97.
5. Janet M. Conrad, Michael H. Shaevitz and Tim Bolton, *Precision Measurements with High Energy Neutrino Beams*, hep-ex/9707015, submitted to Rev. Mod. Phys. (1997)
6. K.S. McFarland et al. (CCFR/NuTeV Collaboration) *A Precision Measurement of Electroweak Parameters in Neutrino-Nucleon Scattering*, FNAL-Pub-97/001-E. B.J.

King, Columbia University Ph.D. Thesis, 1994; Nevis Report: Nevis-283, CU-390, Nevis Preprint R-1500 (1994).
7. R.M. Barnett et al., Physical Review D54, 1 (1996) and 1997 off-year partial update for the 1998 edition available on the PDG WWW pages (URL: http://pdg.lbl.gov/).
8. S. Geer, *The Physics Potential of Neutrino Beams From Muon Storage Rings*, Proc. Workshop on Physics at the First Muon Collider and Front End of a Muon Collider, Fermilab, November 6-9, 1997.
9. I.I Bigi, *Open Questions in Charm Decays Deserving an Answer*, CERN-TH.7370/94, UND-HEP-94-BIG08 (1994). I.I Bigi, *The Expected, The Promised and the Conceivable - on CP Violation in Beauty and Charm Decays.*, UND-HEP-94-BIG11 (1994).
10. Tiehui (Ted) Liu, *The D0-Dobar Mixing Search – Current Status and Future Prospects*, HUTP-94/E021 (1994). Gustavo Burdman, *Charm Mixing and CP Violation in the Standard Model*, FERMILAB-Conf-94/200 (1994).

LOW ENERGY PHYSICS
WITH COLD μ± BEAMS

A $\mu \to e + \gamma$ Experiment with 10^{-14} Sensitivity?

Hans-Kristian Walter

Paul Scherrer Institut, CH-5232 Villigen, Switzerland

Abstract. The motivation for a new experiment to search for the forbidden decay of a muon into an electron and a photon is based on predictions for the decay rate at an experimentally accessible level within the frame of supersymmetric grand unified theories. The present status of the LAMPF MEGA experiment is confronted with new preliminary ideas, how to reach sensitivities to this decay of the order of 10^{-14}, i.e. two orders of magnitude below the expected one for MEGA. Technically a high resolution magnetic spectrometer for the electron, and a high resolution calorimeter for the photon will be discussed as well as the necessary R&D work to be done before a proposal can be made. The πE5 beam at PSI is proposed as the best source for muons with the required quality.

MOTIVATION

New developments in particle physics theory considering supersymmetric grand unification of particles and forces (1-4) suggest several low energy experiments to become extremely important. These are in particular the decay $\mu \to e + \gamma$, muon-electron conversion in nuclei, and the electric dipole moments of the neutron and the electron. The decay $\mu \to e + \gamma$ has been studied at LAMPF over the last 10 years by the MEGA collaboration (5) and if not found an upper limit of $\sim 5 \cdot 10^{-12}$ is expected from this experiment. Muon - electron conversion is being studied at PSI (6) and if not found an upper limit of $\sim 5 \cdot 10^{-14}$ is expected within the next two years. The upper limit on the electric dipole moment of the neutron is $\sim 10^{-25}$ e · cm from experiments in Petersburg and Grenoble. Supersymmetric grand unification theory predicts finite values for these observables at most two orders of magnitude below these limits.

A proposal to study muon - electron conversion with a sensitivity of 10^{-16} has been accepted by Brookhaven National Lab (7). A letter of intent to improve the sensitivity to the electric dipole moment of the neutron by a factor of 250 by a new EDM-collaboration has been submitted to LANSCE (8).

REMAINS $\mu \to e + \gamma$

Several groups from Georgia, Italy, Japan, Poland, Russia and Switzerland consider an experiment searching for the decay $\mu \to e + \gamma$ with a sensitivity of 10^{-14}, which corresponds, according to the above-mentioned theories, to a sensitivity to muon-electron conversion of about 10^{-16}. At a workshop held at PSI in March 1997 the theoretical motivation was summarized and first ideas how to perform the experiment were discussed. Several detector geometries have been proposed. This experiment, at least for a few years to come, can only be done at the Paul Scherrer Institut (PSI) in Switzerland. Only here fluxes of $>10^8$ muons/s at sufficient low energy are available.

First discussions suggest that the two particles, a positron and a photon each with an energy of 52.8 MeV and emitted back to back, should be measured with a high resolution magnetic spectrometer and a high resolution calorimeter, respectively.

A beam of polarized surface muons with intensity of a few 10^8 /s is focussed onto a thin target where it is stopped. 50 MeV positrons are analyzed in a high resolution spectrometer with an axial or solenoidal magnetic field. Very thin tracking detectors measure the position and the angle of the positron, which then is stopped in a thick scintillation counter for triggering and timing. The spectrometer should have a solid angle of 10 - 20 % of 4π, a momentum acceptance of a few % and a momentum resolution of about 0.1 %. The most critical part is the tracking detector and the target, the thicknesses of which influence the verification of the 180° correlation of the two particles.

Photons are detected in a calorimeter shielded from charged particles by a magnetic field. Two options are considered: liquid Xenon and crystals made from fast and highly luminous materials, such as Ce-doped orthosilicates or -aluminates of Yttrium, Lutetium and/or Ytterbium. The LXe calorimeter may consist of a single cell, the surface of which is densely covered with UV sensitive photomultipliers, as proposed by several Japanese groups ("mini- Kamiokande"). Monte Carlo simulations show that a resolution for the photon impact point of a few mm can be obtained by determining the light emission centroid. The impact point is needed in order to establish the 180° correlation between photon and positron. An open problem is how to cope with pile-up in this single cell calorimeter.

The crystal calorimeter can be segmented into single crystals with transverse sizes of the order of the Molière radius. A preshower crystal plate, viewed from the entrance area with densely packed photomultipliers or avalanche photodiodes, may be necessary to obtain the impact point with a few mm accuracy and to give a good timing signal. Because of price reasons only this preconverter may consist of the new crystal material, whereas the back part may consist of CsI doped with Na. Figure 1 shows one possible detector setup, proposed by A. v. d. Schaaf (9).

In order to advance the project and proceed towards a technical proposal a number of open questions remain, requiring R&D work mainly in the following areas:
- Study of the phase space distribution of the πE5 beam at PSI.
- Detailed Monte Carlo simulation and tests of the muon stop distribution in an optimized target, the thickness of which determines the angular resolution between photon and positron.
- Detailed Monte Carlo simulation of the momentum acceptance and resolution of magnetic positron spectrometers.

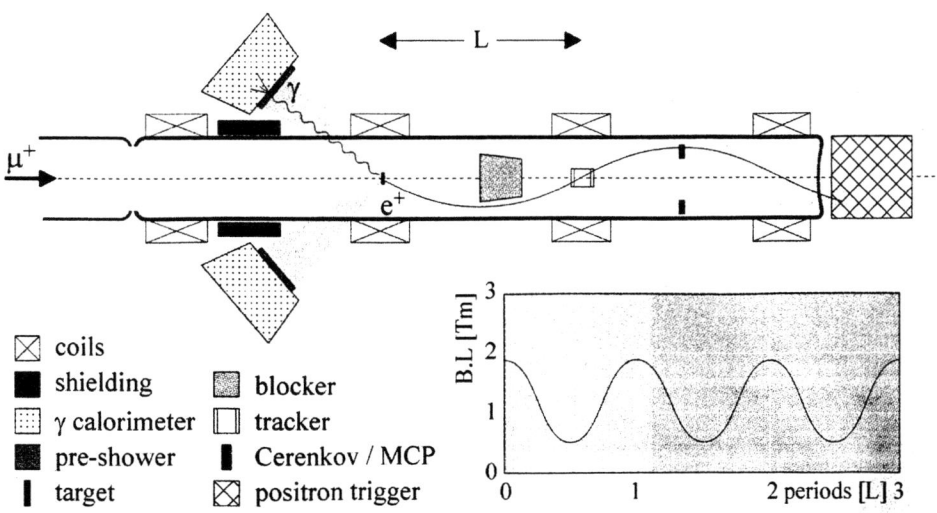

FIGURE 1. One possible setup for a new $\mu \rightarrow e + \gamma$ experiment, proposed by A. v. d. Schaaf (9).

- Studies with prototype LXe calorimeters using UV sensitive readout devices. Measurement of light yield, time resolution, light attenuation, conversion point resolution, pile-up rejection power, purification, stabilisation, etc.
- Production of prototype crystals and active converter plates. Measurement of parameters as above. Price optimization.

A second workshop is held at PSI on March 16-18, 1998, with the objectives to review the R&D progress since last year and to prepare a Letter of Intent to be submitted in April 1998.

REFERENCES

1. Barbieri, R., et al., *Nucl. Phys.* **B445**, 219 (1995).
2. Hisano, J., et al., *Phys. Rev.* **D53**, 2442 (1996).
3. Carlos, B. de, et al., *Phys. Rev.* **D53**, 6398 (1996).
4. Gomez, M. E., and Goldberg H., *Phys. Rev.* **D53**, 5244 (1996).
5. http://bart.phys.uh.edu/~empl/mega/
6. http://www1.psi.ch/www_sindrum2_hn/sindrum2.html
7. http://plato.ps.uci.edu/~meco/
8. http://p2hp2.lanl.gov/edm/edm.html
9. Schaaf, A. v. d., presented at a workshop at PSI in march 1997.

The MECO Experiment to Search for $\mu^- N \to e^- N$ with Sensitivity Below 10^{-16}

William R. Molzon

Department of Physics and Astronomy, University of California, Irvine, CA 92697

Abstract. I review the motivation for and status of searches for violation of muon and electron number conservation in muon initiated processes. I discuss the expected progress in muon conversion experiments, concentrating on the MECO experiment, E940 at BNL. It will improve the sensitivity for the process $\mu^- N \to e^- N$ to below 10^{-16}, roughly 4 orders of magnitude better than the present limit.

INTRODUCTION

Since the discovery of muons and the realization that all differences between muons and electrons are due only to their mass difference, there has been interest in understanding why more than one family of leptons exists and how they are related. An important question is whether or not muon, electron and tau lepton number are conserved. Non-conservation of these numbers is often referred to as lepton flavor violation (LFV).

In principle, LFV could be induced by neutrino mixing, by much the same mechanism that effective flavor changing neutral currents are induced in the quark sector. However, current limits on neutrino mass differences and mixing angles preclude the possibility that experimentally observable LFV effects occur via neutrino mixing. This means that observation of LFV would indicate the existence of new physics processes, and LFV searches are among the most sensitive means at our disposal for exploring physics beyond the Standard Model.

Because there is no theory in which muon and electron conservation follow from invariance under a local gauge transformation, there is strong theoretical prejudice that LFV will be seen. Aside from the underlying motivation to test conservation laws with the best possible sensitivity, there is theoretical motivation derived from the many proposed extensions to the Standard Model which allow LFV. In general, these models are not devised for the purpose of predicting LFV. In many cases, the stringent limits already set restrict the allowed values of parameters within these models. The sensitivity to various scenarios for physics beyond the Standard Model have been discussed, for example by W. Marciano at this conference.

Experimental programs are searching actively for LFV in both muon and kaon induced processes. In the kaon system, the sensitivity is approaching 10^{-12} in branching fraction. In the muon system, sensitivity of 10^{-12} and beyond has been achieved.

A particularly sensitive LFV probe is conversion of muons to electrons in the field of a nucleus. Searches for $\mu^- N \to e^- N$ are conceptually simple. A large flux of μ^- is brought to rest in a thin target. There they either decay or are captured on the nucleus. If they convert to electrons, the signature is an electron originating in the stopping target with energy $\sim m_\mu c^2$. The process is mostly coherent, with the nucleus recoiling without excitation. This process is closely related to $\mu^+ \to e^+\gamma$, if the process is mediated by a photon. In that case, for a given strength interaction, $B(\mu^+ \to e^+\gamma)$ would be $\sim 300 \times R_{\mu e}$ [1], where $R_{\mu e} \equiv \Gamma(\mu^- N \to e^- N)/\Gamma(\mu^- N \to \nu N')$. [1] Even though the branching fraction is smaller, searches for $\mu^- N \to e^- N$ have a significant experimental advantage. Since only one final state particle is observed, the experiments are not limited by accidental backgrounds, as are $\mu^+ \to e^+\gamma$ experiments. The current limit, $R_{\mu e} < 7.8 \times 10^{-13}$, is set by the SINDRUM2 collaboration [2]. That collaboration has proposed to improve their sensitivity to below 10^{-13} using a new muon beam. Further improvement in sensitivity, to below 10^{-16} will require new techniques, and the MECO collaboration has proposed such an experiment at BNL.

THE MECO MUON TO ELECTRON CONVERSION EXPERIMENT

The **M**uon to **E**lectron **Co**nversion collaboration has recently proposed [3] to extend the experimental sensitivity for $\mu^- N \to e^- N$ to below 10^{-16} using a new beam and detector operating at the Brookhaven National Laboratory Alternating Gradient Synchrotron.

One critical aspect of MECO is a very intense μ^- beam which uses the idea from the MELC proposal of Djilkibaev and Lobashev [4,5] to place the production target in a graded solenoidal field and collect π's over essentially 4π solid angle. MECO uses a higher energy proton beam than that proposed for MELC, following the realization by muon collider proponents [6,7] that higher proton energy produce muons with high efficiency. A second critical aspect of MECO is a pulsed beam, as used in an earlier experiments [8], which reduces the prompt backgrounds from π^- and e^- contamination. The basic idea is to stop a pulse of muons and detect conversion electrons only after all π^- and e^- in the beam have either decayed or passed through the detector region. These considerations lead to a beam pulse frequency of ~ 1 MHz and the use of an Aluminum stopping target, in which the μ^- lifetime is 880 ns. A third critical aspect of the MECO experiment is a high precision magnetic spectrometer capable of operating in the high rate environment resulting from the very high μ^- stopping rate.

A schematic drawing of the MECO beam-line and experimental area is shown in figure 1. Pions are produced from a tungsten target in a solenoid. The axial component of the field is graded, resulting in a very high capture probability for π's and μ's, since charged particles emitted away from the μ^- beam are reflected in the field and are captured in the beam with high efficiency. The μ^- resulting from π^- decay are guided to the stopping target in a curved transport solenoid. Particles propagating in helical trajectories in the curved solenoid drift vertically and are sign and momentum selected with appropriate collimators. The μ^- stopping target and the electron detector are located in a solenoid, in which the axial component of the field decreases from 2 T at the entrance to 1 T just after the target. The use of an axially graded field results in very good acceptance and reduces backgrounds from electrons produced in the upstream pole piece of the production solenoid. The heart of the detector is the low mass magnetic spectrometer in which the e^- momentum is measured. The electron trigger detector's is used primarily as a trigger and secondarily to check the e^- energy, help distinguish e^- from other particles, and helps in rejecting cosmic ray induced backgrounds. The detector region is surrounded by a cosmic ray shield.

Physics and Detector Background Sources

Physics backgrounds originate from a variety of sources: μ^- decay in a Coulomb bound orbit, radiative μ^- capture, beam electrons, μ^- decay in flight, π^- decay in flight, radiative π^- capture, $\overline{p}$ induced electrons, and cosmic ray induced electrons. The first two are intrinsic to μ^- stopped in the target; they can be minimized only by improving the measurement of the electron energy. The other sources derive from prompt processes, with the electron detected close in time to the arrival of a particle in the detector, and are reduced with a pulsed beam. Very slow $\overline{p}$'s have a very long transit time in the muon beam-line and arrive at the stopping target essentially continuously. Hence, they are not significantly reduced by using

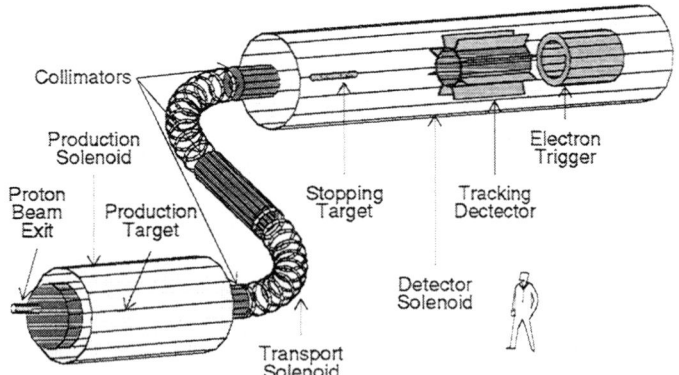

FIGURE 1. Schematic drawing of the MECO beam and apparatus.

a pulsed beam. Cosmic ray background is reduced with appropriate active and passive shielding.

The most important background is from μ^- decay in orbit; the rate is approximately proportional to $(E_{max} - E_e)^5$ near the endpoint [9]. Hence it is extremely sensitive to both the central part of the detector response function and possible high energy tails. Figure 2 shows the signal and background for $R_{\mu e} = 10^{-16}$, calculated in a full GEANT simulation of the MECO experiment [3]. By accepting

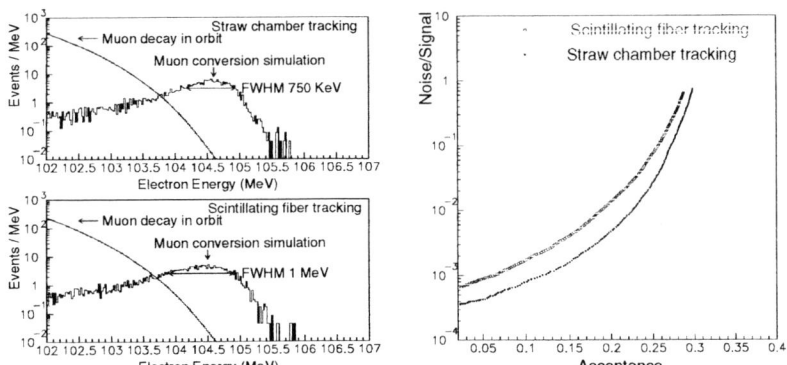

FIGURE 2. The histograms on the left show simulations of the expected signal and background for $R_{\mu e} = 10^{-16}$ for two possible detectors. The normalization of the signal and background curves is for $R_{\mu e} = 10^{-16}$ and a luminosity corresponding to 10^7 seconds running time. The plot on the right is a parametric plot of the background/signal and acceptance as a function of the minimum allowed measured e^- energy.

events between 103.9 MeV and 105.4 MeV, the noise to signal ratio is below 0.05 with large acceptance for signal events.

Other backgrounds have also been studied extensively by the MECO collaboration. The results are summarized in a later section. One potential source of background not discussed in the MECO proposal is due to anti-protons in the beam; it is now being studied actively by the MECO collaboration, and we make some general comments here. The production cross section at low $\bar{p}$ energy is not measured, and depends sensitively on the Fermi momentum in the nucleus. Several means to suppress this source of background exist if it proves to be a problem. One is to simply lower the proton beam energy. The yield as a function of incident proton energy is being calculated, and preliminary results show that the $\bar{p}$ flux entering the transport solenoid can be reduced by orders of magnitude by reducing the incident proton momentum from 8 GeV/c to 5 GeV/c. A second is to place thin absorbers in the muon transport channel. Preliminary results show that a 0.02 cm Be absorber in the middle of the transport solenoid will completely remove $\bar{p}$'s from the beam, and the potential background from annihilation products is being calculated. A third is to sweep the $\bar{p}$'s into a collimator in a crossed electric and magnetic field; an electric field of 2 MV/m in the 2 T transport field would be sufficient.

The high stopping rates in MECO will result in significant detector rates, and the MECO design has a geometry designed to minimize them. The detector gains will be reduced during the stopping pulse when rates are very high. During the detection interval, rates are dominated by p's and γ's from μ capture processes and e^- from μ^- decay in orbit; a full GEANT simulation was used to calculate them. Most electrons from muon decay are contained at radii smaller than the inner radius of the detectors. To reduce the flux of p's and γ's, the detector is located 2 m from the stopping target to reduce the subtended solid angle. Thin absorbers before the chambers reduce the flux of low energy protons hitting them without substantially degrading the momentum resolution. The detector is very low mass to minimize photon conversions.

The total rate in per detector element was calculated to be < 300 kHZ; it results primarily from protons and γ's. The typical detector channel occupancy is under 1%.

The Pulsed Muon Beam

An appropriately pulsed beam is critical to background rejection in MECO. It will use the RF structure of the AGS to pulse the proton beam. The AGS will operate with two filled bunches in the 2.7 μs revolution time, extracting a bunched beam [10].

The properties of a bunched extracted beam were measured [3]. One RF bunch was filled, accelerated to 24 GeV, and extracted. The pulses are $\sim$ 15 ns wide and the extinction between bunches is below 10^{-6} and in unfilled bunches is of order 10^{-4}. These measurements were made with only minimal tuning of the AGS, and substantially improved performance is expected [10]. It may not be possible to reach an extinction below 10^{-8} in the extracted beam, and other means of reducing the off pulse rate have been explored. The preferred solution is a pulsed electric or magnetic kicker [3,11,12] in the proton transport line.

The design of the μ^- production area is based on that of the MELC experiment [4,5] and adopted for the muon collider [6,7] source. Pions are produced in a radiation cooled, 0.8 cm diameter, 16 cm long tungsten target in a high field solenoid. Those with sufficiently small transverse momentum travel in helical trajectories inside the solenoid and decay to μ's. The field is graded with the axial component varying from 3.3 T at the upstream end to 2 T at the muon beam channel entrance. The super-conducting coil is protected with a heat and radiation shield with an inner radius of 0.3 m. At the intensity proposed, the maximum target temperature is below 2450 K, resulting in < 0.1% target evaporation per year. Heat load from the particle spray on the super-conducting solenoid was studied with a GEANT simulation. Less than 50 W is deposited in a 6 cm thick coil pack outside the shield, the maximum instantaneous local heat load is below 0.2

mW/gm and the total radiation load in a 10^7 s run is below 50 MRad.

There is little information on low energy pions produced by protons of a few GeV/c incident on heavy targets. These production cross sections are now being measured by E910 [13] at BNL. Model calculations based on FLUKA, GHEISHA [14], SHIELD, MARS, DPMJET2, and ARC, vary by a factor of 6. The production and transport parameters were optimized, and the resulting pion yield calculated using the GHEISHA model. Subsequent to those calculations, a measurement [15] of low energy pion production by 10 GeV protons on heavy targets was found which indicates [16] the GHEISHA model overestimates the yield by about a factor of two.

Charged particles are transported to the detector solenoid using a curved solenoid, one purpose of which is to decrease the transmission of both high momentum and positive particles. Charged particles of sufficiently low momentum follow helical trajectories centered on magnetic field lines. In a torus, they drift in a direction perpendicular to the plane of the torus. This is exploited to remove > 90% of the positive particles and all negative particles above 100 MeV/c.

The Stopping Target and Experimental Apparatus

The stopping target and detectors are located in a solenoid of diameter $\sim .9$ m, with a graded axial field, varying from 2T at the entrance to 1 T in the region of the detectors. The use of a graded field has two consequences. First, the quantity p_T^2/B_Z is constant, and hence 105 MeV e^-, either in the beam or produced at the upstream end of the solenoid, will have $p_T < 74$ MeV/c at the detector and can be eliminated as background by requiring $p_T > 75$ MeV/c. Second, conversion electrons emitted at angles of $90° \pm 30°$ with respect to the solenoid axis will have trajectories which intercept the tracking detector and which have a restricted range of p_T. Those initially moving away from the detector bounce in the graded field. Sample trajectories illustrating this are shown in figure 3.

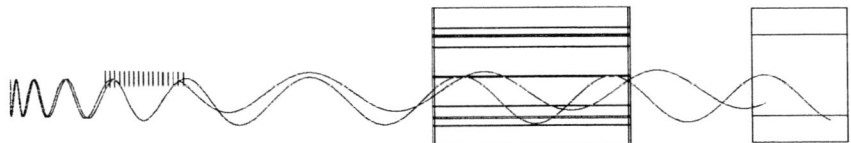

FIGURE 3. A schematic drawing of the MECO detector region with two typical conversion electron trajectories produced by the GEANT simulation superimposed.

The stopping target consists of 17 disks of Aluminum, each 0.02 cm thick and ~ 6 cm radius. The thickness chosen results from a tradeoff between stopping efficiency and energy loss straggling of conversion electrons. Approximately 40% of the μ^- in the beam are stopped in the target, and the energy straggling contributes ~ 200 keV to the energy resolution.

The goal for the tracking detector is to measure with good efficiency the parameters of the helical trajectory of electrons. The precision of the measurement is dominated by multiple scattering. The detector consists of a cylinder and 8 equally spaced *vanes* of tracking detectors, as shown in figure 1. All individual detector elements are oriented in the axial direction. The baseline design of the cylinder and vanes uses three layers of 5 mm diameter, 2 m long straw tubes. The axial coordinate is measured using pads external to the straws, in which signals are induced. The straws will be made of carbon loaded kapton. An alternate design has been studied in which the cylinder consists of 4 layers of 0.5 mm diameter scintillating fibers, arranged in pairs at small angle stereo, to get the axial coordinate. This design is easier to manufacture, but there is more scattering and energy loss in the cylinder, which reduces the precision of the electron energy measurement.

The performance of the detectors was calculated using a full GEANT simulation of the stopping target and detectors. Using the resolution function determined from Monte Carlo simulated events, the level of muon decay in orbit background was calculated by convolving the resolution function and the background electron energy spectrum. Figure 2 shows the noise to signal ratio vs. acceptance for the two detector possibilities, parameterized as a function of the lower edge of the accepted electron energy range. This and other background contributions are summarized in the section on the expected sensitivity of MECO.

The purpose of the electron trigger calorimeter is to detect 105 MeV electrons that have passed through the tracking system. The proposed detector is a scintillator cylinder of outer radius 70 cm and inner radius 41 cm and 1 m in length, segmented in azimuthal and axial directions. The performance was studied with a GEANT simulation, including effects of extra particles in the detector. The trigger rate will be dominated by muon decay in orbit electrons. It has been estimated by convolving the electron energy distribution with the response function of the calorimeter and imposing a minimum energy requirement of 65 MeV. Including the effects of pile-up of photons and neutrons at twice the expected level, the rate is calculated to be 1.7 kHz.

A potential source of background is cosmic ray induced electrons. This source is unique in that the level is proportional to the data collection time, and not the sensitivity. Hence, significant improvement in CR induced background rejection is not required. The background is reduced with a combination of active and passive shielding. It consists of 0.5 m of steel surrounding the detector solenoid (some of which is provided by the return yoke of the magnet), two layers of plastic scintillator, and 2 m of concrete shielding. It is assumed that the probability of not detecting a penetrating charged particle in either layer of scintillator is 10^{-4}. The cosmic ray induced background was calculated using measured cosmic μ^- fluxes and a GEANT simulation of the shielding and detector. From this simulation, the cosmic ray background is predicted to be 0.0035 events in 10^7 seconds of exposure.

Expected Performance and Sensitivity of MECO

The expected MECO sensitivity depends on the running time, proton intensity, number of muons produced and transported to the stopping target per proton, stopping probability, fraction of stopped μ^- which capture, trigger efficiency, accidental cosmic ray veto loss, tracking acceptance, and losses due to analysis inefficiencies and background rejection selection criteria. The values of the acceptance and efficiency for these are given in table 1. Loss of events due to accidental cosmic ray vetos and dead-time losses are expected to be small; losses in pattern recognition

TABLE 1. A summary of the expected MECO sensitivity.

Running time (sec)	10^7
Proton flux (sec^{-1})	4×10^{13}
μ/p entering solenoid	0.006
Stopping probability	0.370
μ capture probability	0.600
Fraction of μ which capture in time window	0.480
Electron trigger efficiency	0.900
Fitting and selection criteria	0.250
Detected events for $R_{\mu e} = 10^{-16}$	5.800

in the tracking detector are also expected to be small but have not yet been estimated. In one year (10^7 s) running time, a few events can be detected at a value of $R_{\mu e} = 10^{-16}$.

The muon yield used in table 1 is different from that in the MECO proposal. We have updated the number based on the measurements of pion yields discussed earlier, which gives a muon flux a factor of 2 lower than the GHEISHA prediction. The proton intensity has been increased with respect to that reported in the proposal by a factor of two based on the new expectations for AGS cycle time when running the machine at 8 GeV. If it is necessary to reduce the proton energy to 5 GeV to eliminate backgrounds from $\bar{p}$'s, the muon yield will decrease by a factor of two, resulting in ~ 3 detected events for a branching fraction of 10^{-16} in a 10^7 second run. The experiment is not limited by backgrounds, and increased running time will improve the sensitivity proportionally.

Table 2 shows the expected background rates for the sensitivity quoted above. The background is dominated by the μ^- decay in orbit contribution. Substantial improvement in discrimination against this source of background can be had with modest loss in acceptance, as shown in figure 2. Many of the background depend on the proton beam extinction, taken to be 10^{-10}. The potential background due to $\bar{p}$'s in the beam is currently under study. At the proposed level, the experiment is not expected to be limited by background.

TABLE 2. A summary of the expected MECO background level at the sensitivity given in table 1.

Source	Events	Comment
μ decay in orbit	0.29	signal/noise = 20 for $R_{\mu e} = 10^{-16}$
Radiative μ capture	$<< 0.05$	
μ decay in flight	< 0.003	without scatter in target
μ decay in flight	0.004	with scatter in target
Radiative π capture	0.007	from proton during detection time
Radiative π capture	0.014	from late arriving π
π decay in flight	$<< 0.001$	
Beam electrons	< 0.002	
Cosmic ray induced	0.004	assuming 10^{-4} CR veto inefficiency
Total background	0.37	

SUMMARY

Experiments to search for LFV have now been done for over 40 years, with ever increasing sensitivity. Current limits are at the level of 10^{-11} to 10^{-12} for the processes $\mu^+ \to e^+\gamma$, $\mu^+ \to e^+e^+e^-$, and $\mu^- N \to e^- N$. These limits place stringent constraints on many scenarios for physics beyond the Standard Model.

Improvements in muon beams and detector technology hold promise for making further significant improvements. The SINDRUM2 experiment is expected to improve the sensitivity to $\mu^- N \to e^- N$ to below 10^{-13} in the next year or two. Further improvement, to a sensitivity below 10^{-16}, is promised by the MECO experiment, now approved at BNL. These very substantial improvements in experimental sensitivity allows some optimism that the first evidence for muon and electron number violation may be found. If these proposed experiments are successfully executed, they will be sensitive to the level of lepton flavor violating signals suggested in many models. In particular, predictions of a class of grand unified supersymmetric models will be confronted directly by experimental measurements.

REFERENCES

1. A. Czarnecki, W. Marciano and K. Melnikov, hep-ph/9801218 (1997).
2. F. Riepenhausen, presented at the *Sixth Conference on the Intersections of Particle and Nuclear Physics*, Big Sky, Montana (1997).
3. M. Bachman, et al., "A Search for $\mu^- N \to e^- N$ with Sensitivity Below 10^{-16}", AGS P940 (1997).
4. R.M. Djilkibaev and V.M. Lobashev, Sov., J. Nucl. Phys. **49(2)**, 384 (1989).
5. V.S. Abadjev, et al., "MELC Experiment to Search for the $\mu^- A \to e^- A$ Process", INR preprint 786/92, November 1992.
6. "Beam Dynamics and Technology Issues for $\mu^+\mu^-$ Colliders", J. Gallardo ed., 1995.

7. "$\mu^+\mu^-$ Collider a Feasibility Study", the $\mu^+\mu^-$ collaboration, ed. J. Gallardo (1996).
8. Badert et al., Nucl. Physics A**377**, 406 (1979).
9. O. Shankar, Phys. Rev. D**25**, 1847 (1982).
10. M. Brennan, private communication.
11. A. Soukas, private communication.
12. A. Soukas, J. Sandberg, W. Meng, "A Pulsed Kicker for Secondary Extinction in a Proton Beam", BNL AGS memo, (1997).
13. I. Chemakin et al., P910 to BNL AGS (1995).
14. H.C. Fesefeldt, "Simulation of Hadronic Showers, Physics and Applications", PITHIA Report, Aachen 85-02 (1985).
15. D. Artmutliski, et al., Sov. J. Nucl. Phys. **48**, 161 (1988).
16. R. Djilkibaev, MECO internal note meco-018, "MECO Muon Yield Simulation Using Experimental Data" (1988).

PHYSICS AT HIGGS FACTORY AND PRECISION ELECTROWEAK DATA STUDIES

The Scientific Case for a Higgs Boson $\mu^+\mu^-$ Collider Factory

David B. Cline

*Center for Advanced Accelerators
Department of Physics and Astronomy, Box 951547
University of California, Los Angeles, CA 90095-1547 USA*

Abstract. We discuss the scientific motivation for a Higgs boson factory, including the possible evidence that may accumulate in the next decade to (1) provide convincing evidence that a low-mass Higgs boson exists from electroweak data, (2) the possible information from the LHC, (3) the mass range where the Higgs factory may be essential, (4) and to propose the time period for the actual construction of a Higgs factory in the USA. Tests for new physics at the Higgs factory are also discussed.

1. THE CONCEPT OF A HIGGS BOSON FACTORY $\mu^+\mu^-$ COLLIDER

The concept of a Higgs factory $\mu^+\mu^-$ collider was born at the first dedicated $\mu^+\mu^-$ workshop in Napa, California, December 1992;[1,2] Figure 1 shows a schematic of the scan for the Higgs presented at that meeting.[1] The next workshop also changed the role of a $\mu^+\mu^-$ collider.[3] Subsequently very nice theoretical work on this issue has been carried out by Barger, Berger, Gunion, and Han,[4] and then a paper (which included Table 1) defining the Higgs particle factory was presented at the Snowmass DPF meeting in 1996 by the author.[5] At that 1996 DPF meeting, a first pass design of a $\mu^+\mu^-$ collider was presented by the Muon Collider Consortium.[6] We will make use of all of these materials and of more recent work in this brief article, as well as proceedings from meetings we have organized[2,3,7] and other recent workshops.

When the Higgs factory was envisioned in 1992, there seemed to be little scientific support. At this conference, we see the first real evidence that the Higgs mass may be low and a possible justification for this machine.

The major purpose of the Higgs factory is to find the exact Higgs mass (or masses) and then measure the important parameters, such as the width(s) and the common and rare branching fractions. This concept is based mainly on a relatively low-mass Higgs (below 300 GeV). In the low-mass region (below 150 GeV), the Higgs could well be supersymmetric (SUSY), and the width measurement will be crucial. Above 150 GeV, the Higgs could be more of a standard-model type. However, this will once again lead to the issue of what keeps the scalar system stable, which might be answered by the study

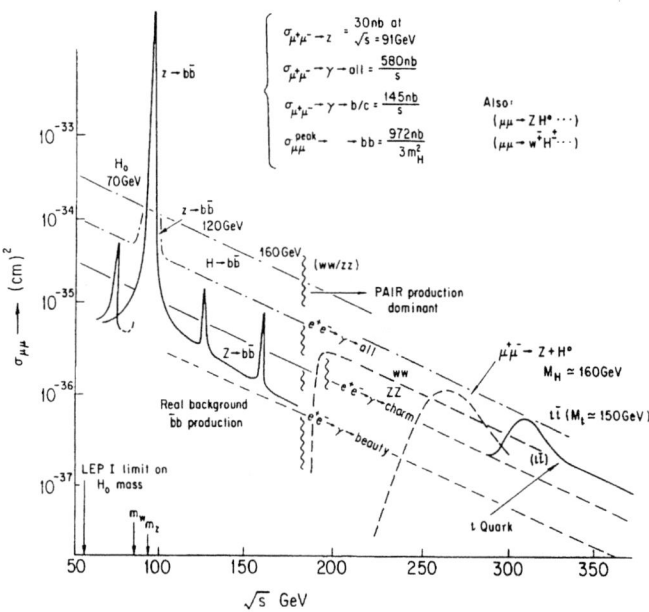

FIGURE 1. The first concept of a Higgs factory $\mu^+\mu^-$ collider from the Napa Workshop.[1]

TABLE 1. Arguments for a Higgs-Factory $\mu^+\mu^-$ Collider[1,5]

1. The $m_\mu \backslash m_e$ ratio gives coupling 40,000 times greater to the Higgs particle. In the SUSY model, one Higgs $m_h < 120$ GeV!!
2. The low radiation of the beams makes precision energy scans possible.
3. The cost of a "custom" collider ring is a small fraction of the $\mu^\pm$ source.
4. Feasibility report to Snowmass established that $\mathcal{L} \sim 10^{33}$ cm^{-2} s^{-1} is feasible.

of rare decays of the Higgs particles (in progress). In the near future, there could be evidence for the Higgs mass obtained from precise electroweak parameter measurements and later from the LHC. This will be a crucial input for the development of the Higgs factory. In addition, if Nature is supersymmetric, there will be additional SUSY-Higgs particles to study and, thus, the Higgs factory concept will include the search for and study of the SUSY Higgs (H,A ...). This is an experimental issue – theory can only take us just so far!

From all we now know about elementary particle physics, the scalar or SUSY scalar sector is the key to future understanding. A complete understanding of this sector is really the goal of the Higgs factory and of nearly all elementary particle physics these days.

The Higgs factory is designed to first give the exact Higgs mass using an energy scan and then measure the general properties of the Higgs, such as the field width, largest branching fractions, *etc.* It would produce 10^4 Higgs/yr and could investigate rare branching modes. If there are more Higgs, the Higgs factory would be used to scan and find and study these in detail as well. A $\mu^+\mu^-$ collider provides the Higgs factory, since scalars couple like m_f^2 and the collider has little radiative energy spreading (see Table 2). Some important meetings are listed in Refs. 7-9, and some references to the early concept of a $\mu^+\mu^-$ collider can be found in Ref. 10.

TABLE 2. Logic of Detailed Study of the Higgs Sector

If particles in the scalar sector are ever discovered, it will be essential to determine their properties, which will give direct information about the nature of the particle and the underlying theory. Three simple examples can be cited:

1. Suppose a Higgs-like particle is discovered with mass 110 GeV. This could either be the standard model (SM) Higgs or an MSSM Higgs. A measurement of the width of the state would presumably tell the difference. However, the SM width is 5 MeV - a formidable measurement!

2. Suppose a Higgs-like particle is discovered with a mass of 150 GeV. This is presumably beyond the MSSM bound, but it could be an NMSSM or an SM Higgs. A measurement of the width could presumably resolve the issue.

3. Suppose a Higgs-like particle of mass 165 GeV is discovered. This is presumably even beyond the NMSSM limits. If this is an SM Higgs, can we learn more by the study of the rare decay modes?

2. THE SCALAR ELECTROWEAK SECTION AND SUPERSYMMETRIC HIGGS BOSON

We expect the supercollider LHC to extract the signal from background (*i.e.*, seeing either $h^0 \to \gamma\gamma$ or the very rare $h^0 \to \mu\mu\mu\mu$ in this mass range, since $h \to b\bar{b}$ is swamped by hadronic background). However, detectors for the LHC are designed to extract this signal. Figure 2 gives a picture of the various physics thresholds that may be of interest for a $\mu^+\mu^-$ collider. In this low mass region, the Higgs is also expected to be a fairly narrow resonance and, thus, the signal should stand out clearly from the background from

$$\mu^+\mu^- \to \gamma \to b\bar{b} \to Z_{\text{tail}} \to b\bar{b} \quad . \tag{1}$$

For masses above 180 GeV, the dominant Higgs decay is

$$h^0 \to W^+W^- \quad \text{or} \quad Z^0Z^0 \quad , \tag{2}$$

and the LHC should easily detect this Higgs particle. Thus the $\mu^+\mu^-$ collider is better adapted for the low mass region.

The strongest argument for the low-energy collider comes from the growing evidence that the Higgs should exist in this low-mass range from:

(i) The original works of Cabibbo and colleagues, which shows that, when $m_t > M_Z$ and assuming a grand unification theory (GUT), $M_H < 2 M_Z$.[11]
(ii) Fits to LEP data imply that a low mass h^0 could be consistent with $m_t > 150$ GeV.[12]

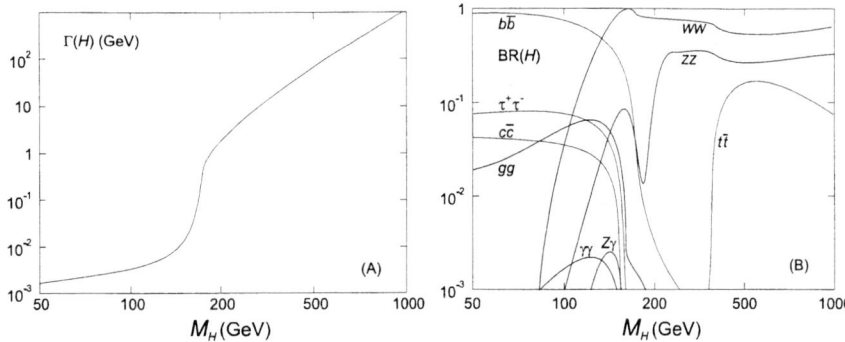

FIGURE 2. (A) The width and (B) branching fractions for the standard model Higgs boson as a function of mass.

(iii) The extrapolation to the GUT scale that is consistent with SUSY also implies that one of the Higgs should have a low mass, perhaps below 130-150 GeV.[12]
This evidence implies the exciting possibility that the Higgs mass is just beyond the reach of LEP II and in a range that is very difficult for the LHC to detect.[3] Other arguments for the detailed study of the scalar sector are given in Table 2.

3. FINDING AND STUDYING THE HIGGS AT A HIGGS FACTORY $\mu^+\mu^-$ COLLIDER

In this section, we assume for the sake of argument that the CMS detector at the LHC has barely detected signal at $m \sim 130$ GeV ($h^0 \to \gamma\gamma$) and at an experimental width of ~8 GeV (Step 1, illustrated in Fig. 3). The questions now are:
(i) Is this a Higgs boson or not?
(ii) Is it the standard model Higgs or a SUSY Higgs?
We envision the next step would be to construct the $\mu^+\mu^-$ collider operating between the energies of $E_{\mu^+\mu^-} \sim m_{h^0}$ (COM) and $E_{\mu^+\mu^-} \sim m_Z + m_{h^0}$ (COM) or the use of the NLC to observe $e^+e^- \to Z^0 h^0$.[13] We build the $\mu^+\mu^-$ collider (after already having built a $\mu^\pm$ source), and for Step 2 operate near the $Z^0 + h^0$ (CMS) threshold to determine m_{h^0} and Γ_{h^0} to ~ 1 GeV. (See Fig. 4 for the cross sections). For Step 3, we envision an energy scan of the mass region by varying the $\mu^+\mu^-$ energy.[3,14] At some point, the mass and width are determined and then used to distinguish between the standard model Higgs and a SUSY Higgs (Fig. 5).

The final step is to measure the branching fractions for different decay modes.[11,12] Figure 2 shows the expectations for the standard model Higgs; the observation of rare Higgs decay could also be important (Fig. 6).

We show in Fig. 2 the logic of the Higgs factory scan. Different methods (LHC, NLC, $\mu\mu$ collider) are used to narrow the energy range that must be scanned.[3,14]

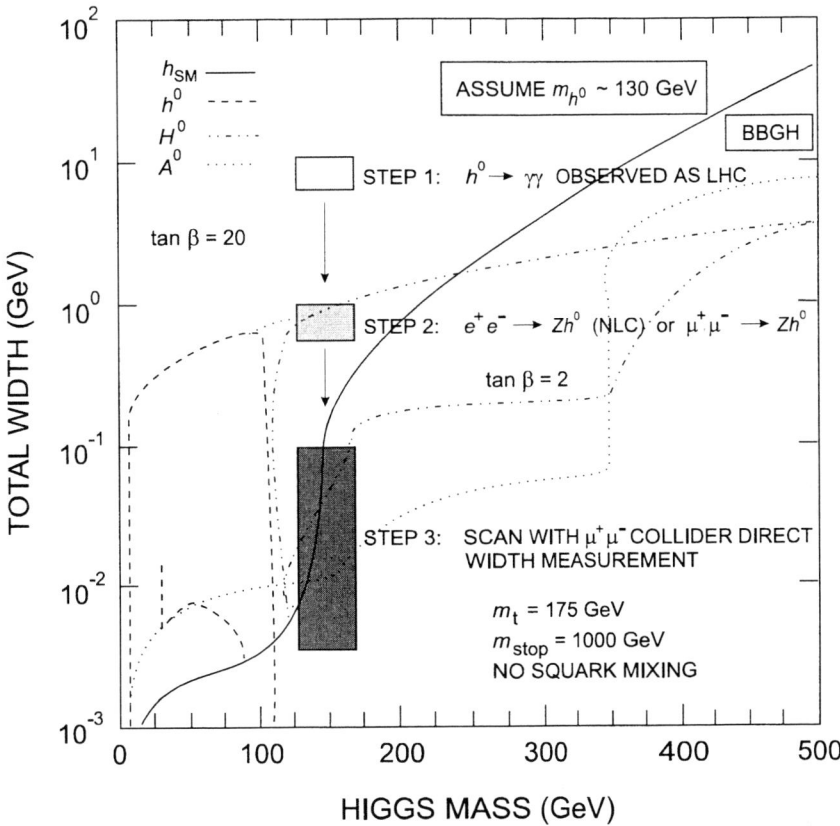

FIGURE 3. Higgs-factory μ⁺μ⁻ collider concept. The Higgs is discovered at the LHC (CMS) and the width further reduced at the NLC or at a μ⁺μ⁻ collider. The final stage is to scan for the Higgs at the μ⁺μ⁻ collider. Existing models can be distinguished by their widths. [Adapted from Refs. 4 (BBGH = Barger, Berger, Gunion, Han) and 5.]

There are several ways to determine the approximate mass of the Higgs boson in the future.[13] Suppose it is expected to be at a mass of 135 ± 2 GeV, the energy spread of a μ⁺μ⁻ collider can be matched to the expected width. An energy scan could yield a strong signal to background especially with polarized μ⁺μ⁻ in the scalar configuration.[8,15] Once the Higgs is found, the following could be carried out:

(i) Measurement of width – to separate the SM Higgs from SUSY or other Higgs models,[4,11]
(ii) Measurement of the branching fractions – the rare decay will involve loop effects that can sample very high energies (see Fig. 2).

Polarization will play an essential role for any μ⁺μ⁻ collider![16,17]

Some of the key branching fractions for the Higgs for two masses of 130 GeV/c^2 and 200 GeV/c^2 are given in Fig. 2.

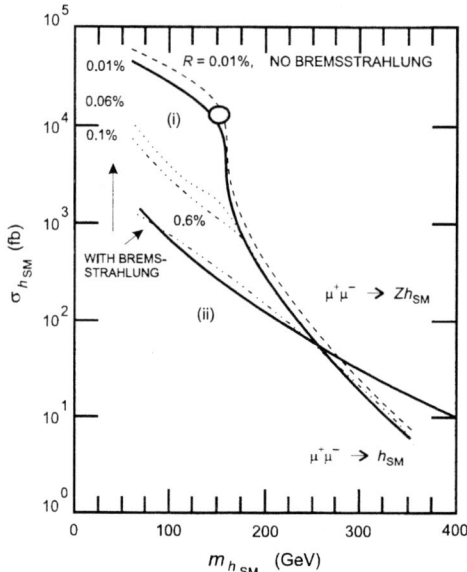

FIGURE 4. Cross sections versus $m_{h_{SM}}$ for inclusive standard model Higgs production: (i) the s-channel σ_h for $\mu^+\mu^- \to h_{SM}$ with $R = 0.01\%$, 0.06%, 0.1%, and 0.6%; and (ii) $\sigma(\mu^+\mu^- \to Zh_{SM})$ at $\sqrt{s} = m_Z + \sqrt{2}\, m_{h_{SM}}$. Also shown is the result for $R = 0.01\%$ if bremsstrahlung effects are not included. (Adapted from Ref. 4.)

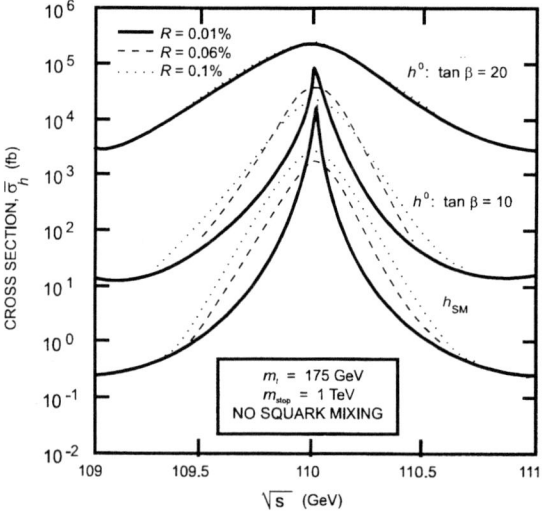

FIGURE 5. The effective cross section, $\bar{\sigma}_h$, obtained after convoluting σ_h with the Gaussian distributions for $R = 0.01\%$, 0.06%, and 0.1%, is plotted as a function of $\sqrt{s}$ taking $m_h = 110$ GeV. Results are displayed in the cases h_{SM}, h^0 with $\tan\beta = 10$ and $= 20$. In the MSSM h^0 cases, two-loop/RGE-improved radiative corrections have been included for Higgs masses, mixing angles, and self-couplings assuming $= 1$ TeV and neglecting squark mixing. The effects of bremsstrahlung are not included in this figure. (Adapted from Ref. 4.)

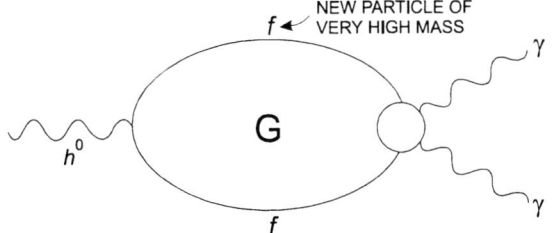

FIGURE 6. Possible probe of new physics in a rare Higgs decay mode.

4. STUDY OF THE SUSY HIGGS SECTOR

In the SUSY model, we expect several scalar and pseudo-scalar particles,[4] and the reports in Refs. 3, 7, and 8 give a very good overview of the detection and possible study of such scalars at $\mu^+\mu^-$ colliders. There are recent fits to the electroweak parameter that suggest that the Higgs mass is below 200 GeV (Table 3). Note that there are constraints on the Higgs mass related to the energy level of new physics (see Fig. 7). However, a Higgs factory is only needed for certain mass ranges, as discussed in Table 4.

TABLE 3. Recent Fits for the Higgs Mass

$m_h = 129 ^{+103}_{-92}$ GeV	Davier and Höchen[18]
$m_h = 122 ^{+134}_{-77}$ GeV	Erlon & Longacher[18]
$m_h = 116 ^{+115}_{-65}$ GeV ; $m_h < 400$ GeV 95% CL[18]	

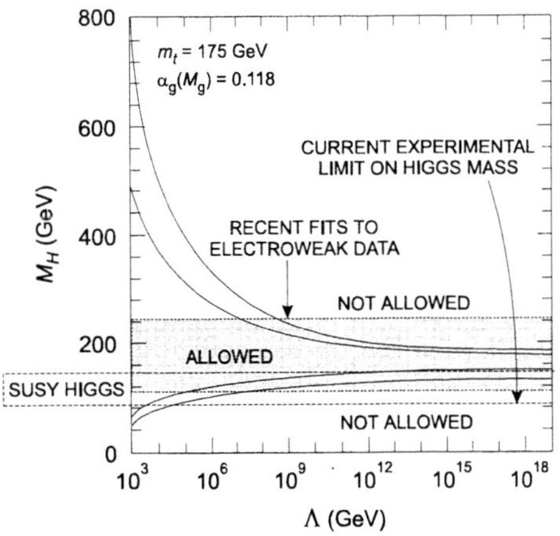

FIGURE 7. Limits of stability for a SM Higgs boson plotted vs the energy, Λ, where the model breaks down (presumably due to new physics). We also show the preferred region for the Higgs mass from recent fits and the region of SUSY Higgs. (Adapted from Ref. 19.)

Recently a schematic of a possible Higgs factory at FNAL has been prepared (Fig. 8). While it is premature, it shows that such a machine is a small addition to the FNAL complex.

ACKNOWLEDGMENTS

I wish to thank V. Barger and J. Gunion, as well as members of the µµ Consortium, for discussions.

Table 4. When a Higgs Factory is Needed and Why

m_h (GeV)	Standard Model (SM) or SUSY	µµ Collider HF Needed	Comments
~88* – 150**	SUSY-Higgs mass range – Higgs Γ very small	Yes	However, if m_{SUSY} is low, LHC may be flooded with $h \to b\bar{b}$ (CMS study)
~150 – ~200	Outside of SUSY Higgs – but self-coupling stability should be explored	Yes!!	In difficult mass range to study at LHC
200 – 300	Outside of SUSY range – but can be observed at LHC with CMS detector	No? for h^0 Yes for H,A	If low-mass SUSY observed, and even if LHC flooded with h^0, H,A difficult to study at LHC
> 300	For SM Higgs For H,A ... SUSY Higgs	No for h^0 Yes for H,A	

*LEP II limit.   **NMSSM model limit

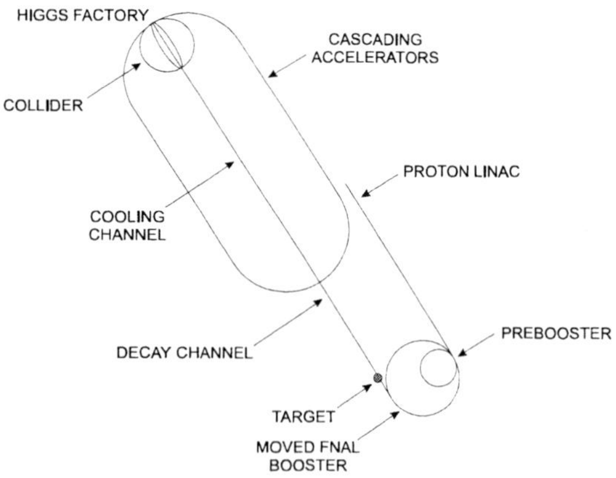

FIGURE 8. Schematic of a possible µµ collider Higgs factory in the FNAL complex..

REFERENCES

1. Cline, D., *Nucl. Instrum. Methods* **A350**, 24 (1994).
2. See collection of papers from the First Workshop on the Physics Potential and Development of $\mu^+\mu^-$ Colliders (Napa, CA, 1992), in *Nucl. Instrum. Methods* **A350**, 24-56 (1994).
3. *Physics Potential and Development of $\mu^+\mu^-$ Colliders* (Proc., 2nd Wksp., Sausalito, CA, 1994) AIP Conf. Proc. 352, Cline, D. B., ed., Woodbury, NY: American Institute of Physics, , 1996.
4. Barger, V., Berger, M. S., Gunion, J. F., and Han, T., U. Wisconsin-Madison report MAD-PH-963/hep-ph/9702334 (1997); also *Phys. Rev. Lett.* **78**, 3991 (1997); also *ibid.* **75**, 1462 (1995).
5. Cline, D.B., "A Higgs Factory $\mu^+\mu^-$ Collider," in *1996 DPF/DPB Summer Study on New Directions for High Energy Physics* (Snowmass, CO, June 25-July 12, 1996) p. 593.
6. *1996 DPF/DPB Summer Study on New Directions for High Energy Physics* (Snowmass, CO, June 25-July 12, 1996).
7. *Physics Potential & Development of $\mu^+\mu^-$ Colliders* (Proc. 3rd Intl. Conf., San Francisco, Dec. 1995), Cline, D. B., ed., *Nucl. Phys. B* (PS) **51A** (1996).
8. *Beam Dynamics and Technology Issues for $\mu^+\mu^-$ Colliders* (Proc., 9th Advanced ICFA Beam Dynamics Workshop, Montauk, LI, NY, Oct. 1995), AIP Conf. Proc. 372, Gallardo, J. C., ed., Woodbury, NY: American Institute of Physics, 1996.
9. *Future High Energy Colliders* (Proc., Symposium, Santa Barbara, CA, Oct. 1996), AIP Conference Proceedings 397, Parsa, Z., ed., Woodbury, NY: American Institute of Physics, 1997.
10. Early references for $\mu\mu$ colliders are: Perevedentsev, E. A. and Skrinsky, A. N., in *Proc., 12th Int. Conf. on High Energy Accelerators*, Cole, R. T. and R. Donaldson, R., eds. (Madison, WI, 1983), p. 485; Neuffer, D., *Part. Accel.*, **14**, 75 (1984); Neuffer, D., in *Advanced Accelerator Concepts*, AIP Conference Proceedings 156, New York: American Institute of Physics, 1987, p. 201.
11. For important references, see Dawson, S., Gunion, J. F., Haber, H. E., and Kane, G. L., *The Physics of the Higgs Bosons: Higgs Hunter's Guide*, Menlo Park, CA: Addison Wesley, 1989.
12. Barger, V., et al., "Particle Physics Opportunities at $\mu^+\mu^-$ Colliders" in *Physics Potential & Development of $\mu^+\mu^-$ Colliders, Nucl. Phys. B* (PS), **51A**, 13-31 (1996).
13. CMS Proposal for the LHC, CERN (1994) unpublished.
14. Report of the $\mu^+\mu^-$ Consortium, distributed at Snowmass '96.
15. Cline, D., "Physics Potential and Development of $\mu^+\mu^-$ Colliders," UCLA preprint CAA-115-12/94 (1994).
16. Norum, B. and Rossmanith, R., in *Physics Potential & Development of $\mu^+\mu^-$ Colliders, Nucl. Phys. B* (PS), **51A**, 191-200 (1996).
17. Cline, D., in *Beam Dynamics and Technology Issues for $\mu^+\mu^-$ Colliders* (Proc., 9th Advanced ICFA Beam Dynamics Workshop, Montauk, LI, NY, Oct. 1995), AIP Conf. Proc. 372, Gallardo, J. C., ed., Woodbury, NY: American Institute of Physics, 1996, p. 279.
18. See references in Renton, P., these proceedings.
19. Adapted from CERN report T4/97-367.

Precision Electroweak Data: Present Status and Future Prospects

Peter B. Renton*

*Nuclear Physics Lab., University of Oxford, Oxford OX2 9PU UK;
e-mail: p.renton@physics.ox.ac.uk

Abstract. The most recent precision electroweak data from LEP, SLC and FNAL are reviewed. The mass of the Higgs boson is extracted from global electroweak fits and the influence of those measurements which are not very compatible with the Standard Model (SM) is discussed. Future expectations are also discussed.

I   PRECISION ELECTROWEAK DATA

The accuracy of the precision electroweak data has improved substantially in the last five years such that the data are now precise enough to give an indirect measurement of the mass of the Higgs boson, with the framework of the Standard Model. During this period the top quark has been discovered at the FNAL Tevatron, and its mass has been precisely measured to give $M_t = 175.6 \pm 5.5$ GeV [1,2]. The mass and width of the Z boson have been measured with impressive accuracy at LEP. The high values of longitudinal polarisation ($P_e \simeq 80$ %) which have been achieved at the SLC have allowed the SLD experiment to make an extremely precise measurement of $A_{\rm LR}$. The individual Z-fermion couplings to both leptons and heavy-quarks have also been measured with good accuracy at both LEP and the SLC. The mass of the W boson is also being measured at the Tevatron and LEP with ever increasing precision.

Recent reviews of the electroweak data, the formalism and the methods used to extract electroweak quantities can be found in [3,4]. In this review the results from the 1997 Summer Conferences, many still preliminary, are also included (see [5]); to which the reader is referred for more details and references.

II   HEAVY QUARK COUPLINGS

Extracting electroweak results from heavy flavor data is a rather involved procedure. This is because knowledge is required of the various c-quark and b-quark hadron lifetimes, multiplicities, branching ratios and fragmentation properties. The quantities of interest are $R_b = \Gamma_b/\Gamma_{had}$, $R_c = \Gamma_c/\Gamma_{had}$, and the pole forward-backward asymmetries for b and c quarks, $A_{FB}^{0,b}$ and $A_{FB}^{0,c}$. At the SLD the left-right-forward-backward asymmetry is also measured for $b\bar{b}$ and $c\bar{c}$ final states and these give direct measurements of $\mathcal{A}_b$ and $\mathcal{A}_c$ respectively.

For R_b and R_c the most reliable and accurate methods exploit double tags. The number of single and double tags for a b (or c) quark is found and this can be used to determine both the tagging efficiency and R_b (or R_c) (see [5] for details). For $A_{FB}^{0,b}$, the two main methods are the lifetime tag plus jet-charge and lepton analyses. The current values of these, plus other electroweak quantities are given in Fig.1.

An alternative approach in trying to understand the possible implications of the heavy flavor results is to extract the individual quark couplings [3,4]. The measurements used are $R_b^0 = \Gamma_b/\Gamma_{had}$ (which, using Γ_{had} from the lineshape, gives $(v_b^2 + a_b^2)$), R_c^0 $(v_c^2 + a_c^2)$, $\mathcal{A}_e$ from LEP/SLD (v_e/a_e), $A_{FB}^{0,b}$ $(v_b/a_b, v_e/a_e)$, $\mathcal{A}_b$ (v_b/a_b), $A_{FB}^{0,c}$ $(v_c/a_c, v_e/a_e)$ and $\mathcal{A}_c$ (v_c/a_c). The constraint $\alpha_s(M_Z) = 0.120 \pm 0.005$ is imposed. The results for v_b and a_b are given in Table 1. Note that there is a strong anti-correlation between v_b and a_b.

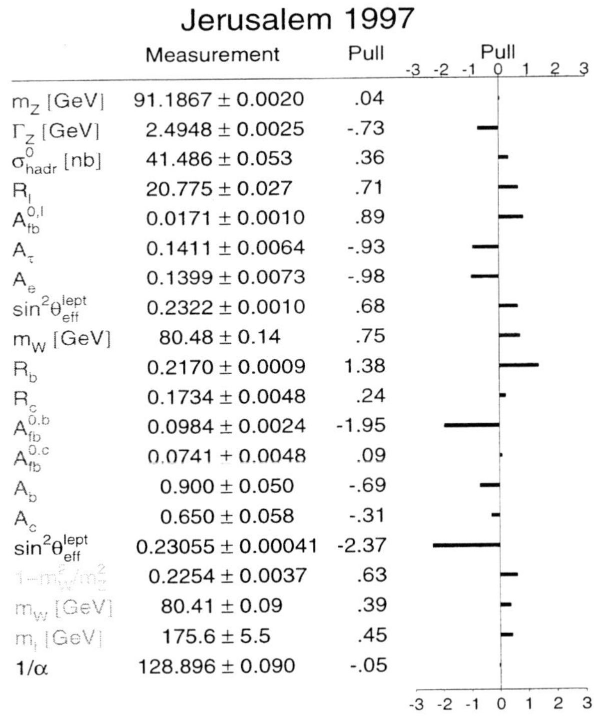

Jerusalem 1997

	Measurement	Pull
m_Z [GeV]	91.1867 ± 0.0020	.04
Γ_Z [GeV]	2.4948 ± 0.0025	-.73
σ_{hadr}^0 [nb]	41.486 ± 0.053	.36
R_l	20.775 ± 0.027	.71
$A_{fb}^{0,l}$	0.0171 ± 0.0010	.89
A_τ	0.1411 ± 0.0064	-.93
A_e	0.1399 ± 0.0073	-.98
$\sin^2\theta_{eff}^{lept}$	0.2322 ± 0.0010	.68
m_W [GeV]	80.48 ± 0.14	.75
R_b	0.2170 ± 0.0009	1.38
R_c	0.1734 ± 0.0048	.24
$A_{fb}^{0,b}$	0.0984 ± 0.0024	-1.95
$A_{fb}^{0,c}$	0.0741 + 0.0048	.09
A_b	0.900 ± 0.050	-.69
A_c	0.650 ± 0.058	-.31
$\sin^2\theta_{eff}^{lept}$	0.23055 ± 0.00041	-2.37
$1 - m_W^2/m_Z^2$	0.2254 ± 0.0037	.63
m_W [GeV]	80.41 ± 0.09	.39
m_t [GeV]	175.6 ± 5.5	.45
$1/\alpha$	128.896 ± 0.090	-.05

FIGURE 1. Electroweak data and pulls with respect to SM fit to all data.

TABLE 1. Results, plus correlation matrix, of a fit to the vector and axial-vector couplings of b and c quarks.

parameter	fitted value	v_b	a_b	v_c	a_c	SM
v_b	-0.309±0.012	1.00	-0.98	-0.20	0.07	-0.343
a_b	-0.523±0.007		1.00	0.20	-0.04	-0.499
v_c	0.188±0.012			1.00	-0.39	0.192
a_c	0.505±0.009				1.00	0.502

The b-quark couplings can also be expressed in terms of the left-handed $\ell_b = (v_b + a_b)/2$ and right-handed $r_b = (v_b - a_b)/2$ couplings. The results are shown in fig. 2. It can be seen that, whereas the c-quark couplings are compatible with the SM, those for the b-quark, in particular the right-handed coupling, are are in poor agreement with the SM expectations. The fitted values of v_b and a_b (or ℓ_b and r_b) give a value of R_b greater than the SM value and a value of $\mathcal{A}_b$ (and $A_{FB}^{0,b}$) less than the SM value. In this sense, the b-quark data are internally compatible with the observed deviations from the SM.

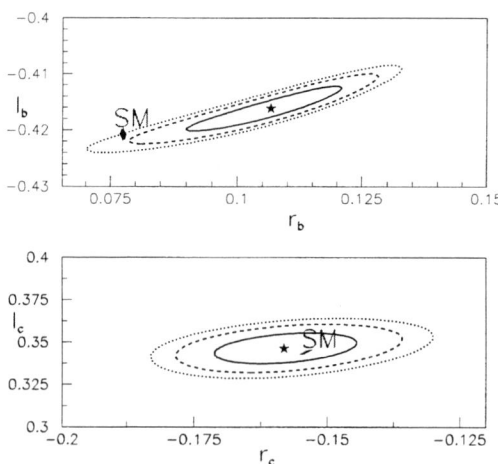

FIGURE 2. Results of a fit to the b-quark left and right-handed couplings. The contours are for the 70, 95 and 99% confidence limits.

TABLE 2. Results of global SM fits and the derived values of m_W.

	LEP (inc. m_W)	All but m_W, m_t	All data
m_t / GeV	158^{+14}_{-11}	157^{+10}_{-9}	173.1 ± 5.4
m_H / GeV	83^{+168}_{-49}	41^{+64}_{-21}	115^{+116}_{-66}
$\log m_H$	$1.92^{+0.48}_{-0.39}$	$1.62^{+0.41}_{-0.31}$	$2.06^{+0.30}_{-0.37}$
$\alpha_s(M_Z)$	0.121 ± 0.003	0.120 ± 0.003	0.120 ± 0.003
χ^2/dof	8/9	14/12	17/15
m_W / GeV	80.298 ± 0.043	80.329 ± 0.041	80.375 ± 0.030

III GLOBAL ELECTROWEAK FITS

Global electroweak fits to the data, within the context of the SM, have been performed. The results are given in Table 2. The input data used in these fits, and the pull distributions with respect to the SM fit to all data, are shown in Fig.1. It can be seen that the two measurements which give the largest χ^2 values are $A_{FB}^{0,b}$ and the SLD measurement of $\sin^2\theta_{eff}^{lept}$ from A_{LR}.

A comparison of the direct and indirect measurements of M_t and M_W is made in Fig.3. There is agreement at the current level of precision, but more accurate measurements of both M_t and M_W are important in searching for deviations from the SM. The distribution of χ^2, as a function of M_H, for the fit to all data is shown in Fig.4. The uncertainty on M_H is essentially logarithmic. The theoretical uncertainty is also shown, as is the limit from direct searches of about 77 GeV. The uncertainty in $\alpha(M_Z)$ is rather large, and gives an error of 0.2 on $\log(M_H)$. At the 95% confidence level M_H is less than 420 GeV.

IV IS THE HIGGS BOSON LIGHT ?

The results of the fits to all electroweak data give a central value for M_H of 115 GeV and a 95% c.l. upper limit of 420 GeV. It is of great importance, particularly in the construction of new accelerators, to understand if these values are reliable. One can adopt two approaches to these fits

1) The overall χ^2 of 17/15 d.f. for the fit to all data is reasonably good. The distribution of the pulls has a mean value of -0.1 ± 0.4 and an rms of 1.1 ± 0.3 and so is compatible with the expected Gaussian distribution. The two measurements with the largest χ^2's ($A_{FB}^{0,b}$ and A_{LR}) are just the expected "tails" of the distribution.

2) The quantities which are most sensitive to M_H are, in order of current sensitivity, A_{LR}, Γ_Z, $A_{FB}^{0,b}$, M_W, $A_{FB}^{0,\ell}$ and P_τ. These 6 quantities contribute 13 to the χ^2. The central value for M_H is sensitive to which data are included. For example, if a fit is performed without the inclusion of A_{LR}, then $M_H = 220^{+185}_{-109}$

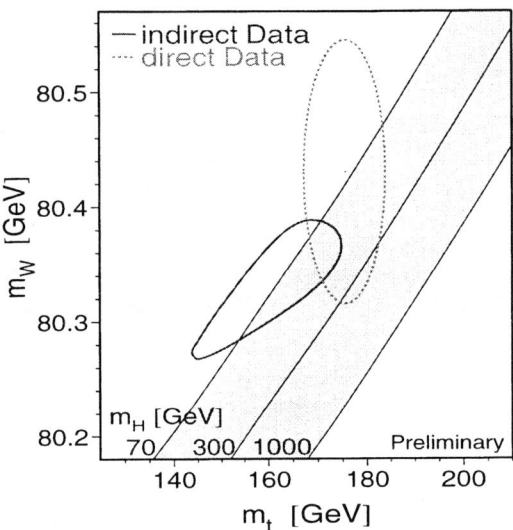

FIGURE 3. Direct and indirect determinations of M_t and M_W.

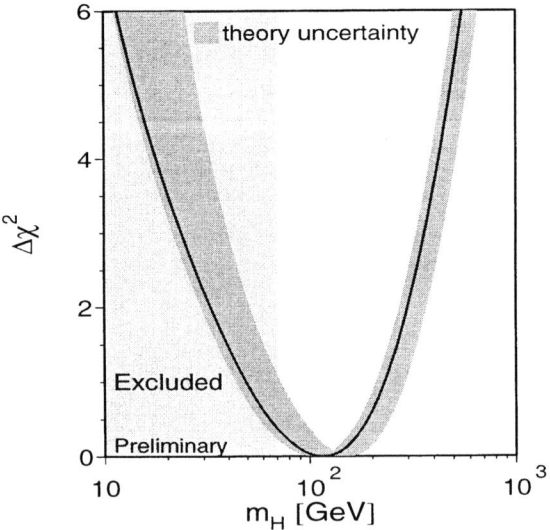

FIGURE 4. $\Delta\chi^2$ v. M_H.

	$M_H = 100$ GeV	$= 300$ GeV	$= 500$ GeV
Stand. assump.	100^{+49}_{-35}	300^{+122}_{-89}	500^{+199}_{-143}
$\delta M_W = 20$ MeV	100^{+46}_{-34}	300^{+108}_{-83}	500^{+175}_{-131}
$\delta \sin^2\theta_{\text{eff}}^{\text{lept}} (A_{LR})$ = 0.00015	100^{+45}_{-33}	300^{+111}_{-83}	500^{+182}_{-134}
$\delta M_t = 0.5$ GeV	100^{+39}_{-30}	300^{+90}_{-71}	500^{+146}_{-114}
$\delta \alpha^{-1} = 0.09$	100^{+62}_{-42}	300^{+141}_{-101}	500^{+229}_{-159}

TABLE 3. Estimated future precison on M_H.

GeV, and $M_H \lesssim 715$ GeV at the 95% c.l.. However if $A_{FB}^{0,b}$ (alone) is excluded from the fit then $M_H = 47^{+87}_{-32}$ GeV and $M_H \lesssim 280$ GeV at the 95% c.l.. One possibilty to take into account the large χ^2 for the 6 most sensitive quantities is to rescale their errors using the standard PDG recipe (approx. factor 1.5). This gives $M_H = 188^{+152}_{-91}$ GeV, and a 95% c.l. that $M_H \lesssim 590$ GeV.

In summary, the best estimate is that the Higgs is relatively light. However, the data are not fully compatible, so some caution in intrepreting the data is necessary.

V FUTURE PROSPECTS

Details of the expected future errors and the assumptions used in estimating these can be found in [4]. Although the LEP 1 data taking is complete some further improvements in precision are expected as the analyses are completed. For SLD the additional data should lead to a final error on $\sin^2\theta_{\text{eff}}^{\text{lept}}$ of about 0.00025. Improvements on δM_W from LEP 2 and then Fermilab should reduce this error to about 30 MeV. Also δM_t is assumed to get reduced to 3 GeV. The error on $\alpha(M_Z)^{-1}$ is assumed to go down to 0.05 (from 0.09). The expected errors on M_H, for different values of M_H, are given in Table 3 for these "standard" assumptions. Also the values are given when a specific error assumption is changed. This illustrates the important in reducing the error on $\alpha(M_Z)^{-1}$ by making low energy e^+e^- measurements which are needed to achieve this.

REFERENCES

1. CDF Collaboration: J. Lys, Top Mass Measurement at CDF, Proc. ICHEP96, Warsaw, 25-31 July 1996, 1196.
2. D0 Collaboration: S. Abachi *et al. Phys. ReV. Lett.* **79** (1997) 1197.
3. P.B. Renton, Proc. of 17th Int. Symp. on Lepton-Photon Interactions, Beijing, China, 10-15 Aug 1995, p35; World Scientific, Singapore.
4. P.B. Renton, International Journal of Modern Physics **A12** (1997) 4109.
5. A Combination of Preliminary Electroweak Measurements and Constraints on the Standard Model, The LEP Collaborations, CERN-PPE/97-154.

Resonant Higgs Enhancement At The First Muon Collider[*]

Basim Kamal, William J. Marciano and Zohreh Parsa

Physics Department, Brookhaven National Laboratory, Upton, New York 11973

Abstract. The effect of beam polarization on Higgs resonance signals and backgrounds ($b\bar{b}$, $\tau\bar{\tau}$, $c\bar{c}$) at the First Muon Collider is studied. Angular distributions (forward-backward charge asymmetries) are examined. The resulting effective enhancement of the Higgs signal relative to the background is investigated as is the reduction in scan time required for Higgs "discovery".

If the Higgs boson has a mass $\lesssim 160$ GeV (i.e. below the W^+W^- decay threshold), it will have a very narrow width and can be resonantly studied in the s-channel via $\mu^-\mu^+ \to H$ production at the First Muon Collider (FMC) [1,2]. A strategy for "light" Higgs physics studies would be to first find the Higgs particle at LEPII, the Tevatron, or the LHC and then thoroughly scrutinize its properties on resonance at the FMC. There, one would hope to precisely determine the Higgs mass, width, and primary decay rates [3].

The FMC Higgs resonance program would entail two stages: 1) "Discovery" via an energy scan which pinpoints the precise resonance position and (perhaps) determines its width. Since pre-FMC efforts may only determine the Higgs mass to $\sim \pm 0.2$–1 GeV and its width is expected to be narrow $\mathcal{O}(1\sim 30$ MeV) for $m_H \lesssim 160$ GeV, the resonance scan may be very time consuming [3]. 2) Precision measurements of the primary Higgs decay modes. Deviations from standard model expectations could point to additional Higgs structure or elucidate the framework of supersymmetry [3]. (Expectations for $m_H = 110$ GeV are illustrated in Table 1.)

The Higgs resonance "discovery" capability and scan time will depend on $N_S/\sqrt{N_B}$ (the scan time is proportional to N_B/N_S^2), where N_S is the Higgs signal and N_B is the expected background. The precision measurement sensitivity will be determined by $N_S/\sqrt{N_B + N_S}$. For both, it will be extremely important to enhance the signal and suppress backgrounds as much as possible. To that end, one should employ highly resolved $\mu^+\mu^-$ beams with a very small energy spread. The

---

[*] Supported by U.S. Department of Energy contract number DE-AC02-76CH00016.

TABLE 1. Expected signals and backgrounds (fully integrated) for a standard model Higgs with $m_H = 110$ GeV, $\Gamma_H \simeq 3$ MeV. Muon collider resonance conditions with no polarization, $\Delta E/E \simeq 3 \times 10^{-5}$, and $L = 0.5$ fb^{-1} are assumed. The total number of Higgs scalars produced is $\sim 30,000$. Realistic efficiency and acceptance cuts are likely to dilute signal and backgrounds for $b\bar{b}$ and $c\bar{c}$ by a 0.5 factor.

$H \to$	$b\bar{b}$	$c\bar{c}$	$\tau\bar{\tau}$
N_S (events)	24,000	2,100	2,700
N_B (events)	25,200	24,160	9,450
$\pm\sqrt{N_S + N_B}/N_S$	± 0.009	± 0.08	± 0.04

proposed $\Delta E/E \simeq 3 \times 10^{-5}$ is well matched to the narrow Higgs width. It allows $N_S/N_B \sim \mathcal{O}(1)$ for the primary $H \to b\bar{b}$ mode (see Table 1). Unfortunately, high resolution is accompanied by luminosity loss. The original on-resonance goal of $\mathcal{L}_{ave} \simeq 5 \times 10^{30}$cm^{-2}s^{-1} was judged in [4] to be too low. Hence, we have assumed in Table 1 and throughout this paper that an additional order of magnitude increase in luminosity to 5×10^{31}cm^{-2}s^{-1} is attainable while maintaining outstanding beam resolution, otherwise the values in Table 1 must be scaled down accordingly.

In this paper, we describe two additional ways of potentially enhancing the Higgs signal to background ratio: beam polarization and final state angular distributions. The Higgs signal $\mu^-\mu^+ \to H \to f\bar{f}$ results from left-left (LL) or right-right (RR) beam polarizations and leads to an isotropic (i.e. constant) $f\bar{f}$ signal in $\cos\theta$ (the angle between the μ^- and f). Standard model backgrounds $\mu^-\mu^+ \to \gamma^*$ or $Z^* \to f\bar{f}$ result from LR or RL initial state polarizations and give rise to $(1 + \cos^2\theta + \frac{8}{3}A_{FB}\cos\theta)$ angular distributions. Similar statements apply to WW^* and ZZ^* final states, but those modes will not be discussed here.

To illustrate the difference between signal, $\mu^-\mu^+ \to H \to f\bar{f}$, and background, $\mu^-\mu^+ \to \gamma^*$ or $Z^* \to f\bar{f}$, we give the combined differential production rate with respect to $x \equiv \cos\theta = 4\mathbf{p}_{\mu^-} \cdot \mathbf{p}_f/s$ for polarized muon beams and fixed luminosity

$$\frac{dN(\mu^-\mu^+ \to f\bar{f})}{dx} = \frac{1}{2}N_S(1 + P_+P_-) \qquad (1)$$
$$+ \frac{3}{8}N_B[1 - P_+P_- + (P_+ - P_-)A_{LR}](1 + x^2 + \frac{8}{3}xA_{eff}).$$

$P_+(P_-)$ is the $\mu^+(\mu^-)$ polarization with $P = -1$ pure left-handed, $P = +1$ pure right handed, and $P = 0$ unpolarized. N_S is the fully integrated $(-1 < x \leq 1)$ Higgs signal and N_B the integrated background for the case of unpolarized beams, $P_+ = P_- = 0$. In that general expression,

$$A_{LR} \equiv \frac{\sigma_{LR \to LR} + \sigma_{LR \to RL} - \sigma_{RL \to RL} - \sigma_{RL \to LR}}{\sigma_{LR \to LR} + \sigma_{LR \to RL} + \sigma_{RL \to RL} + \sigma_{RL \to LR}}, \qquad (2)$$

where, for example, $LR \to LR$ stands for $\mu_L^-\mu_R^+ \to f_L\bar{f}_R$. The effective forward-backward asymmetry is given by

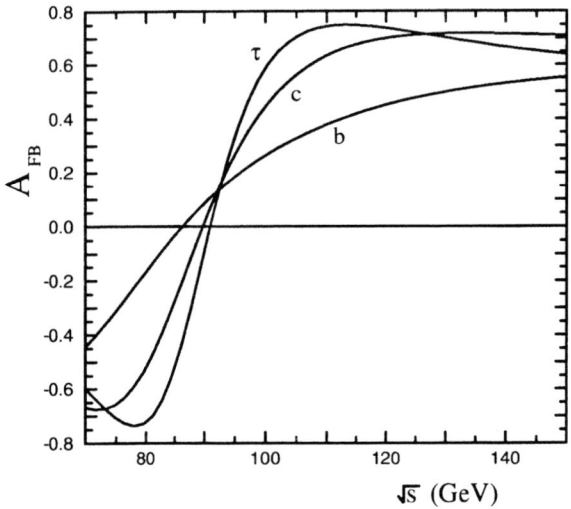

FIGURE 1. Forward-backward asymmetry for $\mu^-\mu^+ \to f\bar{f}$.

$$A_{eff} = \frac{A_{FB} + P_{eff}A_{LR}^{FB}}{1 + P_{eff}A_{LR}}, \qquad (3)$$

with

$$P_{eff} = \frac{P_+ - P_-}{1 - P_+P_-}, \qquad (4)$$

$$A_{FB} = \frac{3}{4}\frac{\sigma_{LR\to LR} + \sigma_{RL\to RL} - \sigma_{LR\to RL} - \sigma_{RL\to LR}}{\sigma_{LR\to LR} + \sigma_{RL\to RL} + \sigma_{LR\to RL} + \sigma_{RL\to LR}}, \qquad (5)$$

$$A_{LR}^{FB} = \frac{3}{4}\frac{\sigma_{LR\to LR} + \sigma_{RL\to LR} - \sigma_{LR\to RL} - \sigma_{RL\to RL}}{\sigma_{LR\to LR} + \sigma_{RL\to LR} + \sigma_{LR\to RL} + \sigma_{RL\to RL}}. \qquad (6)$$

and the $\mu_i^-\mu_j^+ \to f_{i'}\bar{f}_{j'}$, cross sections $(i \neq j)$ are to lowest order

$$\sigma_{ij\to i'j'} = (N_C)\sigma_0\left[1 - \frac{s}{m_Z^2}\left(1 + \frac{(T_{3\mu_i} - Q_\mu\sin^2\theta_W)(T_{3f_{i'}} - Q_f\sin^2\theta_W)}{Q_\mu Q_f \sin^2\theta_W \cos^2\theta_W}\right)\right]^2,$$

$$T_{3\mu_L} = T_{3\tau_L} = T_{3b_L} = -T_{3c_L} = -1/2, \qquad (7)$$

$$T_{3f_R} = 0, \quad Q_\mu = Q_\tau = 3Q_b = -\frac{3}{2}Q_c = -1 \qquad (N_C = 3 \text{ for } f = b,c).$$

Realistic cuts, efficiencies, systematic errors etc, will not be considered. They are likely to dilute the $b\bar{b}$ and $c\bar{c}$ event rates by a factor of 0.5. In addition, we ignore the radiative Z production tail under the assumption such events are vetoed.

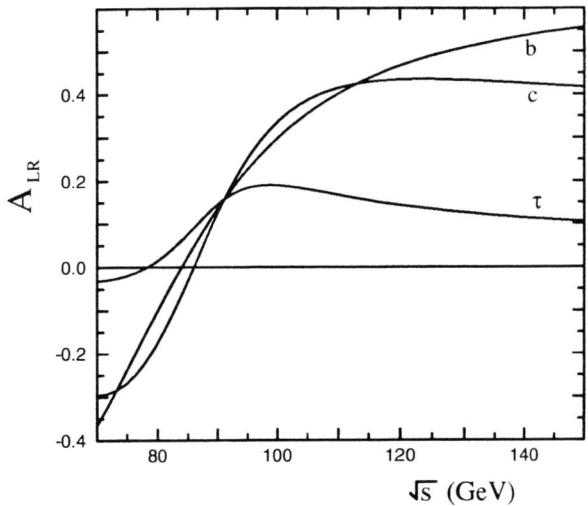

FIGURE 2. Left-right asymmetry for $\mu^-\mu^+ \to f\bar{f}$.

The (unpolarized) forward-backward asymmetries are illustrated in Fig. 1. Note that A_{FB} is large (near maximal) for $\tau\bar{\tau}$ and $c\bar{c}$ in the region of interest. As we shall see, that feature can help in discriminating signal from background.

In principle, large polarization in both beams can be important for enhancing "discovery" and precision measurement sensitivity for the Higgs. From Eq. (1), we find for fixed luminosity that $N_S/\sqrt{N_B}$ is enhanced (for integrated signal and background) by the factor

$$\kappa_{\rm pol} = \frac{1 + P_+ P_-}{\sqrt{1 - P_+ P_- + (P_+ - P_-)A_{LR}}}, \tag{8}$$

where the A_{LR} are shown in Fig. 2. That result generalizes the $P_+ = P_-$ case [5]. For natural beam polarization [1], $P_+ = P_- = 0.2$ (assuming spin rotation of one beam), the enhancement factor is only 1.06. For larger polarization, $P_+ = P_- = 0.5$, one obtains a 1.44 enhancement factor (statistically equivalent to about a factor of 2 luminosity increase). Similarly, $P_+ = P_- = 0.7$ leads to a factor of 2 enhancement or equivalently a factor of 4 scan time reduction. Unfortunately, obtaining even 0.5 polarization simply by muon energy cuts reduces each beam intensity [1] by a factor of 1/4, resulting in a luminosity reduction by 1/16. Such a tradeoff is clearly unacceptable. Polarization will be a useful tool in Higgs resonance "discovery" and studies only if high polarization is achievable with little luminosity loss. Ideas for increasing the polarization are still being explored [1,6]. Tau final state polarizations can also be used to help improve the $H \to \tau\bar{\tau}$ measurement.

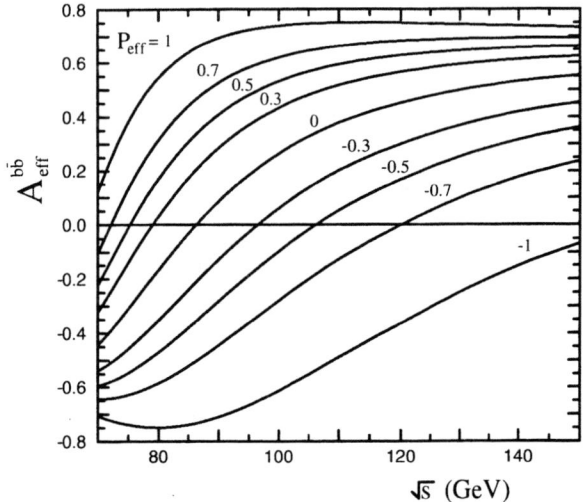

FIGURE 3. Effective forward-backward asymmetry for $\mu^- \mu^+ \to b\bar{b}$.

Some "discovery" or sensitivity enhancement can also be obtained from angular discrimination. A proper study would include detector acceptance cuts and maximum likelihood fits. Here, we wish to only approximate the gain. For that purpose, we assume perfect (infinitesimal) binning and obtain a (maximal) measurement sensitivity enhancement factor

$$\frac{1}{2}(1 + P_+ P_-)\sqrt{N_S + N_B} \left[\int \frac{dx}{dN/dx} \right]^{1/2}, \tag{9}$$

which becomes, from Equations (1) and (8),

$$\kappa_{\text{pol}} \sqrt{\frac{2}{3}} \sqrt{\frac{N_S + N_B}{N_B}} \left(\frac{\tan^{-1}\left(\frac{2}{\zeta}\sqrt{1 - \frac{16}{9}A_{\text{eff}}^2 + \zeta}\right)}{\sqrt{1 - \frac{16}{9}A_{\text{eff}}^2 + \zeta}} \right)^{1/2}, \quad \zeta \equiv \frac{4}{3}\frac{N_S}{N_B}\frac{\kappa_{\text{pol}}^2}{1 + P_+ P_-}. \tag{10}$$

The actual gain from angular information will be less than that idealized factor. For $A_{\text{eff}} \simeq 3/4$, $\zeta \simeq 0.38$ (which roughly applies to $\tau\bar{\tau}$) and $P_+ = P_- = 0$, one finds a sensitivity enhancement of 1.33. That means the ±4% statistical error in Table 1 would be reduced to ±3%; not a significant improvement. Similar sensitivity enhancements apply to $c\bar{c}$. In the case of $H \to b\bar{b}$, the primary discovery mode, $A_{\text{eff}} \simeq 0.4$ and one finds only a 3% enhancement. One can increase the effective $b\bar{b}$ forward-backward asymmetry via $P_{\text{eff}} \neq 0$ (see Fig. 3). However, one must again confront the issue of luminosity loss.

In the case of "discovery", high polarization and/or a near maximal forward-backward asymmetry can significantly reduce the scan time. For the idealized coverage and binning assumed above, the time is reduced by the factor

$$\frac{1}{\kappa_{\text{pol}}^2} \frac{3}{\pi} \sqrt{1 - \frac{16}{9} A_{eff}^2} \ . \tag{11}$$

Of course, that naive formula must be corrected for realistic acceptances, efficiencies, etc.; so, it should not be taken too literally (particularly for $A_{eff} \simeq 3/4$). Also, discovery will entail the detection of some minimal signal (perhaps 5 events); so, the scan time cannot be reduced beyond some level. Nevertheless, applying it to the $b\bar{b}$ discovery mode with "natural" $P_+ = P_- = 0.2$ and $A_{eff} \simeq 0.37$ gives a scan reduction time factor of 0.74. If $P_+ = P_- = 0.7$ were achievable without loss of luminosity, the scan time would be significantly reduced by a factor of 0.19.

The $H \to \tau\bar{\tau}$ "discovery" time is about 15 times longer than that of the $b\bar{b}$ (with efficiencies) for fully integrated signals. Employing $A_{FB} \simeq 0.743$ and assuming tau detection down to about 15° from the beams and $P_+ = P_- = 0.2$, reduces that time by about a factor of $6 \sim 7$, making it somewhat less than $1/2$ as effective as $b\bar{b}$. Using both along with all background angular information should, therefore, reduce the scan time by almost a factor of 2 compared to using the integrated $b\bar{b}$ signal alone. Such a reduction would be quite welcome, particularly if the luminosity is less than expected.

In conclusion, we have shown that polarization is potentially useful for Higgs resonance studies, but only if the accompanying luminosity reduction is not significant. Large forward-backward asymmetries can also be used to enhance the Higgs "discovery" signal or improve precision measurements, particularly for $\tau\bar{\tau}$. However, to make the s-channel Higgs "factory" a compelling facility, one must focus on attaining the outstanding beam resolution assumed here and maintaining the highest luminosity possible.

REFERENCES

1. Muon Collider Feasibility Study, BNL Report BNL-52503 (1996).
2. Cline, D., "The Problems and Physics Prospects for a $\mu^+\mu^-$ Collider", in *Future High Energy Colliders*, edited by Z. Parsa, AIP Conference Proceedings **397**, 1997, pp. 203–218.
3. Barger, V., Berger, M.S., Gunion, J.F., and Han, T., "The Physics Capabilities of $\mu^+\mu^-$ Colliders", in *Future High Energy Colliders*, edited by Z. Parsa, AIP Conference Proceedings **397**, 1997, pp. 219–233; *Phys. Rep.* **286**, 1–51 (1997); *Phys. Rev. Lett.* **75**, 1462–1465 (1995).
4. Kamal, B., Marciano, W., Parsa, Z., BNL Report BNL-65193 (1997), hep-ph/9712270.
5. Parsa, Z., $\mu^+\mu^-$ Collider and Physics Possibilities (1993) (unpublished).
6. Skrinsky, A., *these proceedings*.

$\mu^+\mu^-$ COLLIDER STUDIES

MUON COLLIDER DESIGN

R. B. Palmer for the Muon Collider Collaboration[1]

Physics Department
Brookhaven National Laboratory,
Upton, NY 11973-5000, USA

Abstract. Parameters are given of machines with center-of-mass (CoM) energies of 3 TeV and 400 GeV but, besides a comment on neutrino radiation, the paper concentrates on progress on the design of a machine to operate at a light Higgs mass, assumed, for this study, to be 100 GeV (CoM). This article reviews the same material covered in the report [1].

INTRODUCTION

The possibility of muon colliders was introduced by G. I Budker [2], Skrinsky et al. [3] and Neuffer [4]. More recently, a collaboration of over 100 members, lead by BNL, FNAL, LBNL, BNIP, University of Mississippi, Princeton University and UCLA has been formed to coordinate studies on specific designs. The studies on the particle physics that could be done at the collider is lead by the University of Wisconsin, UCD and Indiana University. Work has been done on designs at a 3-4 TeV, 0.4-0.5 TeV and $\approx$100 GeV [5–9]. Tb. 1 gives the parameters of such colliders, and Figs. 1 and 2 show possible outlines of the 3 TeV and 100 GeV machines.

The original motive for considering muon colliders was the effective energy advantage of any lepton collider over hadron machines, together with the fact that muons, unlike electrons, generate negligible synchrotron radiation. As a result, a muon collider can be circular and much smaller than the current designs of linear electron colliders, and also much smaller than a hadron machine with the same *effective* energy.

In addition, a $\mu^+\mu^-$ collider would have some unique physics advantages over an e^+e^- collider:

1. The direct coupling of a lepton-lepton system to a Higgs boson has a cross section that is proportional to the square of the mass of the lepton. As a

---

[1] Members of the Collaboration can be found at **http://www.cap.bnl.gov/mumu/**

TABLE 1. Parameters of Collider Rings

(CoM) energy	TeV	3	0.4	0.1		
p energy	GeV	16	16	16		
p's/bunch	10^{13}	2.5	2.5	5		
bunches/fill		4	4	2		
rep rate	Hz	15	15	15		
p power	MW	4	4	4		
μ/bunch	10^{12}	2	2	4		
μ power	MW	28	4	1		
wall power	MW	204	120	81		
collider circ	m	6000	1000	300		
depth	m	500	100	10		
rms $\frac{\Delta p}{p}$	%	.16	.14	.12	.01	.003
6D ϵ_6	$10^{-12}\,(\pi m)^3$	170	170	170	170	170
rms ϵ_n	π mm mrad	50	50	85	195	280
β^*	cm	0.3	2.3	4	9	13
σ_z	cm	0.3	2.3	4	9	13
σ_r spot	μm	3.2	24	82	187	270
tune shift		0.043	0.043	0.05	0.02	.015
Luminosity	$cm^{-2}sec^{-1}$	$5\,10^{34}$	10^{33}	$1.2\,10^{32}$	$2\,10^{31}$	10^{31}
(CoM) $\frac{\Delta E}{E}$	10^{-5}	80	80	80	7	2
Higgs/year	$10^3\,year^{-1}$			1.6	4	4

result, the cross section for direct Higgs production from the $\mu^+\mu^-$ system is 40,000 times that from an e^+e^- system.

2. Because of the lack of beamstrahlung, a $\mu^+\mu^-$ collider can be operated with an energy spread of as little as 0.003 %. Furthermore, with the naturally occurring polarization it would be possible, by observing g-2, to determine the absolute energy to an accuracy of 10^{-6} or better [10]. It should thus be possible to use a $\mu^+\mu^-$ collider to make precision measurements of masses and direct measurements of the Higgs width (assumed to be $\approx$ 2 MeV), that would be otherwise impossible, with an e^+e^- collider.

Machines with energies higher than 3-4 TeV, have significant beam current constraints from off site neutrino radiation limits. If the required luminosities are to be reached without unacceptable hazards, then significant improvements in muon emittance over the current base line values are needed. There is however reason to believe such improvements are achievable, and machines with a center of mass energy of 10 TeV and luminosities of 10^{35} $cm^{-2}s^{-1}$ and above may be possible [11]. For energies below 3 TeV, for fixed muon currents, this radiation falls as the energy cubed, and it should be little problem for machines with energies of 1.5 TeV or less.

Recent work in the collaboration has concentrated on the lowest energy machine ($\approx$ 100 GeV), whose energy is taken to be representative of the possible

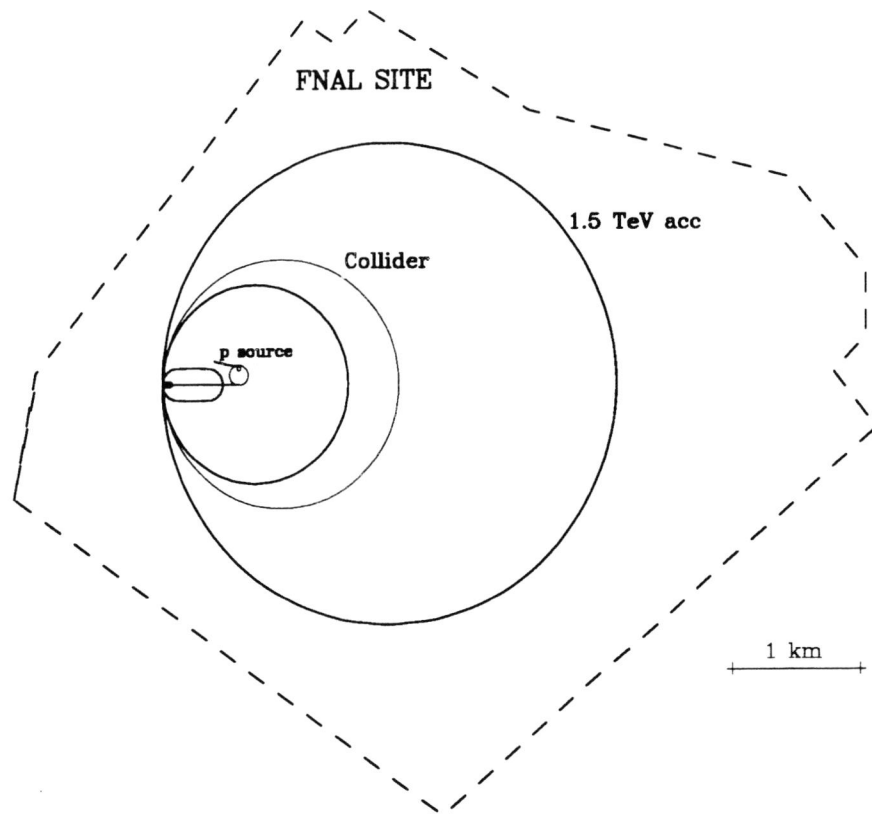

FIGURE 1. Plan of a 3 TeV Muon Collider.

mass of a light Higgs particle. Such a machine would serve as a demonstration of Muon Collider technology, a needed step before high energy machines can be considered, and as a unique physics tool to make and study, if they exist, Higgs particles in the S-channel.

PROTON DRIVER

π production rises approximately linearly with proton energy up to about 10 GeV after which it continues rising more slowly, but the requirement of very short bunches sets an effective minimum proton energy of about 16 GeV. The baseline specification used in Tb.2 is for a 16 GeV proton with a repetition rate of 15 Hz, 10^{14} protons per cycle in 2 or 4 bunches (depending on the collider energy), each with an rms bunch length of 1-2 ns. The total beam power is 4 MW. A design worked out at FNAL [12] would involve: a) An upgraded linac (0.4 → 1 GeV); b) higher energy booster (8 → 16 GeV) and

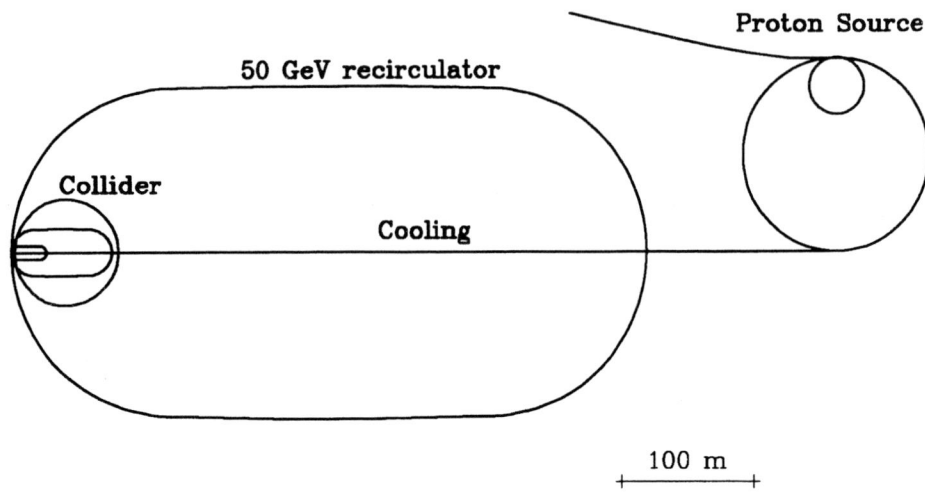

FIGURE 2. Plan of a 100 GeV Muon Collider.

c) new pre-booster. Some parameters are given in Tb. 2.

TABLE 2. Proton Driver Specifications

		Linac	Pre-Booster	Booster
Final energy	GeV	1.0	4.5	16
Protons/bunch			$5\ 10^{13}$	$5\ 10^{13}$
No of bunches			2	2
Rep. freq	Hz	15	15	15
Circumference	m		180.6	474.2
Norm. 95% emit.	$\pi\ mm\ mrad$		200	240
sp ch tune shift			.39	.39
Final field	T		1.3	1.3

Another study had been done at BNL [13] that, while it did not quite reach the same beam power, involved far less upgrade: a) upgraded linac ($0.2 \rightarrow 0.6$ GeV); b) increased AGS rep rate: 2.5 Hz.

In order to reduce the cost of the muon phase rotation section and for minimizing the final muon longitudinal phase space, it appears now that the final proton bunch length should be 1-2 ns.

An experiment [14] at the AGS has tested a method to generate such short bunches by rapidly bringing the tune of the machine near transition and allowing a strong phase rotation to occur. Bunches were shortened from 8 ns rms to 2.2 ns with initial longitudinal phase space similar to that specified in the above design. Shorter bunches are expected in later experiments with better control.

Another experiment [15] has used variable inductors to reduce the longitudinal space charge effects.

Target and Pion Capture

π production is maximized by the use a well focused proton beam, small diameter target and a high Z target material. Tungsten, platinum or lead would be good, but the heating could not be easily removed and shock damage could be a problem. The use of a rapidly flowing liquid can solve the heating problem, but the shock could damage the enclosure, if one is used. We are thus considering the use of an open liquid jet. Such a jet has been tested [16] using mercury, although this was never exposed to a beam, and the jet did not move in a strong magnetic field, as required in our case. Theoretical studies of liquid metal flow in magnetic fields are underway [17,18], and the possibilities of using insulating liquids (e.g. PtO_2, Re_2O_3) and slurries (e.g. Pt in water) are being considered.

If the axis of the target is coincident with that of the solenoid field, then there is a relatively high probability that pions produced at the start of the target will reenter, interact again later and be lost. The probability for such interactions is reduced, and the overall production rate increased (by about 60 %) if the target and proton beam are set at an angle (10-15°) with respect to the field axis [19,20].

Three different codes [21-23] have been used to estimate π yields and, despite detailed differences between them, overall μ production was very similar. In addition, the collaboration is involved in an AGS experiment [24] to measure the π yields. The production is peaked at a relatively low pion momentum ($\approx$ 200 MeV/c), but has a very wide distribution: $\frac{\Delta E}{E}$ rms $\approx$ 100 %. The pion multiplicity, per 16 GeV proton, is about 2. At these low energies, the transverse momenta are of the order of 200 MeV/c. If a substantial fraction of these pions are to be captured, a very wide band system is required. A 20 T solenoid, 16 cm inside diameter is found to capture about half of all produced pions, and with target efficiency included, about 0.6 pions per proton emerge from the solenoid end [25]. Such a solenoid is well within the parameters of existing magnets [26]. It would have a superconducting outsert, and an 8 MW water cooled copper insert [27] (see Fig. 3).

After capture, the 20 T solenoid is matched [28] into a decay channel with 5 T fields and diameter of 30 cm.

Phase Rotation Linac

The pions, and the muons into which they decay, have an energy spread with an rms value of approximately 100 %. It would be difficult to handle such a wide spread in any subsequent system. A linac is thus introduced along the decay channel, with frequencies and phases chosen to deaccelerate the fast particles and accelerate the slow ones; i.e. to phase rotate the muon bunch. Tb.3 gives an example of parameters of such a linac. It is seen that the lowest

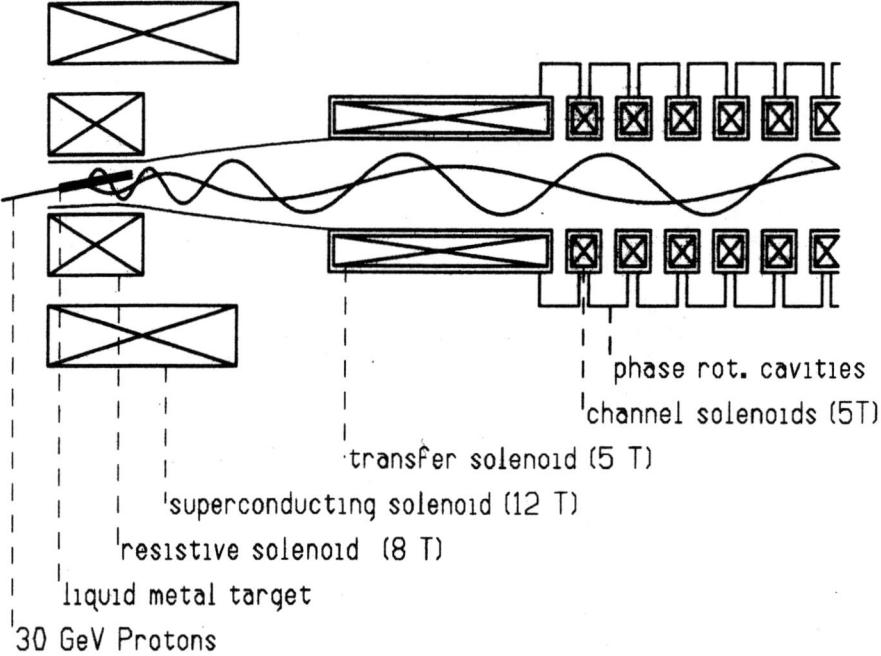

FIGURE 3. Schematics of the front end: skewed target, high field solenoid and decay and phase rotation channel

frequency is 30 MHz, a low but not impossible frequency for a conventional structure.

TABLE 3. Parameters of Phase Rotation Linacs

Linac	Length m	Frequency MHz	Gradient MeV/m
1	3	60	5
2	29	30	4
3	5	60	4
4	5	37	4

Fig. 4 shows the energy vs. ct at the end of the decay and phase rotation channel. A bunch is defined with mean energy 150 MeV, rms bunch length 1.7 m, and rms momentum spread 20 % (95 %, $\epsilon_L = 3.2$ eVs) in the Monte Carlo study [8]. The number of muons per initial proton in this selected bunch was 0.38, which can be compared with a value of 0.3 assumed in the baseline parameters.

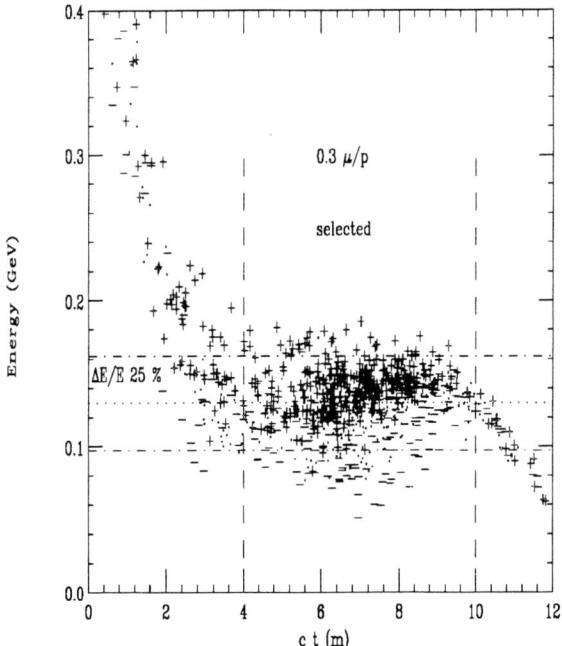

FIGURE 4. Energy vs. ct of μ's at end of decay channel with phase rotation.

Use of Both Signs

Protons on the target produce pions of both signs, and a solenoid will capture both, but the required subsequent phase rotation rf systems will have opposite effects on each. The baseline solution is to use two proton bunches, aim them at the same target one after the other, and adjust the rf phases such as to act correctly on one sign of the first bunch and on the other sign of the second.

A second possibility would be to separate the charges into two channels, and phase rotate them separately. However, the separation, probably using a bent solenoid, is not simple and would not be fully efficient. Whether a gain in overall efficiency could be achieved is not yet known.

Polarization

Polarized Muon Production

In the center of mass of a decaying pion, the outgoing muon is fully polarized (-1 for μ^+ and +1 for μ^-). In the lab system the polarization depends [29] on the decay angle θ_d and initial pion energy. For pion kinetic energy larger than the pion mass, the average is about 20 %, and if nothing else is done, the polarization of the captured muons and phase rotated by the proposed system is approximately this value.

If higher polarization is required, some selection of muons from forward pion decays ($\cos\theta_d \to 1$) is required. Fig. 4, above, showed the polarization of the phase rotated muons. The polarization P$> \frac{1}{3}$, $-\frac{1}{3} < P < \frac{1}{3}$, and P$< -\frac{1}{3}$ is marked by the symbols +, . and − respectively. If a selection is made on the minimum energy of the muons, then greater polarization is obtained. The tighter the cut, the higher the polarization, but the less the fraction F_{loss} of muons that are selected. Fig. 5 gives the results of a Monte Carlo study.

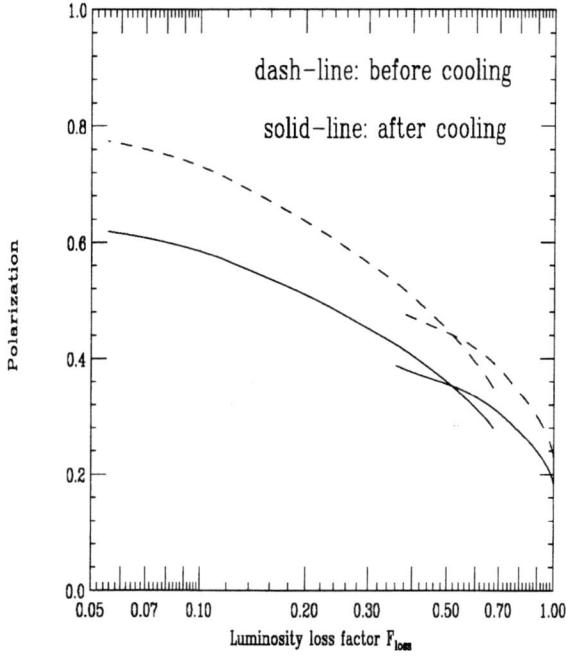

FIGURE 5. Polarization vs F_{loss} of μ's accepted.

If this selection is made on both beams, and if the proton bunch intensity is maintained, then naturally the muon bunch is reduced by the factor F_{loss} and the luminosity would fall by F_{loss}^2. But if, instead, proton bunches are merged so as to obtain half as many bunches with twice the intensity, then the muon bunch intensity is maintained and the luminosity (and repetition rate) falls only as F_{loss}.

One also notes that the luminosity could be maintained at the full unpolarized value if the proton source intensity could be increased. Such an increase in proton source intensity in the unpolarized case would be impractical because of the resultant excessive high energy muon beam power, but this restriction does not apply if the increase is used to offset losses in generating polarization.

Polarization Preservation

A paper [30] has discussed the preservation of muon polarization in some detail. During the ionization cooling process the muons lose energy in material and have a spin flip probability $\mathcal{P}$,

$$\mathcal{P} \approx \int \frac{m_e}{m_\mu} \beta_v^2 \frac{dE}{E} \tag{1}$$

where β_v is the muon velocity divided by c, and $\frac{\Delta E}{E}$ is the fractional loss of energy due to ionization loss. In our case the integrated energy loss is approximately 3 GeV and the typical energy is 150 MeV, so the integrated spin flip probability is close to 10 %. The change in polarization $\frac{\Delta \mathcal{P}}{\mathcal{P}}$ is twice the spin flip probability, so the reduction in polarization is approximately 20 %. This loss is included in Fig. 5.

During circulation in any ring, the muon spins, if initially longitudinal, will precess by $(g-2)/2\gamma$ turns per revolution; where $(g-2)/2$ is 1.166 10^{-3}. A given energy spread $\frac{\Delta \gamma}{\gamma}$ will introduce variations in these precessions and cause dilution of the polarization. But if the particles remain in the ring for an exact integer number of synchrotron oscillations, then their individual average γ's will be the same and no dilution will occur.

In the collider, bending can be performed with the spin orientation in the vertical direction, and the spin rotated into the longitudinal direction only for the interaction region. The design of such spin rotators appears relatively straightforward, but long. This might be a preferred solution at high energies but is not practical for instance, in the 100 GeV machine. An alternative is to use such a small energy spread, as in the Higgs factory, that though the polarization vector precesses, the beam polarization does not become significantly diluted.

COOLING

For a collider, the phase-space volume must be reduced within a time of the order of the μ lifetime. Cooling by synchrotron radiation, conventional stochastic cooling and conventional electron cooling are all too slow. Optical stochastic cooling [31], electron cooling in a plasma discharge [32] and cooling in a crystal lattice [33] are being studied, but appear difficult. Ionization cooling [34] of muons seems relatively straightforward.

Ionization Cooling Theory

In ionization cooling, the beam loses both transverse and longitudinal momentum as it passes through a material medium. Subsequently, the longitudinal momentum can be restored by coherent reacceleration, leaving a net loss of transverse momentum.

The approximate equation for transverse cooling (with energies in GeV) is

$$\frac{d\epsilon_n}{ds} = -\frac{dE_\mu}{ds}\frac{\epsilon_n}{E_\mu} + \frac{\beta_\perp (0.014)^2}{2\, E_\mu m_\mu\, L_R}, \qquad (2)$$

where ϵ_n is the normalized emittance, $\beta_\perp$ is the betatron function at the absorber, dE_μ/ds is the energy loss, and L_R is the radiation length of the material. The first term in this equation is the coherent cooling term, and the second is the heating due to multiple scattering. This heating term is minimized if $\beta_\perp$ is small (strong-focusing) and L_R is large (a low-Z absorber).

The equation for energy spread (longitudinal emittance) is:

$$\frac{d(\Delta E)^2}{ds} = -2\, \frac{d\left(\frac{dE_\mu}{ds}\right)}{dE_\mu} < (\Delta E_\mu)^2 > + \frac{d(\Delta E_\mu)^2_{\text{straggling}}}{ds} \qquad (3)$$

where the first term is the cooling (or heating) due to energy loss, and the second term is the heating due to straggling.

Energy spread can be reduced by artificially increasing $\frac{d(dE_\mu/ds)}{dE_\mu}$ by placing a transverse variation in absorber density or thickness at a location where position is energy dependent, i.e. where there is dispersion. The use of such wedges can reduce energy spread, but it simultaneously increases transverse emittance in the direction of the dispersion. Six dimensional phase space is not reduced.

Cooling Components

We require a reduction of the normalized transverse emittance by almost three orders of magnitude (from 1×10^{-2} to 5×10^{-5} m-rad), and a reduction of the longitudinal emittance by one order of magnitude. This cooling is obtained in a series of cooling stages. In general, each stage consists of two components:

1. material in a strong focusing (low $\beta_\perp$) environment alternated with linac accelerators. These components will cool the transverse phase space.

2. lattice that generates dispersion, with absorbing material wedges introduced to interchange longitudinal and transverse emittance.

Simulations have been performed on examples of each component using the program ICOOL [35] which includes Vavilov distributions (with Landau and Gaussian limits) for dE/dx, and Moliere scattering distributions (with Rutherford limit). The only effects which are not yet included are space-charge and wake-field effects. Analytic vacuum calculations indicate that these effects are not overwhelming. A correct simulation must be done before we are assured that no real problems exist.

Transverse Cooling

The baseline solution for the first component involves the use of liquid hydrogen absorbers in strong solenoid focusing fields, interleaved with short linac sections. The solenoidal fields in succesive absorbers must be reversed to avoid build up of the canonical angular momentum. Fig. 6 shows the cross section of one cell of such a system. The top plot in Fig. 7 shows the reduction of transverse emittance in 10 such cells (20 m); the middle one shows the increase in longitudinal emittance induced by straggling and the adverse dependence of dE/dx with energy; while the bottom one shows the overall reduction in 6-dimensional emittance. This simulation has been confirmed, with minor differences by the codes double precision GEANT [36] and PARMELA [37].

Using 30 T solenoids at the end of a cooling sequence can attain a transverse emittance of 190 mm mrad and a six dimensional emittance of 30×10^{-12} m^3 (cf. 280 mm mrad and 170×10^{-12} m^3, respectively required for a Higgs factory).

Other solutions, e.g. rapidly alternating solenoids and LiH absorbers [38] and current carrying Li rods have been and will continue to be studied, but do not appear to be required to meet the baseline parameters (see below).

Linac

The linacs used in the above simulations has a frequency of 805 MHz and required an accelerating gradient (peak phase) of 24 MV/m. The current designs use cavities separated by thin Be foils, $\frac{2\pi}{3}$ or $\frac{2\pi}{4}$ phase advanced per cavity, and powered in 3 of 4 separate interleaved side-coupled standing-wave systems [39,40]. In order to reduce power source requirements the cavities may be operated at liquid nitrogen temperatures.

Longitudinal-Transverse Exchange

The exchange of longitudinal and transverse emittance requires dispersion in a large acceptance channel. One way of achieving this is in a bent solenoid. Fig. 8 shows transverse positions vs. their momenta: a) before the bend, b) after the bend, and c) after hydrogen wedges. The *rms* momentum spread in

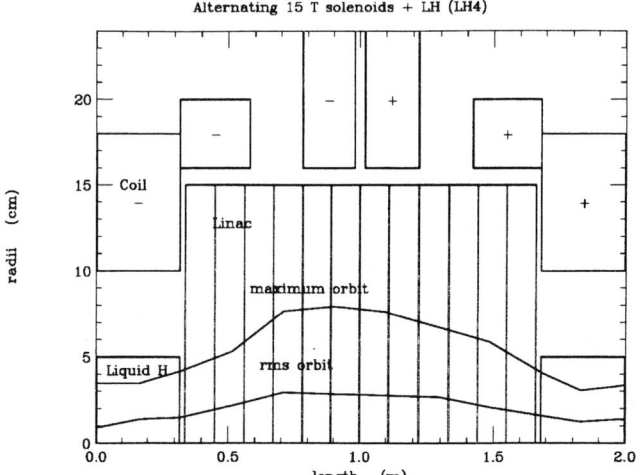

FIGURE 6. The cross section of one cell of an alternating solenoid cooling system

this example is reduced from 8 MeV/c to 4.6 MeV/c with an accompanying approximate equivalent increase in the x-y emittance.

Emittance exchange in solid wedges in the presence of ideal dispersion has also been simulated using SIMUCOOL [41]. Dispersion generation by weak focussing spectrometers [42] and dipoles with solenoids [43] are also studied.

Cooling System

The required total 6 dimensional cooling is about 10^6. Since a single stage, as illustrated above, gives a factor of 2 reduction, about 20 such stages are required. The total length of the system would be of the order of 500 m, and the total acceleration would be of the order of 6 GeV. The fraction of muons remaining at the end of the cooling system is estimated to be $\approx 60\%$.

In a few of the later stages, current carrying lithium rods might replace item (1) above. In this case the rod serves simultaneously to maintain the low $\beta_\perp$, and attenuate the beam momenta. Similar lithium rods, with surface fields of 10 T, were developed at Novosibirsk (BINP) and have been used as focusing elements at FNAL and CERN [44,45]. Cooling in beam recirculators could lead to reduction of costs of the cooling section [42].

ACCELERATION

Following cooling and initial bunch compression, the beams must be rapidly accelerated. A sequence of linacs would work, but would be expensive, some form of circulating acceleration is preferred. At lower energies, the acceleration

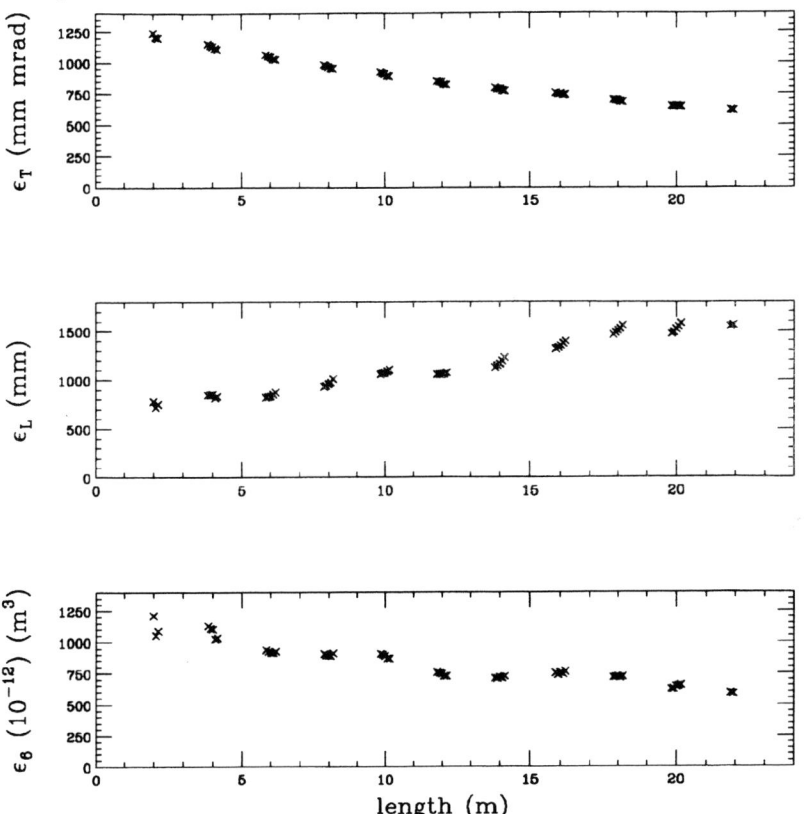

FIGURE 7. Emittance vs. length in 10 alternating solenoid cells; top: transverse emittance; middle: longitudinal emittance; and bottom: 6-dimensional emittance

time is so short that any form of magnet ramping is probably impractical. The conservative option is to use a sequence of recirculating accelerators (similar to that used at TJNL), but fixed frequency alternating gradient acceleration (FFAG) is also being studied [46]. At higher energies, it is probably more economical to use fast rise time pulsed magnets in more conventional synchrotrons [47].

Scenarios

Tbs. 4 and 5 give an example of possible sequences of accelerators for a 100 GeV Higgs Factory and a 3 TeV collider. In both cases, following initial linacs, recirculating accelerators are used. Designs [48] have been made of multiple aperture superconducting magnets for use in recirculating acceleration. The use of such magnets was not assumed in the scenarios, but they would reduce

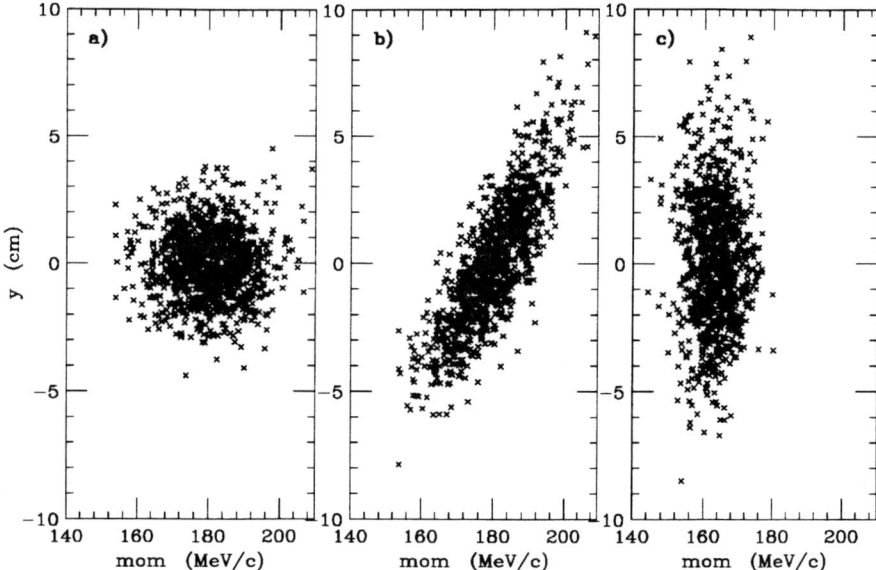

FIGURE 8. Transverse trajectory positions vs. their momenta: a) before the bend, b) after the bend, and c) after hydrogen wedges.

the diameter of the recirculating accelerator ring and lower particle loss from decay.

In the high energy case, the final three stages use pulsed magnet synchrotrons. If only pulsed magnets were used the power consumed by a ring would be high and its circumference large, but a hybrid ring with alternating pulsed warm magnets and fixed superconducting magnets appears practical. In the example, the last two such rings are located in the same tunnel with differing ratios of pulsed to fixed magnets. The fixed magnets are superconducting at 8 T; the pulse d magnets are warm with fields that swing from - 2 T to + 2 T.

In both cases, except for the earliest stages, superconducting rf is employed. The reason for this, in the earlier stage, is that the instantaneous acceleration power requirement is very high, and the use of superconducting cavities allows a longer rf fill time and a reduced rf power source requirement. For the higher energy accelerators the use of superconducting cavities is dictated by the need to achieve high wall to beam efficiency.

A study [49] tracked particles through a similar sequence of recirculating accelerators and found a dilution of longitudinal phase space of the order of 10% and negligible particle loss.

TABLE 4. Parameters Higgs Factory (100 GeV) Accelerators

acc type rf type		linac sledCu	linac sledCu	recirc sledCu	recirc sledCu	recirc SC Nb	sums
E_{init}	(GeV)	0.10	0.20	0.70	2	7	
E_{final}	(GeV)	0.20	0.70	2	7	50	
circ	(km)	0.04	0.07	0.06	0.18	1.21	1.57
turns		1	1	8	10	11	
decay loss	(%)	2.31	3.98	6.74	7.77	9.88	27.29
decay heat	W/m	0.85	1.88	10.50	12.39	12.14	
B_{fixed}	(T)			2	2	2	
pipe width	cm			30.66	21.22	10.44	
pipe ht	cm			10	8	4.30	
rf freq	(MHz)	90	90	120	170	400	
acc/turn	(GeV)	0.20	0.40	0.17	0.50	4	
acc time	(μs)			1	6	43	
acc Grad	(MV/m)	8	8	8	10	15	
grad sag	%			13.08	16.82	27.15	
rf time	ms	0.55	0.56	0.37	0.24	2.04	
peak rf /m	(MW/m)	2.72	2.56	2.21	4.40	0.20	
ave rf power	MW	0.61	1.10	0.28	0.88	1.99	4.88
total wall p	MW	4.71	8.50	1.67	5.20	5.87	25.94
beam power	MW	0.00	0.01	0.03	0.12	0.92	1.08
wall-beam eff	%	0.06	0.15	1.93	2.22	15.62	4.16

COLLIDER STORAGE RING

After acceleration, the μ^+ and μ^- bunches are injected into a separate storage ring. The highest possible average bending field is desirable, to maximize the number of revolutions before decay, and thus maximize the luminosity. Collisions would occur in one, or perhaps two, low-β^* interaction areas. Parameters of the rings were given earlier in Tb. 1.

Lattice Design

In order to maintain the required short bunches, without excessive rf, approximately isochronous Flexible Momentum Compaction lattices [50] would be used.

In the high energy cases, the required betas at the intersection point are very small (e.g. $\beta^* = 3$ mm for 4 TeV), and the quadrupoles needed to generate them are large (20-30 cm diameter). At 100 GeV, the betas are not so small and the quadrupoles are more conventional, but in both cases it has been found that local chromatic correction is essential [51].

Preliminary lattices have been designed for both 4 TeV and 0.5 TeV machines [52], and several designs now exist for the 100 GeV case. Fig. 9 gives the dynamic aperture of one such lattice [53] for the required 1000 turns.

TABLE 5. Parameters 3 TeV Collider Accelerators

acc type magnet type rf type		linac sledCu	recirc warm sledCu	recirc warm sledCu	recirc warm SC Nb	pulsed warm SC Nb	pulsed hybrid SC Nb	pulsed hybrid SC Nb	sums
E_{init}	(GeV)	0.10	0.70	2	7	50	200	1000	
E_{final}	(GeV)	0.70	2	7	50	200	1000	1500	
circ	(km)	0.07	0.12	0.25	1.16	4.65.	11.30	11.36	28.93
turns		2	8	10	11	15	27	17	
decay loss	(%)	6.11	12.11	10.38	9.53	10.68	10.07	2.65	47.68
decay heat	W/m	3.46	14.20	16.03	15.49	19.44	30.97	18.09	
pulsed B_{max}	(T)					2	2	2	
B_{fixed}	(T)		0.70	1.20	2	2	8	8	
ramp freq	(kHz)	900	109	40.02	7.99	1.43	0.33	0.53	
sig beam	cm	0.59	0.51	0.39	0.25	0.11	0.08	0.06	
sig width	cm		3.03	3.65	3.85	1.36	0.59	0.18	
mom compactn	%		-1	-2	-2	-1	-1	-1	
pipe width	cm		30.31	36.49	38.53	13.63	5.86	3	
pipe ht	cm		10	8	4.30	3	3	3	
rf freq	(MHz)	90	50	90	200	800	1300	1300	
acc/turn	(GeV)	0.40	0.17	0.50	4	10	30	30	
acc time	(μs)		3	8	41	232	1004	631	
eta	(%)	0.73	0.22	0.33	0.44	10.15	14.37	12.92	
acc Grad	(MV/m)	8	8	10	15	15	25	25	
synch rot's		0.54	0.82	1.91	9.16	27.07	76.78	31.30	
phase slip	deg		6.90	4.62	5.35	1.64			
cavity rad	(cm)	122	220	134	76.52	19.13	11.77	11.77	
loading	%	4.23	6.22	11.98	16.54	210	527	296	
grad sag	%		3.16	6.18	8.65				
rf time	msec	0.56	1.35	0.59	2.04	0.40	1.25	0.96	
peak rf /m	(MW/m)	2.56	3.43	6.05	0.81	0.91	0.56	0.50	
ave rf power	MW	1.11	1.54	2.84	7.20	6.32	21.91	15.07	55.99
rf wall	MW	8.50	9.05	16.69	21.18	18.59	44.72	30.76	149
magnet ps	MJ						34.31	13.19	47.51
magnet wall	MW						3.7	1.4	5.1
total wall	MW	8.50	9.05	16.69	21.18	18.59	48.4	32.2	155
beam power	MW	0.02	0.04	0.15	1.17	3.68	17.54	9.86	32.47
wall-beam eff	%	0.26	0.49	0.91	5.51	19.81	39.23	32.06	21.72

Scraping

Collimation schemes have been designed [54] for colliders at both high and low energies. At low energies, as in the Higgs Factory, tungsten collimators have been shown to be effective. At higher energies, the muons are scattered, but not stopped, by such collimators. For this case it has been shown that electrostatic septa followed by sweeping magnets could effectively extract the tail muons. Lattices [52] have been designed incorporating these systems.

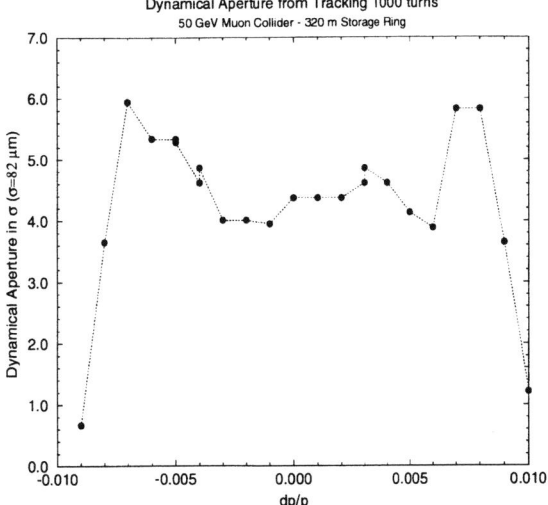

FIGURE 9. Dynamical aperture after 1000 turns.

Instabilities

The studies [55,56] have considered beam emittance growth due to beam-beam tune shift, and both, although some assumptions were made, predict negligible effects in 1000 tunes at the values shown in Tb. 1.

A study [57] has examined the resistive wall impedance longitudinal instabilities in rings at several energies. At the higher energies and larger momentum spreads, solutions were found with small but finite momentum compaction, and moderate rf. For the special case of the Higgs Factory, with its very low momentum spread, a solution was found with no synchrotron motion, but rf provided to correct the first order impedance generated momentum spread. The remaining off momentum tails, that would not affect the luminosity, but which might generate background, could be removed by a higher harmonic rf correction.

Given the very slow, or nonexistent synchrotron oscillations, the transverse beam breakup instability is significant. But this instability can be stabilized using rf quadrupole [58] induced BNS damping. For instance, in the 3 TeV case, to stabilize the resistive wall instability, the required tune spread, calculated [59] using the two particle model approximation, for a 1 cm radius aluminum pipe, is only 1.58×10^{-4}.

However, this application of the BNS damping to a quasi-isochronous ring, and other head-tail instabilities due to the chromaticities ξ and η_1, needs more careful study.

Bending Magnet Design

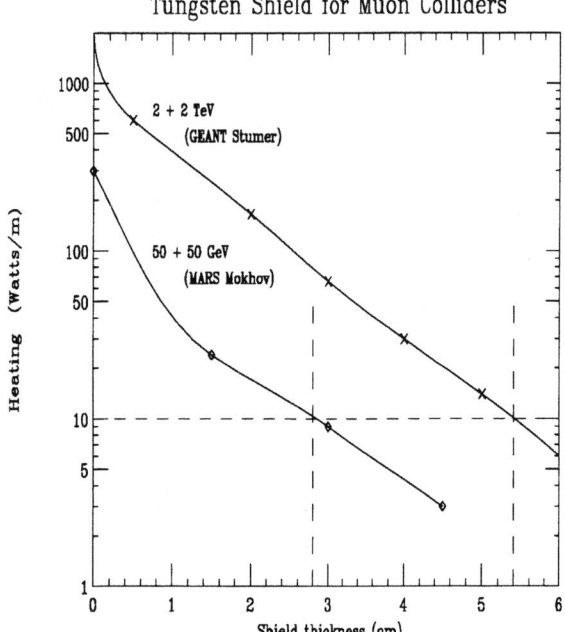

FIGURE 10. Power penetrating tungsten shields vs. their thickness for a) 4 TeV, and b) 100 GeV, collider.

The magnet design is complicated by the fact that the μ's decay within the rings ($\mu^- \to e^- \overline{\nu}_e \nu_\mu$), producing electrons whose mean energy is approximately 0.35 that of the muons. With no shielding, the average power deposited per unit length would be about 2 kW/m in the 4 TeV machine, and 300 W/m in the 100 GeV Higgs factory. Fig. 10 shows the power penetrating tungsten shields of different thickness. One sees that 3 cm in the low energy case, or 6 cm at high energy would reduce the power to below 10 W/m which can reasonably be taken by superconducting magnets.

The quadrupoles could use warm iron poles placed as close to the beam as practical. The coils could then be either superconducting or warm, placed at a greater distance from the beam and shielded from it by the poles.

NEUTRINO RADIATION

Bruce King [62] has shown that the surface radiation dose $D_B(Sv)$ in a time $t(s)$, in the plane of a bending magnet of field B(T), in a circular collider with beam energy $E(TeV)$, average bending field $<B(T)>$, at a depth $d(m)$ (assuming a spherical earth), with muon current (of each sign) of I(muons/sec/sign) is given by:

$$D_B \approx 4.4 \; 10^{-24} \frac{I_\mu \; E^3 \; t}{d} \frac{<B>}{B} t$$

and that the dose D_S at a location on the surface, in line with a high beta straight section of length ℓ is:

$$D_S \approx 6.7 \; 10^{-24} \frac{I_\mu \; E^3 \; t}{d} \ell <B> t$$

The first formula has been confirmed by a Monte Carlo simulation using the MARS code [63]. In all cases it is assumed that the average divergence angles satisfy the condition: $\sigma_\theta << \frac{1}{\gamma}$. This condition is not satisfied in the straight sections approaching the IP, and these regions, despite their length, do not contribute a significant dose.

For the 3 TeV parameters given in Tb. 1 (muon currents $I = 6 \times 10^{20} \; \mu^-/yr$, $<B> = 6\,T$, $B = 10\,T$, and taking the federal limit on off site radiation Dose/year, D_{Fed} to be 1 mSv/year (100 mrem/year), then the dose D_B per year (defined as 10^7 s), in the plane of a bending dipole is:

$$D_B = 1.07 \; 10^{-5} \; (\text{Sv}) \approx 1\% \; D_{Fed}$$

and for a straight section of length 0.6 m is:

$$D_B = 9.7 \; 10^{-5} \; (\text{Sv}) \approx 10\% \; D_{Fed}$$

which may be taken to be a reasonable limit.

Special care will be required in the lattice design to assure that no field free region longer than this is present. But it may be noted that the presence of a field of even 1 T over any length, is enough to reduce the dose to the 10 % Federal limit standard. For machines above 3 TeV, the muon current would probably have to be reduced.

For lower energy machines, the requirements get rapidly easier: a 0.5 TeV at 500 m depth could have 130 m straight sections, or if at 100 m depth 25 m lengths, for the same surface dose. For a 100 GeV machine the doses are negligible.

DETECTOR BACKGROUND

There will be backgrounds in the detector from the decay of muons in the ring and approaching the IP. A recent study [60] of electromagnetic, hadronic and muon components of the background has been done using the GEANT codes [61]. This study:

TABLE 6. Detector backgrounds from μ decay

Radius	cm	5	10	20	100
Photons hits	cm^{-2}	26	6.6	1.6	.06
Neutrons hits	cm^{-2}	0.06	0.08	0.2	0.04
Charged hits	cm^{-2}	8	1.2	0.2	0.01
Total hits	cm^{-2}	34	8	2	0.12
Pixel size	μm^2	60x150	60x150	300x300	300x300
Occupancy all	%	0.6	0.14	0.4	0.02
Occupancy charged	%	0.14	0.02	0.04	0.002

- followed shower neutrons and photons down to 40 keV and electrons to 25 KeV.

- used a tungsten shield over the beam, extending in to within 14 cm of the intersection point, and extending outward to an angle of 20 degrees from the axis.

- inside this shield, between its smallest aperture 1 m from the IP, and its tip, the inner surface is shaped into a series of rising collimating steps and slopes, designed so that, the detector could not *see* any surface directly illuminated by the initial decay electrons, whether seen in the forward or backward (albedo) directions.

- from the aperture point of minimum to a few meters (2.5 m for Higgs Factory) upstream, the inside forms another series of stepped collimators placed at $\pm$ 4 σ_{θ_0} (where σ_{θ_0} is the rms divergence of the beam).

- further upstream, prior to the first quadrupole (from 2.5 to 4 m in the Higgs case) an 8 T dipole, with collimators inside, is used to sweep decay electrons before the final collimation.

Tb. 6 gives the hit density for the Higgs factory from the various sources and the occupancy of pixels of the given sizes. In all cases the numbers are given per bunch crossing. The hit density for the higher energy machines are found to be somewhat lower than these, due to the small decay angles of the electrons. The radiation damage by the neutrons on a silicon detector has also been estimated. In the Higgs case, at 5 cm from the vertex, the number of hits from neutrons above 100 KeV is found to be $1.8\,10^{13}$ per year ($10^7\,s$). This is an order of magnitude less than that expected at the LHC. The damages for silicon detectors in the higher energy machines are of the same order.

This study also found a significant flux of muons, with quite high energies, from μ pair production in electromagnetic showers (Bethe Heitler). Their most serious effect appears to arise when they make deeply inelastic interactions and deposit spikes of energy in the electromagnetic and hadronic calorimeters. This is not serious in the Higgs case, when the fluxes and cross sections are low, but at higher energies timing and/or longitudinal calorimeter segmentation appears necessary to identify and remove the problem.

An ealier study using the code MARS [64], using less sophisticated shielding, gave results qualitatively in agreement with those from Geant [60].

CONCLUSION

Motive

- Because they can be circular, muon colliders appear to be far smaller than hadron or e^+e^- machines of similar *effective* energy.

- It is thus hoped that muon colliders could have a lower cost per TeV than other options.

- Their smaller size would allow machines up to 3 TeV effective energy (roughly equivalent to a 30 TeV hadron machine) to fit on existing laboratory sites.

- The low synchrotron and beamstrahlung radiation with muons could allow energy spreads as small as 0.003 % (3×10^{-5}).

- By measuring g-2 of the muon it should be possible to determine the energy to even greater precision.

- The above, plus the large cross section for s-channel production, could make a muon collider into a precision tool to study Higgs particles and their decays.

Progress

- The theory of operation of all components of a muon collider are now well understood.

- Simulations of examples of all components of a baseline design have now been performed. The simulated performances of many of these components has exceeded the baseline specifications. All known effects have been included except space-charge in the cooling, whose effect, calculated analytically, appears not to be too large.

What is needed?

- More detailed simulations of all components, including space-charge in the cooling channel.

- Complete scenarios of the cooling stages and acceleration.

- An experimental study of the target.

- The construction and test of one or more of the cooling stages.

- Technical development of components: 1) a large high field solenoid for capture; 2) low frequency rf linacs; 3) multi-beam or pulsed magnets for acceleration; 4) warm bore shielded high field dipoles for the collider; 5) muon collimators and background shields, etc..

  But none of these components can be described as *exotic*, and their specifications are not beyond what has been demonstrated.

ACKNOWLEDGMENTS

This research was supported by the U.S. Department of Energy under Contract No. DE-ACO2-76-CH00016 and DE-AC03-76SF00515.

REFERENCES

1. R. B. Palmer, *Muon Collider: Introduction and Status*, submitted to the Proceedings of the Workshop on Physics at the First Muon Collider and at the Front End of a Muon Collider, Fnal, Oct. 1997.
2. G. I. Budker, *Accelerators and Colliding Beams*, Proc. of the 7th International Acc. Conference (Erevan, 1969); presentation at the International High Energy Conference (Kiev, 1970).
3. V. V. Parkhomchuk and A. N. Skrinsky, Proc. 12th Int. Conf. on High Energy Accelerators, F. T. Cole and R. Donaldson, Eds., (1983) 485; A. N. Skrinsky and V.V. Parkhomchuk, Sov. J. of Nucl. Physics **12**, (1981) 223; *Early Concepts for $\mu^+\mu^-$ Colliders and High Energy μ Storage Rings, Physics Potential & Development of $\mu^+\mu^-$ Colliders. 2nd Workshop*, Sausalito, CA, Ed. D. Cline, AIP Press, Woodbury, New York, (1995).
4. D. Neuffer, Fermilab Note FN-319, July 1979; Proc. 12th Int. Conf. on High Energy Physics (1983) 481; *Principles and Applications of Muon Cooling*, Part. Acc. **14** 75 (1983)
5. *Proceedings of the Mini-Workshop on $\mu^+\mu^-$ Colliders: Particle Physics and Design*, Napa CA, Nucl Inst. and Meth., **A350** (1994) ; Proceedings of the Muon Collider Workshop, February 22, 1993, Los Alamos National Laboratory Report LA-UR-93-866 (1993) and *Physics Potential & Development of $\mu^+\mu^-$ Colliders 2nd Workshop*, Sausalito, CA, Ed. D. Cline, AIP Press, Woodbury, New York, (1995); Proceedings of the 9th Advanced ICFA Beam Dynamics Workshop, Ed. J. C. Gallardo, AIP Press, Conference Proceedings 372 (1996).

6. R. B. Palmer et al., *Monte Carlo Simulations of Muon Production, Physics Potential & Development of $\mu^+\mu^-$ Colliders 2nd Workshop*, Sausalito, CA, Ed. D. Cline, AIP Press, Woodbury, New York, pp. 108 (1995); R. B. Palmer, et al., *Muon Collider Design*, in Proceedings of the Symposium on Physics Potential & Development of $\mu^+\mu^-$ Colliders, Nucl. Phys B (Proc. Suppl.) **51A** (1996); R. B. Palmer and J. C. Gallardo, *Muon-Muon and other High Energy Colliders, Techniques and Concepts of High Energy Physics IX*, Phys. vol. 365, Ed. T. Ferbel, pp. 183, Plenum Pub. (1997).
7. R. B. Palmer and J. C. Gallardo, *High Energy Colliders*, Proceedings of 250th Anniversary Conference on Critical Problems in Physics, Princeton University, Eds. V. Fitch, D. Marlow, M. Dementi, pp. 247 (1997), Princeton Press.
8. R. B. Palmer, *Progress on $\mu^+\mu^-$ Colliders*, submitted to the Proceedings of the PAC97, Vancouver, Canada, June 1997.
9. $\mu^+\mu^-$ *Collider, A Feasibility Study*, BNL-52503, FermiLab-Conf-96/092, LBNL-38946, Proceedings of the 1996 DPF/DPB Summer Study on High-Energy Physics, Snowmass'96. For updated information, see the Muon Collider Collaboration WEB page: http://www.cap.bnl.gov/mumu/.
10. R. Raja and A. Tollestrup, *Calibrating the energy of a 50 x 50 GeV muon collider using spin precession*, LANL preprint archive, hep-ex/9801004; submitted to Phys. Rev. D.
11. B. King, private communication
12. C. Ankenbrandt and B. Noble, *Summary of the Accelerator Working Group*, submitted to the Proceedings of Workshop on Physics at the First Muon Collider and at the Fron End, FNAL, Nov. 1997.
13. T. Roser, *AGS Performance and Upgrades: A Possible Proton Driver for a Muon Collider*, Proceedings of the 9th Advanced ICFA Beam Dynamics Workshop, Ed. J. C. Gallardo, AIP Press, Conference Proceedings 372 (1996).
14. C. Ankenbrandt, K-Y. Ng, J. Norem, M. Popovic, Z. Qian, L. Ahrens, M. Brennan, V. Mane, T. Roser, D. Trbojevic, W. van Asselt, *Bunching Near Transition in the AGS*, Fermilab Pub-98-006, submitted to Phys. Rev. D.
15. J. E. Griffin, K.Y. Ng, Z.B. Qian and D. Wildman, *Experimental Study of Passive Compensation of Space Charge Potential Well Distortion at the Los Alamos National Laboratory Proton Storage Ring*, Fermilab Report, FN-661, Nov. 1997.
16. C. Johnson, *Solid and Liquid Targets Overview*, presentation at the Mini-Wokshop: Target and Muon Collection Magnets and Accelerators, Oxford, MI, Jan. 1997.
17. C. Lu, K. T. McDonald, *Low-Melting-Temperature Metals for possible Use as Primary Targets at a Muon Collider Source*, Princeton/$\mu\mu$/97-3, Revised Dec. 1997, unpublished.
18. R. Weggel, *Deceleration of Conductor by Magnetic Field: 1) Paraxial; 2) Perpendicular*, unpublished
19. M. Green and R. Palmer, *A $\mu - \mu$ collider capture solenoid system for pions froma tilted target*, submitted to the Proceedings of PAC97, May 1997.
20. N.V. Mokhov and A. Van Ginneken, *Pion Production and Targetry at $\mu^+\mu^-$*

Colliders, Fermilab-Conf-98/041 (1998), submitted to Proc. of the 4th Int. Conf. on Physics Potential and Development at $\mu^+\mu^-$ Colliders, San Francisco, CA, December 10-12, 1997

21. D. Kahana, et al., *Proceedings of Heavy Ion Physics at the AGS-HIPAGS '93*, Ed. G. S. Stephans, S. G. Steadman and W. E. Kehoe (1993); D. Kahana and Y. Torun, *Analysis of Pion Production Data from E-802 at 14.6 GeV/c using ARC*, BNL Report # 61983 (1995).
22. N. V. Mokhov, *The MARS Code System User's Guide*, version 13(95), Fermilab-FN-628 (1995).
23. J. Ranft, DPMJET Code System (1995).
24. Experiment E-910 at AGS, BNL, private communication.
25. N.V. Mokhov and S.I. Striganov, *Towards Reliable Prediction of Particle Production for 6-120 GeV Proton Beams*, presentation at the Workshop on Physics at the First Muon Collider and at the Front End of a Muon Collider, Nov. 1997.
26. J. R. Miller, M. Bird, S. Bole et al., *An Overview of the 45 T Hybrid Magnet System for the National High Field Magnet Laboratory*, IEEE Transactions on Magnetics 30, pp. 1563 (1994).
27. R. Weggel, *4-MW Hollow-Conductor Magnets for 20 T Hybrid Systems to Collect Pions for a Muon Collider*, presentation at the Mini-Wokshop: Target and Muon Collection Magnets and Accelerators, Oxford, MI, Jan. 1997.
28. N. Mokhov, R. Noble and A. Van Ginneken, *Target and Collection Optimization for Muon Colliders*, Proceedings of the 9th Advanced ICFA Beam Dynamics Workshop, Ed. J. C. Gallardo, AIP Press, Conference Proceedings 372 (1996).
29. K. Assamagan, et al., Phys. Lett. **B335**, 231 (1994); E. P. Wigner, Ann. Math. **40**, 194 (1939) and Rev. Mod. Phys., **29**, 255 (1957).
30. B. Norum and R. Rossmanith, *Polarized Beams in a Muon Collider*, in Proceedings of the Symposium on Physics Potential & Development of $\mu^+\mu^-$ Colliders, Nucl. Phys B (Proc. Suppl.) **51A** (1996).
31. A. A. Mikhailichenko and M. S. Zolotorev, Phys. Rev. Lett. **71**, (1993) 4146; M. S. Zolotorev and A. A. Zholents, SLAC-PUB-6476 (1994).
32. A. Hershcovitch, Brookhaven National Report AGS/AD/Tech. Note No. 413 (1995).
33. Z. Huang, P. Chen and R. Ruth, SLAC-PUB-6745, *Proc. Workshop on Advanced Accelerator Concepts*, Lake Geneva, WI , June (1994); P. Sandler, A. Bogacz and D. Cline, *Muon Cooling and Acceleration Experiment Using Muon Sources at Triumf*, Physics Potential & Development of $\mu^+\mu^-$ Colliders 2nd Workshop, Sausalito, CA, Ed. D. Cline, AIP Press, Woodbury, New York, pp. 146 (1995).
34. Initial speculations on ionization cooling have been variously attributed to G. O'Neill and/or G. Budker see D. Neuffer in [5]; D. Neuffer, in Advanced Accelerator Concepts, AIP Conf. Proc. 156, 201 (1987); see also [3–5]; R. C. Fernow and J. C. Gallardo, *Muon Transverse Ionization Cooling: Stochastic Approach*, Phys. Rev. **E52** 1039 (1995).
35. R. Fernow, **ICOOL**, fortran program to simulate muon ionozation cooling.
36. P. Le Brun, *Alternate solenoid in DPGeant*, presented at the Mini-Workshop

on Cooling, BNL Jan. 1998, unpublished.
37. H. Kirk, *Parmela modeling of alternating solenoids* presented at the Mini-Workshop on Cooling, BNL Jan. 1998, unpublished.
38. R. C. Fernow, J. C. Gallardo and R. B. Palmer, *Ionization cooling using a FOFO lattice*, BNL Report BNL #64493, submitted to PAC97, Vancouver, Canada, 1997.
39. Y. Zhao, *The preliminary simulation of $\frac{2\pi}{3}$-mode interleaved side coupled standing wave structures*, presented at the Mini-Workshop on Cooling, BNL Jan. 1998, unpublished.
40. A. Moretti, *Rf Update*, presented at the Mini-Workshop on Cooling, BNL Jan. 1998, unpublished.
41. D. Neuffer and A. Van Ginneken, *Recent Cooling Simulation Studies*, presented at the Mini-Workshop on Cooling, BNL Jan. 1998, unpublished.
42. V. Balbekov and A. Van Ginneken, *Ring Cooler for Muon Collider*, presented at the Mini-Workshop on Cooling, Fermilab Oct. 1997, unpublished.
43. D. Neuffer and W. Wan, *COSY transport for μ cooling*, presented at the Mini-Workshop on Cooling, Fermilab Oct. 1997, unpublished.
44. G. Silvestrov, Proceedings of the Muon Collider Workshop, February 22, 1993, Los Alamos National Laboratory Report LA-UR-93-866 (1993); B. Bayanov, J. Petrov, G. Silvestrov, J. MacLachlan, and G. Nicholls, Nucl. Inst. and Meth. **190**, (1981) 9; C. D. Johnson, Hyperfine Interactions, **44** (1988) 21; M. D. Church and J. P. Marriner, Annu. Rev. Nucl. Sci. **43** (1993) 253.
45. G. Silvestrov, *Lithium Lenses for Muon Colliders*, Proceedings of the 9th Advanced ICFA Beam Dynamics Workshop, Ed. J. C. Gallardo, AIP Press, Conference Proceedings 372 (1996).
46. F. Mills and C. Johnstone, presentation at the 4th Int. Conference on Physics Potential and Development of $\mu - \mu$ Colliders, San Francisco, CA, Dec. 1997.
47. D. Summers, presentation at the 9th Advanced ICFA Beam Dynamics Workshop, Montauk 1995, unpublished.
48. G. Morgan, presentation at the 9th Advanced ICFA Beam Dynamics Workshop, Montauk 1995, unpublished.
49. D. Neuffer, *Acceleration to Collisions for the $\mu^+\mu^-$ Collider*, Proceedings of the 9th Advanced ICFA Beam Dynamics Workshop, Ed. J. C. Gallardo, AIP Press, Conference Proceedings 372 (1996).
50. S.Y. Lee, K.-Y. Ng and D. Trbojevic, FNAL Report FN595 (1992); Phys. Rev. **E48**, (1993) 3040; D. Trbojevic, et al., *Design of the Muon Collider Isochronous Storage Ring Lattice*, Micro-Bunches Workshop, BNL Oct. (1995), AIP Press, Conference Proceedings 367 (1996).
51. K. L. Brown and J. Spencer, SLAC-PUB-2678 (1981) presented at the Particle Accelerator Conf., Washington, (1981) and K.L. Brown, SLAC-PUB-4811 (1988), Proc. Capri Workshop, June 1988 and J.J. Murray, K. L. Brown and T.H. Fieguth, Particle Accelerator Conf., Washington, 1987; Bruce Dunham and Olivier Napoly, *FFADA, Final Focus. Automatic Design and Analysis*, CERN Report CLIC Note 222, (1994); Olivier Napoly, it CLIC Final Focus System: Upgraded Version with Increased Bandwidth and Error Analysis, CERN

Report CLIC Note 227, (1994).

52. A. Garren, C. Johnstone, *Lattice Design for a 100 GeV Muon Collider*, presentation at the 4th Int. Conference on Physics Potential and Development of $\mu - \mu$ Colliders, San Francisco, CA, Dec. 1997; A. Garren and C. Johnstone, *Progress on a Lattice for a 2 TeV Muon Collider*, submitted to the Proceedings of the PAC97, Vancouver, Canada, June 1997.

53. D. Trbojevic and K.-Y. Ng, submitted to Proc. of the 4th Int. Conf. on Physics Potential and Development at mu+mu- Colliders, San Francisco, CA, December 10-12, 1997

54. A. Drozhdin, C. Johnstone and N. Mokhov, *Muon Collider Beam Collimation System*, unpublished.

55. M. Furman, *The Classical Beam-Beam Interaction for the Muon Collider: A First Look*, BF-19/CBP-Note-169/LBL-38563, April 1996.

56. P. Chen, *Beam-Beam interaction at $\mu^+\mu^-$ Colliders*, in Proceedings of the Symposium on Physics Potential & Development of $\mu^+\mu^-$ Colliders, Nucl. Phys B (Proc. Suppl.) **51A** (1996);

57. W.-H. Cheng, A. M. Sessler and J. Wurtele, *Studies of Collective Instabilities in Muon Collider Rings*, Proceedings of the 9th Advanced ICFA Beam Dynamics Workshop, Ed. J. C. Gallardo, AIP Press, Conference Proceedings 372 (1996).

58. A. Chao, *Physics of Collective Beam Instabilities in High Energy Accelerators*, John Wiley & Sons, Inc, New York (1993).

59. K.Y. Ng, *Beam Stability Issues in a Quasi-Isochronous Muon Collider*, Proceedings of the 9th Advanced ICFA Beam Dynamics Workshop, Ed. J. C. Gallardo, AIP Press, Conference Proceedings 372 (1996).

60. I. Stumer et al., *Study of Detector Backgrounds in a $\mu^+\mu^-$ Collider*, Proceedings of the 1996 DPF/DPB Summer Study on High-Energy Physics, Snowmass'96.

61. *Geant Manual*, Cern Program Library V. 3.21, Geneva, Switzerland, 1993.

62. B. King, presentation at the Muon Collider Mini-Workshop: Lattice and Background, UCLA, Feb. 1997 and private communication.

63. N. V. Mokhov and A. Van Ginneken, *Muon Collider Neutrino Radiation*, presentation at the Muon Collider Collaboration Meeting, Orcas Is., Washington (1997); C.J. Johnstone and N.V. Mokhov, *Shielding the Muon Collider Interaction Region*, presented at the PAC97, Vancouver, Canada, 1997.

64. G. W. Foster and N. V. Mokhov, *Backgrounds and Detector Performance at 2 + 2 TeV $\mu^+\mu^-$ Collider, Physics Potential & Development of $\mu^+\mu^-$ Colliders 2nd Workshop*, Sausalito, CA, Ed. D. Cline, AIP Press, Woodbury, New York, pp. 178 (1995).

An Isochronous Lattice Design for a 50 on 50 GeV Muon Collider

C. Johnstone, A. Drozhdin, N. Mokhov, and W. Wan

Fermi National Accelerator Laboratory, Batavia, IL 60510[1]

A. Garren

Lawrence Berkeley National Laboratory, Berkeley, CA 94720

Abstract. Using local chromatic correction techniques, a lattice for a 50 on 5-GeV muon collider has been developed which can serve as a broad-band (broad momentum acceptance) or a high-resolution (narrow momentum acceptance) Higgs factory. To reach design luminosities of 13^{32} and $10^{31}\,\mathrm{cm}^{-2}\mathrm{s}^{-1}$, a short bunch length, minimal ring circumference, and a β^* of 4 cm and 13 cm must be realized in the broad-band and high-resolution machines, respectively. In the broad-band machine, local chromatic correction of the Interaction Region is required to provide adequate momentum acceptance. However, local chromatic correction conflicts with demands for extreme compactness and isochronicity, making the lattice design challenging.

INTRODUCTION

As designs for muon colliders progress at 4-TeV, 500-GeV, and 100-GeV (center of mass), the concept of a Higgs factory became the dominant influence in the design of the low-energy machine. At this time two cases are being considered for a 50 on 50-GeV collider: a ring with a broad momentum acceptance and high luminosity and one with a much narrower momentum acceptance and lower luminosity but capable of resolving the Higgs mass to high precision. Based on an injected charge per bunch of 4.2×10^{12} muons and an normalized emittance of 90π mm $-$ mr (315π mm $-$ mr) for the broad-band(narrow-band) machine, the lattice must fulfill several criteria to attain the design luminosity, $8.5 \times 10^{31}\,\mathrm{cm}^{-2}\,s^{-1}(7.4 \times 10^{30}\,\mathrm{cm}^{-2}\,s^{-1})$. One of the most important requirements is an Interaction Region (IR) with a very low

---

[1] Work supported by the Universities Research Association, Inc., under contract DE-AC02-76CH00300 with the U. S. Department of Energy.

value of β^* at collision. For the case of broad momentum acceptance, the β^* must be 4 cm and for narrow momentum acceptance, 13 cm. In addition, the bunch length must be held comparable to the value of β^* to avoid dilution from the hour-glass effect. To prevent the bunch from spreading in time using only a modest rf system, the constraint of isochronicity is further imposed on the lattice. Finally, the circumference must be kept as short as possible to minimize luminosity degradation due to muon decay. All of these conditions have a significant impact on the lattice design.

For example, the small value of β^* leads to large peak beta values in the final-focus quadrupoles and correspondingly large linear chromaticities in the Interaction Region (IR). For the broad-band machine, local correction of the linear part of the IR chromaticity is required to achieve adequate momentum acceptance. Efficient chromatic correction in turn requires large positive values of dispersion in the correction sextupoles. Because of the short circumference restriction, high dipole packing fractions must be imposed not only in the arcs, but in the local Chromatic Correction Section (CCS) of the 100-GeV collider as well. One consequence of the high dipole concentration in the (CCS) is that a small momentum compaction, or isochronicity, becomes difficult to maintain as a result of the large number of dipoles in regions of high positive dispersion. The following sections will discuss a base ring design which approaches the limit of how short a 50-GeV lattice can be made under the required isochronous conditions.

OVERVIEW

The ring design is a racetrack, with two circular arcs separated by an experimental insertion and a utility insertion for injection, extraction, and beam scraping. The experimental insertion includes the Interaction Region followed by a section optimized to correct the ring's linear chromaticity; a chromaticity which is almost completely generated by the IR. Fig. 1 shows half of the IR, CCS, matching sections, and one of the only three arc modules. The total circumference in the broad-band machine is currently 345 m. The arc modules account for only about a quarter of the ring circumference.

THE INTERACTION REGION

Because of the dynamics of the cooling process, μ^+ and μ^- emerge from the cooling stage with roughly equal emittances. Initially, unequal β^*s, or elliptical beams, were explored at the collision point. From an optics standpoint, elliptical beams are more manageable and less nonlinear than round beams in the design of Interaction Regions. Using a β^* ratio of 1:4 for the horizontal to vertical (factor of 2 in the relative beam sizes), however, causes a decrease in

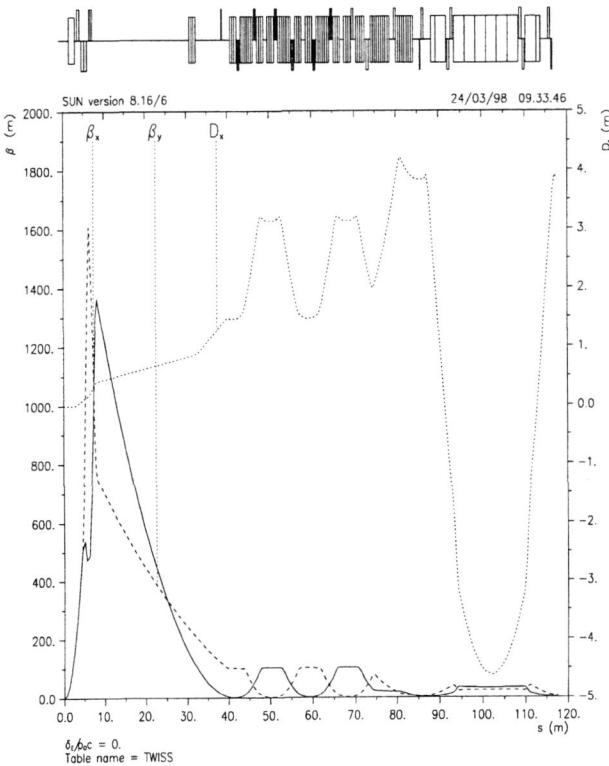

FIGURE 1. The 50-GeV Muon Collider: Half of the IR, local chromatic correction, and one of three arc modules.

the luminosity of a factor of two and this was felt to be unacceptable. Therefore, the condition of round beams at the Interaction Point (IP) has been imposed in all current collider designs.

The β^* at the IP has been fixed to be either 4 cm or 13 cm depending on the desired mode of operation. Since the 4 cm β^* Interaction Region is the more difficult of the two, its properties are emphasized in this report and only occasional references are made to the relaxed beta mode.

The low-beta function values at the IP are mainly produced by three strong superconducting quadrupoles in the Final Focus Telescope (FFT) with poletip fields of 8T. The full IR is symmetric with reflection about the IP. Because of significant, large-angle backgrounds from muon decay, a background-sweep dipole must be included in the FFT and placed it near the IP to protect the detector. It was found that 2.5 m of an 8T field was required to attain sufficient background suppression. Since this dipole is located 2 m from the IP, the first final-focus quadrupole begins 5 m away from the IP. The effect of the dipole has been acounted for in the suppression of dispersion and its first derivative at the IP.

The proximity of the final-focus quadrupoles to the IP determines the maximum beta and this value combined with the quadrupole strengths and lengths determine the natural chromaticity and, ultimately, the nonlinear behavior of the lattice. Unlike the 2-TeV on 2-TeV collider, lattice quadrupole strengths are not a limiting factor in the IR design and lattice performance. With poletip fields reaching 8T, the final-focus triplet in the 100-GeV collider remains short: quadrupole lengths range from .6 to 1.5 m. At 2-TeV, for comparison, 12T poletip fields are needed and, even at these strengths, quadrupole lengths reach 10 m.

With such short quadrupoles, the peak beam size in the 100-GeV machine and, therefore, the natural chromaticity of its Interaction region is almost completely a property of the IP to quadrupole spacing. With the first quadrupole 5 m away from the IP, the beta reaches a maximum of 1.5 km in the FFT when betas are equalized in both planes. At 1.5 km, quadrupole apertures must be at least 15 cm in radius to accomodate 5σ of an 90 π mm-mr, 50-GeV muon beam (normalized rms emittance) plus a 2 cm thick tungsten liner [1].

In the IR shown in Fig. 2, the high-betas are somewhat different, β_{max} is 1680 m vertically and 1300 m horizontally to achieve a more compact IR. Subsequent improvements to the CCS will allow both β_{max}'s to return to the characteristic 1.5 km value. The demagnification in both planes is determined by the relative strengths in the high-beta triplet, the inter-magnet spacing between the two high-beta quadrupoles, especially. These parameters were carefully adjusted to match directly into a chromatic correction section. A smooth transition [2] has been designed into all the collider lattices so as to place the first chromatic correction sextupole at the same phase as the high-beta point.

The optimum design of a very low-beta IR is to make the imaging as point to parallel as is practical to soften chromatic aberrations. The less the applied chromatic correction, the larger, in general, is the dynamic aperture. In the 100-GeV machine, circumference constraints require the IP to be imaged in a short distance; implying stronger than optimal focussing from the high-beta triplet. The IP image distance can be reduced by as much 35 meters on either side of the IP; or about a 30% decrease in the ring circumference. The stronger quadrupole strengths do increase the linear chromaticity of the IR from about 60 to 85 in the vertical with little effect on the horizontal (assuming the triplet powering is FDF). In practice, the demagnification is about halfway between a compact and an optimal or soft-focussing IR. Some deterioration in dynamic aperture is evident with stronger focussing, although studies of high-order and phase dependencies are underway and careful tuning

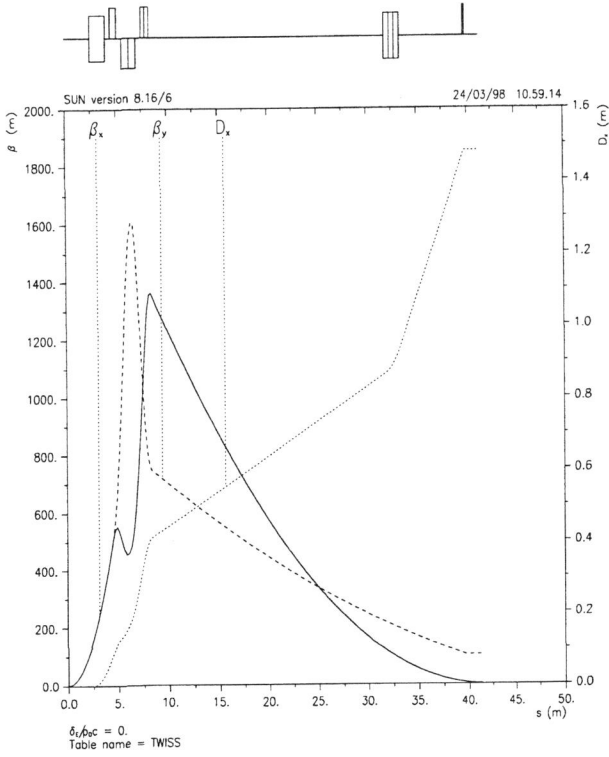

FIGURE 2. The 50-GeV Muon Collider: The $\beta^* = 0.04$ m interaction region.

appears to meloriate these effects.

Initially, the powering of the triplet was chosen such that the vertical apertures in the near dipoles was minimized. This requires placing the vertical high-beta peak at the center of the triplet, so that the triplet sequence is FDF. This has the disadvantage in that the local chromatic correction is not as efficient for the higher the dispersion, the more efficient the chromatic correction. Higher values of dispersion are usually obtained at peaks in the horizontal beta function than in vertical beta peaks. The plane corrected first should be the one with the highest chromaticity; in this case the vertical. If the dispersion is lower, then the chromatic correction, even with π-pairs of sextupoles, is not as efficient and generates stronger nonlinearities. These nonlinearities propagate and appear to be enhanced by the sextupoles of the opposite plane and can be correlated to an observed decrease in the dynamic aperture in this plane.

In a recent lattice, powering of the triplet was reversed to DFD out of concern for the dynamic aperture. The plane with the highest chromaticity and the highest achievable dispersion at the sextupoles was corrected closest to the source, effecting a more efficient chromatic correction. Nonlinear terms were amplified less by sextupoles in the opposing plane as was evidenced by a slight improvement in dynamic aperture. A questionable consequence of installing the horizontal, high-dispersion peak nearest the IP was the unavoidable application of reverse bends to create a dispersion plateau (D'=0) after a defocussing quadrupole. (These reverse bends are not needed if vertical chromaticity is performed first and the dispersion plauteau follows a focussing quadrupole.) The net increase in the circumference due to reverse bends and less efficient dipole packing in general brought the circumference up by at least 50 m; making the circumference more than 400 m when injection and scraping are included. The loss in muon lifetime was felt to outweigh the advantage to the optics of the ring. The final triplet powering remains as FDF with vertical chromaticity being corrected closest to the IP.

LOCAL CHROMATIC CORRECTION

Local chromatic correction of all muon collider interaction regions is required to achieve broad momentum acceptance. The standard approach developed by Brown [3], is used: a pair of sextupoles in each plane separated by a phase advance of π with each pair $(2n+1)\pi/2$ in phase from the IP. Momentum acceptance and dynamic aperture depend sensitively on lattice characteristics and the interaction between certain parameters. Transverse dynamic aperture, for example, is adversely affected by high betatron amplitudes and strong fields in the sextupoles; with sextupole strengths, in turn, being proportional to dispersion. Effective chromatic correction and whether higher orders of chromaticity are generated in the process are greatly determined by the following phase advances: the phase from the IP to the sextupole pairs, and the inter-phasing of the sextupoles comprising a pair. The length of the sextupole contributes to high-order chromatic terms (mainly third-order) and cross correlations depend on the ratio of the betas between planes. Global tune plays an important role in both acceptances.

An innovative module was developed specifically for chromatic correction and implemented first in the 4-TeV muon collider [2]. Its characteristics include a high-dispersion, high-beta plateau in one plane which is consistent with a deep minimum in beta in the opposite plane. When sextupoles are centered in this region, an enhanced β ratio is obtained ($> 10,000$ and about $150 - 200$ for the 4-TeV and 100-GeV colliders, respectively) and provides a clean chromatic correction with minimal cross correlation between planes—

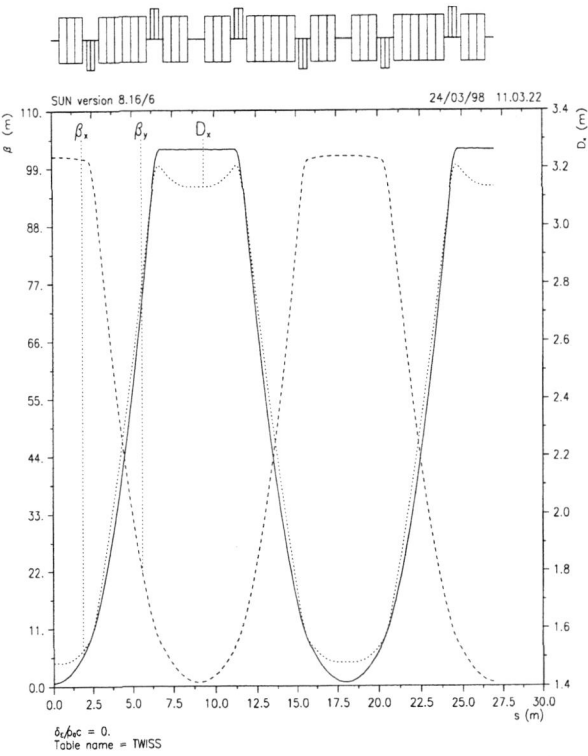

FIGURE 3. The 50-GeV Muon Collider: The Chromatic Correction Module.

unlike previous strong local chromatic corrrection schemes. The high-beta plateaus alternate between planes, with the intervening deep minima establishing a π or near π phase advance between plateaus in the same plane. Thus, sextuple pairs are separated by a π phase advance which is necessary for the almost aberration-free correction of linear chromaticity. Furthermore, the large beta ratio allowed sextupoles in opposite planes to be interleaved without impacting significantly lattice nonlinearity. In point of fact, interleaving improved lattice performance due to a shortening of the CCS, which is accompanied by a lowering of its contribution to overall chromaticity. A string of sharp minima is usually unavoidably chromatic and the extended noninterleaved versions reduced the overall dynamic aperture.

This module, specifically optimized to perform chromatic correction, is particularly powerful in that it can accomodate long sextupoles without beta and phase changes taking place in the plane being corrected. However, because of finite element lengths and changes in the phase advance between sextuples as a function of energy, a tuneshift with amplitude is unavoidable and depends most sensitively on the beta amplitude in the sextupole, but also on the length of the sextupole and the tune of the ring. (The phase advance between sextupoles deviates from π with energy, particularly for shallow minima.) Ultimately, a tuneshift with amplitude constricts the dynamic aperture and a tradeoff exists between momentum acceptance and transverse dynamic aperture. Lattice parameters, in particular, the beta values at the sextupoles and the phase advance around the ring, must be carefully tuned to optimize both acceptances simultaneously.

The Chromatic Correction Section is based on the 4-TeV collider design and was found to make for a lengthy correction section of up to 40% of the ring circumference in the 100-GeV collider ring. The latest CCS design (Fig. 3) was optimized to be the shortest possible module with a β_{max} of only 100 m and a β_{min} less than 1 m ($\beta_{min} = 0.7$ m) coupled to high dispersion at the sextupoles. This results in a β_{ratio} between planes of about 150 without compromising aperture with a large tuneshift with amplitude. Since the momentum compaction of the CCS cannot be made negative at 50 GeV (a consequence of current beta and length requirements), it necessarily contributes a large positive momentum compaction which must be offset by the arcs in order to make the ring isochronous. Optimization of the CCS for length in order to minimize it contribution to momentum compaction resulted in matching minima which were more relaxed than the more compact IR designs ($\beta_{min} = 0.25$ m). The relaxed minima decreased the demagnification required and therefore the focussing strength of the FFT. With the relaxed conditions, the β_{max} can be readjusted to the characteristic 1.5 km in both planes.

If a 12T dipole is considered, the CCS can be considerably shortened because the dispersion can be manipulated more quickly in a FODO-like manner where first a low β_y then a low β_x is inserted to produce the needed π phase advance. With 12T dipole fields, it is anticipated that the circumference of the 100-GeV collider will be under 300 m.

For the narrow band acceptance local chromatic correction; i.e. the sextupoles, are turned off. The momentum acceptance narrows to about a $\delta p/p$ of $\pm.2\%$ while the transverse dynamic aperture increases rapidly to over 10σ at the central momentum.

THE ARC

The arc module Fig. 4 is a new Flexible Momentum Compaction (FMC) design that has the small beta functions characteristic of FODO cells, yet a large, almost separate, variability in the momentum compaction of the module.

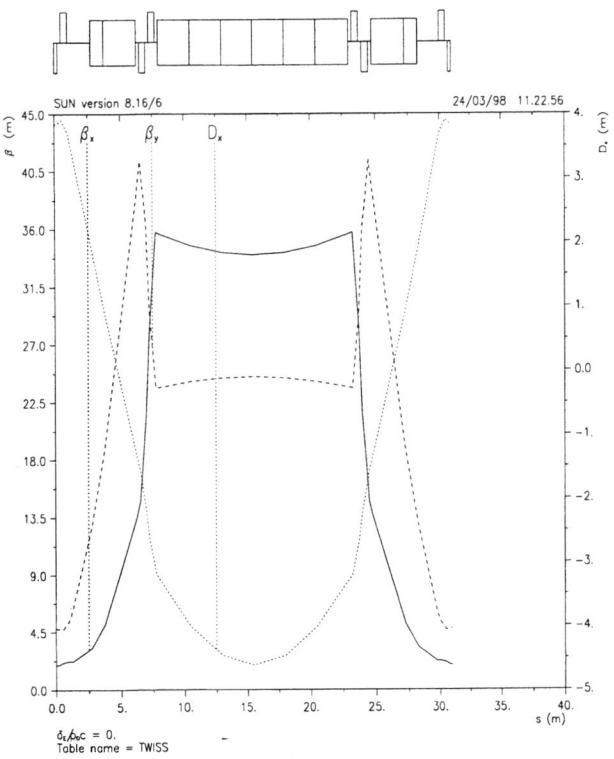

FIGURE 4. The 50-GeV Muon Collider: A New Flexible Momentum Compaction Arc Module.

The smallish beta functions are achieved through the use of a doublet focussing structure which produces a low beta simultaneously in both planes. At the dual minima, a strong focussing quadrupole is placed to control the derivative of dispersion with little impact on the beta functions. (The center defocussing quadrupole is used only to clip the point of highest dispersion.) Ultimately a derivative can be generated which is negative enough to drive the dispersion negative through the doublet and the intervening waist. Extremely negative values of momentum compaction have been achieved, $\alpha = -0.13$ and $\gamma_t = 2i$, with modest values of the beta function. The large negative momentum compaction offsets the positive momentum compaction generated in the CCS and matching sections.

The entire ring was desinged with an eye to controlling momentum

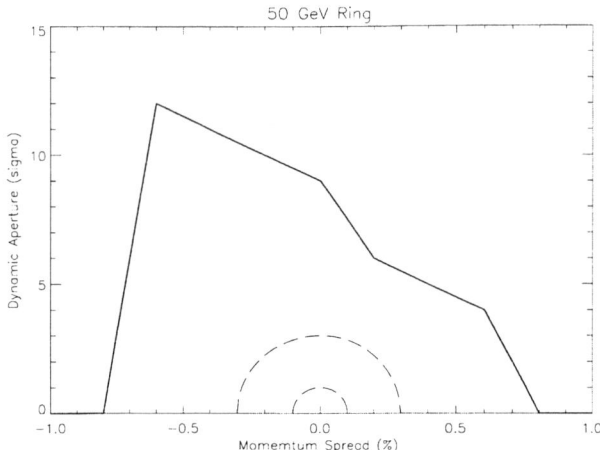

FIGURE 5. The 50-GeV Muon Collider: A preliminary dynamic aperture.

compaction–it was minimiized even in the matching section which links the CCS to the arc. This careful attention to momentum compaction in all parts of the ring design has resulted in a circumference which is just under the 350 m, as opposed to greater 400 m which was characteristic of initial closed-ring designs.

The momentum compaction of the IR, CCS, and matching sections is about 0.04. Since the length of this part is 173 m, the 93 m arc length must have a negative momentum compaction of about −0.09 in order to offset the positive portion of the ring. The arc module was able to generate this amount of negative momentum compaction with beta functions of 40 m or less. The peak dispersion is somewhat high, 4.6 m, although this can probably be reduced in future iterations to under 4 m.

DYNAMIC APERTURE AND MOMENTUM ACCEPTANCE

Studies of the dynamic aperture and momentum acceptance of sextupole dominated lattices are ongoing. The ideal situation is of course to locate sextupoles in high-dispersion and small betas, with no phase advance across the sextupoles, and with exactly π phase advance to its partner. This is not feasible, so beta amplitudes at the sextupoles are slowly decreased to the lowest practical value (beta ratio plus strength considerations), maintaining a phase advance as near π as possible to the next member of the pair. The lower the peak beta at the sextupole, the deeper the minimum required to generate the necessary phase advance. Best dynamic aperture is obained when these parameters are carefully balanced. In the present CCS, β values between 100 and 200 were found to result in sextupole strengths and beta ratios with tune with amplitude terms which were comparable to the chromatic terms. The relatively shallow minima ($\beta = 0.7\,\text{m}$), however, did not produce an exact π (0.92π) phase advance to the partner sextupole. For dynamic aperture, the shallower minima were found to be more important than an exact π phase advance. A preliminary dynamic aperture [4] without optimization and errors is given in Fig. 5.

REFERENCES

1. N. Mokhov, et al. "Scraping Beam Halo in $\mu^+\mu^-$ Colliders" submitted to *4th International Conference on Physics Potential and Development of Muon Colliders*, San Francisco, California, De. 10-12, 1997.
2. C. Johnstone and A. Garren, "An IR and Chromatic Correction Design for a 2-TeV Muon Collider", *Snowmass 96 Proceedings*.
3. K. Brown, "A Conceptual Design of Final Focus Systems for Linear Colliders", SLAC-PUB-4159, (1987).
4. W. Wan, private communication.

The Lattice for the 50-50 GeV Muon Collider

King-Yuen Ng[1] and D. Trbojevic[2]

[1] *Fermi National Accelerator Laboratory,*[3] *P.O. Box 500, Batavia, IL 60510*
[2] *Brookhaven National Laboratory, P.O. Box 5000, Upton, N.Y. 11973-5000*

Abstract. The lattice design of the 50-50 Gev muon collider is presented. Due to the short lifetime of the 50 GeV muons, the ring needs to be as small as possible. The 4 cm low betas in both planes lead to high betatron functions at the focusing quadrupoles and hence large chromaticities, which must be corrected locally. In order to maintain a low rf voltage of around 10 MV, the momentum-compaction factor must be kept to less than 10^{-2}, and therefore the flexible momentum-compaction modules are used in the arcs. The dynamical aperture is larger than 6 to 7 rms beam size for ± 5 rms momentum offset. Comments are given and modifications are suggested.

I THE 50 GEV LATTICE

The recently proposed 50-50 GeV muon collider has a luminosity of 1×10^{33} cm^{2}s^{-1} [1]. To accomplish this, the intense bunches, each containing 1×10^9 muons, will have an rms length of 4 cm, which is also chosen as the low betatron functions at the interaction point (IP). To reduce the high background in the detector due to the decay of the muons, a dipole of length 1 m and field 9 T must be placed downstream of the last focusing quadrupole and outside the detector. As a result, it has been suggested that the last focusing quadrupole should be placed 4.5 m from the IP. The rms momentum spread of these muon bunches is estimated to be 0.0010 and the rms normalized emittance $\epsilon_{N\,\mathrm{rms}} = 80 \times 10^{-6}$ πm. The transverse excursion of the beam will be large. In addition, to prevent the quenching of the superconducting magnets due to the energy released by the decaying muons, they must be shielded by a 2-cm layer of tungsten. Because of all these, the quadrupoles are required to have rather large aperture, and therefore their field gradients are limited. Due to the above restrictions, the betatron functions in the triplet focusing region will be very high, reaching ~ 1550 m in both transverse planes in our design presented below. High betatron functions bring about high natural chromaticities and local correction by sextupoles will be necessary.

Sample lattices have been designed. One of the designs is shown in Fig. 1 starting from the IP. We see first the background clearing dipole followed by the triplet focusing quadrupoles. The interaction region (IR) stops at about 24 m from the IP and the local correction section begins. The Twiss properties of the 4 correction sextupoles are listed in Table 1. The SX1's are the two horizontal correction sextupoles. They are placed at positions with almost the same betatron functions and dispersion function, and are separated horizontally and vertically by phase advances π so that their nonlinear effect will be confined in the region between the

---

[3] Operated by the Universities Research Association, Inc., under contract with the U.S. Department of Energy.

TABLE 1. Twiss properties of the IR correction sextupoles.

	Distance (m)	Phase Advances ν_x	ν_y	Betatron Functions (m) β_x	β_y	Dispersion (m)
SX2	33.3277	0.52280	0.74952	1.06603	100.00063	2.51592
SX2	57.8132	0.99926	1.24953	1.06338	100.00059	2.51851
SX1	46.5892	0.74950	0.96450	100.00552	0.55826	3.20185
SX1	77.5012	1.24858	1.50666	100.56615	1.06291	3.07869

two sextupoles. Their horizontal phase advances are also integral numbers of π from the triplet focusing F-quadrupole so that the chromaticity compensation for that quadrupole will be most efficient [2]. The SX2's are the two vertical correction sextupoles and are placed similarly at designated locations. Notice that the correction section on each side of the IP spans a distance of roughly 62.4 m.

To design the arc section of the collision ring, let us first analyze the possible rf system required. Each muon bunch has an rms length $\sigma_\ell = 4$ cm and rms momentum offset of $\sigma_\delta = 0.0010$. When placed in a matched rf bucket, the synchrotron tune at small amplitude is $\nu_s = |\eta|\sigma_\delta C/(2\pi\sigma_\ell)$, where η is the slippage factor and C is the circumference of the collider ring. The rf voltage required to setup such a bucket will be

$$V_{\rm rf} = \frac{2\pi E \nu_s^2}{|\eta| h} = \frac{|\eta| E}{2\pi h}\left(\frac{C\sigma_\delta}{\sigma_\ell}\right)^2, \qquad (1)$$

where h is the rf harmonic, E the muon total energy, and the muon velocity has been taken as the velocity of light. We have the bucket-bunch relation

$$\frac{\hat{\delta}}{\sigma_\delta} = \frac{\lambda_{\rm rf}}{\pi\sigma_\ell} = \frac{C}{\pi h \sigma_\ell}, \qquad (2)$$

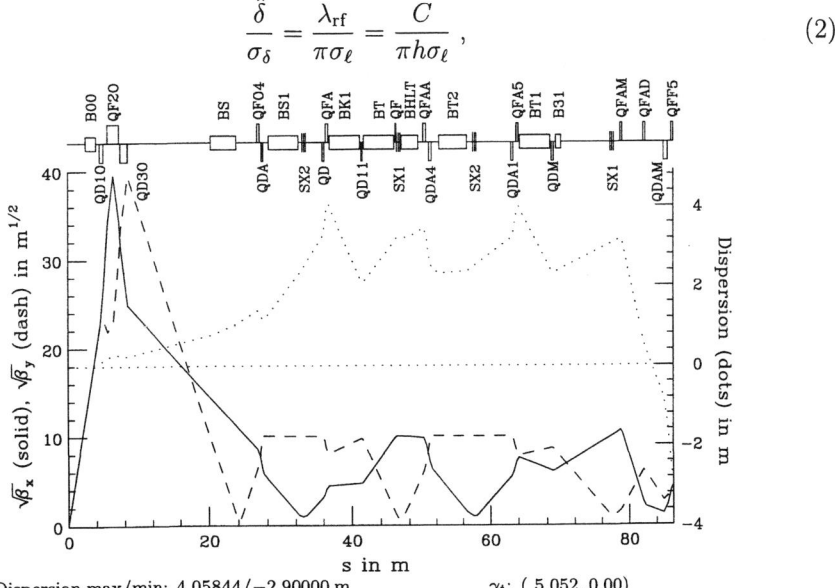

Dispersion max/min: 4.05844/−2.90000 m, γ_t: (5.052, 0.00)
β_x max/min: 1560.92/0.0400 m, ν_x: 1.57703, ξ_x: −41.93, Module length: 86.4056 m
β_y max/min: 1551.07/0.0400 m, ν_y: 2.20717, ξ_y: −42.38, Total bend angle: 1.3888960 rad

FIGURE 1. The lattice structure of the IR including local chromaticity corrections.

where $\hat{\delta}$ is the bucket height and $\lambda_{\rm rf}$ is the rf wavelength. If we take $\hat{\delta}/\sigma_\delta = 5$ and $C = 350$ m, one gets $h = 557$. Then, from Eq. (1), the rf voltage can be solved as $V_{\rm rf} = 1094|\eta|$ MV. In Table 2, the rf voltages corresponding to some possible $|\eta|$ or transition gammas γ_t are tabulated. It is clear that if we want to have a low rf voltage of around 10 MV, we need to limit the momentum-compaction factor to $\sim 1 \times 10^{-2}$. To accomplish that we must use the flexible momentum-compaction (FMC) modules [3] in the arc of the ring. Two such modules will be required for half of the collider ring, one of which is shown in Fig. 2. To close the ring geometrically, there will be a ~ 45.1 m straight section between the two sets of FMC modules. The total length of the collider ring is now only 349.5 m. This is a nice feature, since a small ring allows larger number of collisions before the muons decay appreciably. Note that the IR and local correction sections take up 49.4% of the whole ring.

TABLE 2. Relation between $V_{\rm rf}$ and slippage factor.

$V_{\rm rf}$ (MV)	$|\eta|$	γ_t	ν_s
1	0.000914	33.1	0.00127
5	0.00457	14.8	0.00637
10	0.00914	10.5	0.0127
15	0.01371	8.54	0.0191
20	0.01828	7.40	0.0255
25	0.02285	6.62	0.0318
30	0.02742	6.04	0.0382

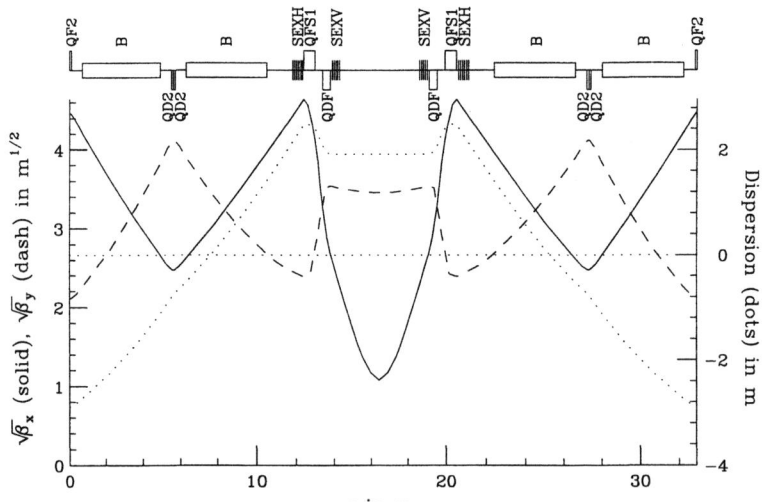

Dispersion max/min: 2.47014/−2.90000 m, γ_t: (0.00, 7.98)
β_x max/min: 21.52/1.15911 m, ν_x: 0.76926, ξ_x: −1.34, Module length: 32.8999m
β_y max/min: 16.95/4.43528 m, ν_y: 0.58984, ξ_y: −0.75, Total bend angle: 0.8763483 rad

FIGURE 2. Lattice structure of the flexible momentum-compaction module.

II DYNAMICAL PROPERTIES

A Dynamical Aperture

Initially 16 particles with the same momentum offset and having vanishing x' and y' are placed uniformly on a circle in the x-y plane. The particles are tracked with the code COSY [4]. The largest radius that provide survival of the 16 particles in 1000 turns is defined here as the dynamical aperture at this momentum offset and is plotted in Fig. 3(a) in units of the rms radius of the beam at the IP. (At the 4 cm low-beta IP, the beam has an rms radius of 82 μm.) We see that the dynamical aperture is not symmetric in momentum offset. It is about 7σ for positive momentum offset, but drops rapidly when the offset becomes negative.

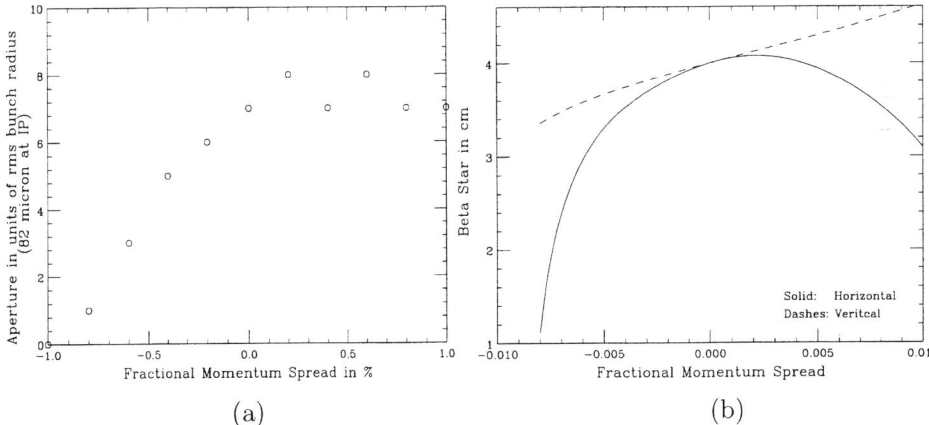

FIGURE 3. (a) Dynamical aperture of the lattice versus momentum offset. (b) Betatron functions at the IP versus momentum offset.

B Betatron Functions at IP with Momentum Offset

We see in Fig. 3(b) that the vertical betatron function at the IP varies almost linearly between ~ 3.4 and ~ 4.6 cm in the range of momentum offset -0.008 and $+0.010$. The horizontal betatron function, on the other hand, varies very nonlinearly between ~ 1.1 and ~ 4.1 cm for the same range of momentum offset. This large and rapid variation is an unpleasant feature of the lattice, because it can easily lead to incorrect computation of the luminosity.

C Tune dependence on Momentum offset

The horizontal and vertical tunes as functions of momentum offset are shown in Fig. 4(a). Although the variation of the vertical tune is small, the variation of the horizontal tune is much larger.

Next we look at the chromaticities of the lattice as functions of momentum offset in Fig. 4(b). Although the chromaticities in both planes have been corrected to

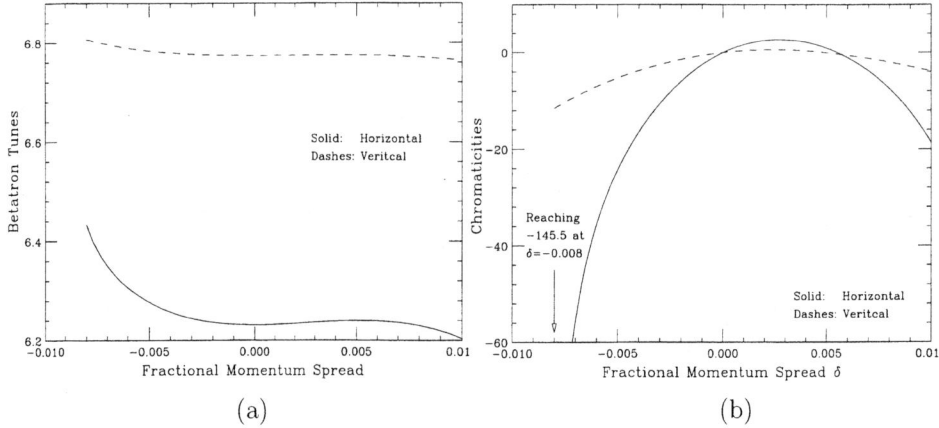

FIGURE 4. (a) Horizontal and vertical tunes versus momentum offset. (b) Horizontal and vertical chromaticities versus momentum offset.

zero, the off-momentum variations are still very large, This is especially true for the horizontal variation which decreases extremely fast at negative momentum offsets. We believe this is the culprit that limits the momentum aperture of the lattice. However, not much can be improved, because we have only one family of correction sextupoles in each plane. These sextupoles, SX1 and SX2, are of length 0.5 m and field strengths 281.9 and -355.2 T/m^2, respectively. Considering a 6-cm aperture, the pole tip fields are 1.01 and 1.27 T, respectively. The sextupoles inside the FMC module only correct the much smaller chromaticities of the FMC module, and have little effect on the chromaticities created by the IR. The on-momentum amplitude-dependent tunes of the ring excluding the straight are

$$\nu_x = 6.231 - 2.40 \times 10^2 \epsilon_x - 3.58 \times 10^3 \epsilon_y ,$$
$$\nu_y = 6.774 - 3.58 \times 10^3 \epsilon_x - 6.65 \times 10^1 \epsilon_y , \qquad (3)$$

where ϵ_x and ϵ_y are the 95% unnormalized horizontal and vertical emittances measured in πm. With the designed rms emittances of 80×10^{-6} πm, the 95% unnormalized emittances are $\epsilon_x = \epsilon_y = 1.014 \times 10^{-6}$ πm (6 times the rms value). Thus the amplitude-dependent tune spreads are only < 0.004 which are small. However, the off-momentum amplitude-dependent tune spreads will be larger due to the larger off-momentum chromaticities.

D Momentum-Compaction Function

The momentum-compaction factor of the whole ring is $\alpha_0 = 0.01114$ and varies rather linearly from 0.0082 to 0.013 when the momentum offset changes from -0.008 to $+0.010$. To further reduce α_0 to ~ 0.001 and hence the rf voltage in operation to $V_{\rm rf} \sim 1$ MV, we must tune the momentum-compaction factor of the FMC module to ~ -0.05, which is very difficult, requiring high dispersion function and high quadrupole gradients [5]. Note that we have already removed the dipoles at the center of the FMC module in Fig. 2 so as to make the momentum-compaction factor more negative.

III PROBLEMS WITH THE LATTICE

(1) What is shown in Fig. 3(a) is the aperture of the *bare* lattice. If we include higher-order field errors, both systematic and random, misalignment errors, synchrotron oscillations, the aperture can be very much reduced. This is a small ring with short magnets and the effects of fringe fields may not be negligible. During the ICFA Frascati Workshop held at Frascati in October of 1997, a lattice for the tau-charm factory was presented by Huang [6]. This machine has a circumference of ~ 380 m. It was pointed out by Koiso that the aperture of the lattice would be very much reduced when fringe fields are included [6]. Therefore, it is very plausible that the momentum aperture of our designed lattice will be reduced to less than ± 5 rms of momentum offset, when all considerations are included.

(2) The horizontal chromaticity changes too abruptly in the negative momentum-offset region, which is the source of the small aperture in that region. Therefore, higher-order chromaticities, like beta-beating, etc, must be corrected. However, this is not possible with only one family of sextupoles in each transverse plane.

(3) Figure 3(b) shows a large variation of the betatron functions at the IP with momentum offset, which is unpleasant. In order to correct for this and higher-order chromaticities, 3 families of sextupoles are required in each transverse plane. Unfortunately, there are only 4 FMC modules in the arc section of ring, and there is not enough space to deploy these sextupoles.

IV SUGGESTIONS FOR IMPROVEMENT
A FODO Cells

If we replace the FMC modules with FODO cells, we will have rooms to install more sextupoles. The 4 FMC modules together with the 45 m straight amount to a total of ~ 177 m. There, we can install 14 FODO cells each of length 12.6 m and phase advance $90°$. Since sextupoles have to be placed π apart to cancel nonlinearity, we can install two more families of sextupoles in each transverse plane. Higher-order chromaticities and the betatron functions at the IP can then be controlled.

The price we need to pay is the more expensive rf system. The horizontal and vertical tunes of the ring will remain about 6 to 8. Since FODO cells are used in the arc region, we expect the transition gamma to be around 6 to 8 as well. According to Table 2, an rf voltage of about 20 to 30 MV will be needed.

B Unequal Betatron Functions at IR

Another suggestion is unequal betatron functions at the IP, for example, $\beta_y^* = 4$ cm, and larger β_x^*. One clear disadvantage is that the luminosity will be reduced almost by half. Also a flat beam will decrease the precision of collision-vertex tagging in the silicon vertex detector. However, there are certainly advantages. First, the final focus can be performed much more easily, since we need to pay attention mostly to the lower β^* only. The chromaticities will be much smaller and can therefore be corrected easily. Second, hopefully, the chromaticities of the IR are so small that local correction will not be necessary. Notice that in the lattice presented earlier, the two local correction sections take up about $\sim 36\%$ of the whole ring. If these sections can be eliminated, more FODO cells can be inserted making the 3-family corrections for the chromaticities and β^* dependence on momentum offset more efficient. Moreover, the collider ring can be designed smaller, allowing for more collisions of the muon bunches per unit time, and we will gain back part of the lost luminosity. If the rms length of the muon bunches can be shortened to 3 cm, we can make $\beta_y^* = 3$ cm and larger for β_x^*. This will also help in gaining back part of the lost luminosity. Finally, the straight section of the lattice presented earlier carries a dispersion up to $\sim \pm 3$ m. On the other hand, dispersion-free straight section can easily be designed with short FODO cells.

In the following we shall present some elementary designs of the IR and estimate their chromaticities using thin-lens theory. We assume the extreme case of $\beta_y^* = 4$ cm and $\beta_x^* = 40$ cm, and use a focusing doublet consisting of quadrupoles of length 1 m each. After the background clearing dipole, the first quadrupole which is defocusing in the x-plane is centered at 5 m from the IP, and the second quadrupole which is focusing is centered at 6 m. Then at $s = 5$ m, $\beta_y \approx s^2/\beta_y^* = 625$ m. Before entering the quadrupole, the *negative-betatron-function-half-slope* is $\alpha_y = -s/\beta_y^* = -125$. In order to bend it to $+125$, the quadrupole is required to have the integrated strength of $k = B'\ell/(B\rho) = 2/s = 0.4$ m^{-1}. The field gradient is therefore $B' = k(B\rho)/\ell = 66.7$ T/m. Taking the rms normalized emittance to be $\epsilon_{N\,\rm rms} = 80 \times 10^{-6}$ πm, the 5σ beam radius is 5.14 cm. Adding 2 cm of shielding, the aperture of the first quadrupole should be 7.14 cm. The pole-tip field required will be 4.8 T only. The vertical chromaticity can be estimated to be

$$\xi_y = -\sum \frac{k\beta_y}{4\pi} \sim -\frac{s}{2\pi\beta_y^*} = -20, \tag{4}$$

where the contribution of only one quadrupole has been included. Thus the total vertical chromaticity of the IR will be around -40, about one half of that in the lattice presented in Sec. I and II. The actual chromaticity should be much smaller, because the design has not been optimized. There will be cancellation of chromaticity between the F- and D-quadrupoles.

We would like to use two examples to give a better illustration. In the first example, the first quadrupole reverses α_y to $-\alpha_y$, and the second quadrupole reverses α_x to $-\alpha_x$. The thin-lens solution is given in Table 3. Whenever two values appear, the first one is at entrance of quadrupole and the second at exit. The integrated strengths of the first and second quadrupoles are $k = +0.40$ and -0.75 m^{-1}, respectively. The chromaticities for the half IR are $\xi_y = +3.98$ and $\xi_x = -7.59$.

In Example 2, we equalize the horizontal and vertical chromaticities and at the same time minimize them. We find that the two quadrupoles have integrated strengths of $k = 0.4330$ and -0.5775 m^{-1}, respectively. The chromaticities for the half IR are $\xi_x = \xi_y = -5.57$. The Twiss properties at the quadrupoles are listed in Table 4. Our results imply that the chromaticities of the whole IR with unequal β^*'s can be as low as ~ -12. Thus local correction at the IR will be unnecessary. We would like to point out, however, that the optimization of chromaticities made in this example does not necessarily be viewed as the best design of the lattice.

TABLE 3. Twiss properties of IR in Example 1.

s (m)	0	5	6
β_y (m)	0.04	625	400
α_y	0	$-125, +125$	$+100, -200$
γ_y (m^{-1})	25	25, 25	25, 99.7
β_x (m)	0.40	62.9	161
α_x	0	$-12.5, -37.7$	$-60.2, +60.2$
γ_x (m^{-1})	2.5	2.50, 22.6	22.6, 22.6

TABLE 4. Twiss properties of IR in Example 2.

s (m)	0	5	6
β_y (m)	0.04	625	358
α_y	0	$-125, +152$	$+115, -91.9$
γ_y (m^{-1})	25	25, 36.9	36.9, 23.6
β_x (m)	0.40	62.9	170
α_x	0	$-12.5, -40.4$	$-66.3, +33.6$
γ_x (m^{-1})	2.5	2.5, 25.9	25.9, 5.91

V CONCLUSION

We have presented a lattice for the 50-50 GeV muon collider that entails a relatively large dynamical aperture. In order to further enlarge the dynamical aperture even in the presence of field errors and fringe-field effects, we propose to replace the FMC modules in the arc with FODO cells and modify the IP by having unequal horizontal and vertical betatron functions. The chromaticities will be much smaller making local correction unnecessary. With the FODO cells, 3 family of sextupoles can be deployed in each plane making the control of variation of β^*'s and chromaticities with momentum offset possible.

REFERENCES

1. R. Palmer for the Muon Collider Collaboration, "Muon Collider: Introduction and Status," to be published in Proc. Workshop on Phys. at First muon Collider and at Front End of a Muon Collider, Fermilab, Batavia, November 6-9, 1997.
2. M. Donald, R. Helm, J. Irwin, H. Moshammer, E. Forest, D. Robin, A. Zholents, and M. Sullivan, "Localized Chromaticity Correction of Low-Beta Insertions in Storage Rings," Proc. 1993 Part. Accel. Conf., Washington, D.C., May 17-20, 1993, pp. 131.
3. S.Y. Lee, K.Y. Ng, and D. Trbojevic, Phys. Rev. **E48**, 3040 (1993).
4. COSY INFINITY, a New Multipurpose Arbitrary Order Beam Dynamics Code, Lawrence Berkeley Laboratory Report LBL-28881, March, 1990. We would like to thank Dr. Weishi Wan to track the dynamical aperture.
5. Ref. 3 shows large dispersion function is required to attain large negative momentum compaction. For the missing-dipole scenario, large quadrupole gradients are required as is illustrated in D. Trbojevic and K.Y. Ng; "A Proton Driver for the Muon Collider Source with a Tunable Momentum Compaction Lattice," to be published in Proc. Part. Accel. Conf., Vancouver, BC, May 12-16, 1997.
6. N. Huang, to be published in Proc. ICFA Workshop on Beam Dynamics Issues for e^+e^- Factories, Frascati, Italy, October 20-25, 1997; H. Koiso, *ibid*.

Calibrating the energy of a 50 × 50 GeV muon collider using spin precession

Rajendran Raja
&
Alvin Tollestrup

Fermi National Accelerator Laboratory
P.O. Box 500
Batavia, IL 60510

Abstract. The neutral Higgs boson is expected to have a mass in the region 90-150 GeV/c^2 in various schemes within the Minimal Supersymmetric extension to the Standard Model. A first generation Muon Collider is uniquely suited to investigate the mass, width and decay modes of the Higgs boson, since the coupling of the Higgs to muons is expected to be strong enough for it to be produced in the s channel mode in the muon collider. Due to the narrow width of the Higgs, it is necessary to measure and control the energy of the individual muon bunches to a precision of a few parts in a million. We investigate the feasibility of determining the energy scale of a muon collider ring with circulating muon beams of 50 GeV energy by measuring the turn by turn variation of the energy deposited by electrons produced by the decay of the muons. This variation is caused by the existence of an average initial polarization of the muon beam and a non-zero value of $g-2$ for the muon. We demonstrate that it is feasible to determine the energy scale of the machine with this method to a few parts per million using data collected during 1000 turns.

THE METHOD

The spin vector $\vec{S}$ of a muon in the muon collider will precess according to the following equation, first derived by Bargmann, Michel and Telegdi [1]

$$\frac{d\vec{S}}{dt} = \vec{\Omega} \times \vec{S} \qquad (1)$$

$$\vec{\Omega} = -\frac{e}{\gamma m_\mu}\left((1+a\gamma)\vec{B}_\perp + (1+a)\vec{B}_\parallel - (a\gamma + \frac{\gamma}{1+\gamma})\vec{\beta} \times \frac{\vec{E}}{c}\right) \qquad (2)$$

where $\vec{B}_\perp$ and $\vec{B}_\parallel$ are the transverse and parallel components of the magnetic field with respect to the muon's velocity $\vec{\beta}c$, e is the electric charge, m_μ the mass of the muon, $a \equiv \frac{g-2}{2}$ is the magnetic moment anomaly of the muon and γ and g are the Lorentz factor and the gyromagnetic ratio of the muon. The value of $a \equiv \frac{g-2}{2}$ for the muon is 1.165924E-3 [2]. In what follows, we will consider the ideal planar collider ring case where $\vec{B}_\parallel$ and $\vec{E}$ are zero. For such a collider ring, $\vec{\Omega}$ is given by

$$\vec{\Omega} = \vec{\Omega}_{cyc}(1 + a\gamma) \tag{3}$$

where $\vec{\Omega}_{cyc}$ is the angular velocity of the circulating beam. From this, it follows that when the beam completes one turn, the spin will rotate by a further $a\gamma \times 2\pi$ radians. We will compute the precision with which γ can be determined by measuring the energy of the electrons produced by muon decay in this ideal case. We will examine the effects of departures from the ideal case in the last section.

It can be shown that the angular distribution of the decay electrons in the muon center of mass is given by the relation [3]

$$\frac{d^2N}{dx\,dcos\theta} = N(x^2(3-2x) - \hat{P}x^2(1-2x)cos\theta) \tag{4}$$

where N denotes the number of muon decays, $x \equiv 2E/m_\mu$ is the electron energy E in the muon rest frame expressed as a fraction of the maximum possible energy ($\approx 0.5m_\mu$), $cos\theta$ is the angle of the electron in the muon rest frame with respect to the z axis which is the direction of motion of the muon in the laboratory and $\hat{P}$ is the product of the muon charge and the z component of the muon polarization. The muon polarization is defined as the average of the individual muon unit spin vectors over the ensemble of muons considered. We note that the distribution is linear in $\hat{P}$.

The average energy $<E>$ and longitudinal momentum $<P_L>$ of the electron in the muon rest frame can be obtained using equation 4 as follows.

$$<E> = \frac{m_\mu}{2} \int\int x \frac{d^2N}{dx\,dcos\theta} dx\,dcos\theta = \frac{7}{10}\frac{m_\mu}{2} \tag{5}$$

$$<P_L> = \frac{m_\mu}{2} \int\int x cos\theta \frac{d^2N}{dx\,dcos\theta} dx\,dcos\theta = \frac{\hat{P}}{10}\frac{m_\mu}{2} \tag{6}$$

These two quantities form the components of a 4-vector, whose transverse components are zero, which may be transformed to the laboratory frame to yield the average electron energy $<E_{lab}>$.

$$<E_{lab}> = \frac{7}{20}E_\mu(1 + \frac{\beta}{7}\hat{P}) \tag{7}$$

where E_μ is the energy of the muon beam. Since the polarization $\hat{P}$ precesses from turn to turn by the amount $\omega = \gamma(g-2)/2 \times 2\pi$ radians, and the number of muons decrease turn by turn due to decay and losses, the total energy $E(t)$ due to decay electrons observed during turn t in an electromagnetic calorimeter will have the following expression

$$E(t) = Ne^{(-\alpha t)}(\frac{7}{20}E_\mu(1 + \frac{\beta}{7}(\hat{P}\cos\omega t + \phi))) \tag{8}$$

where N is the number of muon decays sampled in turn 0, ϕ is an arbitrary phase containing information on the initial direction of polarization and α is the turn by turn decay constant of the muon intensity which in the absence of losses other than decay is given by

$$\alpha = \frac{t_{circ}}{\gamma t_{life}} \tag{9}$$

where t_{circ} is the time taken to circulate around the storage ring and t_{life} is the muon life time.

For a 100% polarized beam, the amplitude of the oscillations is only 1/7 that of the non-oscillating background. It can be seen from equation 4 that the sensitivity to $\hat{P}$ is enhanced by selecting larger values of $\cos\theta$. This implies selecting electrons with higher laboratory energy. Figures 1(a-c) show the deposited electron energy as a function of turn number for polarization $\hat{P} = 1.0$ for individual electron energy ranges of 0-10 GeV, 10-25 GeV and 25-50 GeV respectively as a function of turn number. Figure 2(b) shows very little oscillatory signal, since the electrons in that energy range have small values of $\cos\theta$. Figure 2(d) shows the deposited electron energy with no electron energy cuts. Superimposed is the predicted behavior according equation 8. This serves as a consistency check for our routines. The signal to background ratio increases as we demand electrons with higher value of $\cos\theta$. In what follows, we use electrons with energy greater than 25 GeV during the investigative phase of this analysis and will later optimize this cut. In practice, we can select electrons with energies above a value by momentum analyzing them with a dipole field before they enter the calorimeter.

The method to determine the energy scale of the collider would then entail fitting a functional form of the type

$$f(t) = Ae^{-Bt}(C\cos(D + Et) + F) \tag{10}$$

to the energy observed in the calorimeter. The variables A, B, C, D, E, F are parameters to be fitted. The information on the energy scale is contained in the parameter E.

Parameters of a 50 GeV idealized muon storage ring

In order to arrive at reasonable numbers for α and ω, we consider a storage ring of 50 GeV muons with a uniform bending field of 4 Tesla. This would produce a circular ring with the parameters given in table 1.

It should be noted that for an idealized storage ring with constant B field considered here, α does not depend on γ, since

$$t_{circ} = \frac{m_\mu \gamma}{0.3 B c} \tag{11}$$

$$\alpha = \frac{2\pi m_\mu}{0.3 B c t_{life}} \tag{12}$$

where m_μ is the muon rest mass, B is the bending field of the storage ring and c is the velocity of light. A 100 GeV collider ring will have the same α as a 50 GeV collider ring or a 25 GeV collider ring in this idealized case. As γ changes slightly, t_{circ} changes in proportion, α being the constant used to convert measurements of t_{circ} to γ. Measuring the decay rate of muons also affords a second method to determine γ. The beam circulation time t_{circ} can be measured to precisions of the order of a part in 10^6 and the fractional error in muon lifetime is 1.82E-5 [2]. The fractional error in γ obtainable by observing the rate of decay of the muons will then be dominated by the precision that one can measure α, namely $\delta\gamma/\gamma = \delta\alpha/\alpha$.

Generation of events and fitting for γ

Since equation 4 is linear in $\hat{P}$, the decay distribution of an ensemble of muons depends only on $\hat{P}$, the ensemble average of the z component of the individual muon spin vectors. However, because of the momentum spread of the muons, each individual particle will have a γ slightly different from the average and hence the precession of the spin vector around the ring will be different, leading to a slightly different value of $\hat{P}$ for the next turn. We model the beam by generating an ensemble of 100,000 muons each having its own spin vector and momentum. In an actual collider, it will be possible to sample significantly more decays than this. During each turn, we decay all the beam particles once and record the number and total energy deposited by electrons with individual energies above 25 GeV. Approximately 27% of the decay electrons pass this cut, on average. We decrease the number of decays by the appropriate number expected by muon decay alone for the next turn. At this stage we do not introduce fluctuations in the number of decays from turn to turn, since the 100,000 muons are meant to be representative of a much larger number in the actual ring. We precess the 100,000 spin vectors by their individual precession rates and make them decay again. We repeat this for 1000 turns. We re-use the muons after each turn since the 100,000 muons represent our model of the muon ensemble in the collider.

Generation of muon spin vectors

We generate 4 different samples of events with different ensembles of spin vectors. The z component of the unit spin vector of a muon S_z is allowed to vary from -1 to 1, using a binomial distribution of specified mean. The average value of the distributions are 0.9, 0.74, 0.5 and 0.26 respectively. We study negatively charged muons resulting an initial value of $\hat{P}$ of -0.9,-0.74,-0.5 and -0.26 respectively for these samples. In the absence of momentum spread, the decay distributions would only depend on $\hat{P}$ and not on the details of the distribution of S_z. The angles of the spin vectors are precessed by the individual γ dependent precession rate from turn to turn. In what follows, we assume a beam energy spread of 0.03% for the muons for all samples unless otherwise specified.

Fitting procedure and generation of errors

The energy deposited every turn is fitted to the functional form given by equation 10 using the CERN program MINUIT [4]. In order to study the variation of the fractional error $\delta\gamma/\gamma$ with the number of electrons sampled, we fluctuate the energy observed in the calorimeter E_m by

$$\frac{\sigma_{E_m}^2}{<E_m>^2} \approx \frac{1}{N}(1.03153) \tag{13}$$

where N is the number of electrons sampled. We analyze the case for 41261, 10315, 2579 and 1146 electrons sampled which corresponds to a fractional error in the measured total energy of PERR $\equiv \frac{\sigma_{E_m}}{E_m}$ of .5E-2,1.0E-2,2.0E-2 and 3.0E-2 respectively.

RESULTS

We simulate the muon collider spin precession for a grid of values of $\hat{P}$ =-0.9,-0.74,-0.5 and -0.26 and fractional measurement error for the first turn (PERR) of 0.5E-2, 1.0E-2, 2.0E-2 and 3.0E-2. Figure 2(a) shows the result of the MINUIT fit plotted for 50 turns for $\hat{P}$=-0.26 and PERR=0.5E-2. Figure 2(b) shows the same plot but with the function being plotted only at integer values of the turn number t. A beat is evident in both the theoretical curve and the simulated measurements as a result of sampling the oscillation function at fixed intervals, not connected with the oscillation frequency. The origin of the beat is stroboscopic. Figure 2(c) shows the pulls, defined as $(data - fit)/error$ at each measurement as a function of turn number for 1000 turns. There are no major turn dependent variations in this quantity indicating that the fit converged satisfactorily. Figure 2(d) shows the histogram of the pulls, which approximates a unit Gaussian as desired. Table 2 shows the results of the fit for the grid of values of $\hat{P}$ and PERR. The results presented in table 2 are shown graphically in Figure 3. As an example, for an average polarization $\hat{P} = -0.26$, the fractional error in $\delta\gamma/\gamma$ varies from 5.1E-6 to 1.9E-5 as the fractional error in the electron energy sampled varies from 0.5E-2 to 3.0E-2, corresponding to the number of electrons sampled during the first turn varying from 41261 to 1146. The average number of decays in the muon collider is expected to be 3.2E6 decays per meter for a beam intensity of 10^{12} muons. The error in determining γ is thus going to be dominated by the fluctuations in the number of electrons sampled turn by turn, rather than sampling fluctuations in the calorimeter. We have simulated conditions involving $\approx$ 40,000 decays. It should be possible to go to higher statistical precision than computed here by sampling larger number of electrons.

The results for $\delta\gamma/\gamma$ obtained from the measurement of the turn by turn rate of decay of the electron energy are not competetive with the precession method primarily because of the small value of α (0.8399E-3). This leads to larger fractional errors for γ from this method (which also assumes that the loss of intensity is entirely due to the decay process) by almost three orders of magnitude than from the precession method.

Variation of $\delta\gamma/\gamma$ as a function of muon energy

The spin precession per turn equals 2π for a γ value of 857.689, which corresponds to a muon beam momentum of 90.622 GeV/c. This is the first spin resonance for muons. At this point, the fitting method loses sensitivity completely, since there will be no spin oscillations turn by turn. We now study the error $\delta\gamma/\gamma$ as a function of beam energy for $\hat{P}$=-0.26 and PERR=0.5E-2 (keeping the magnetic field in the idealized storage ring to be 4.0 Tesla) as a function of muon beam energy that straddles the spin resonance. For initial muon collider physics, the interesting beam energies are 45.5 GeV (half the Z mass), 80.3 GeV

Parameter	Value	Parameter	Value
Muon Energy	50 GeV	γ	473.22
spin precession in one turn	3.4667 radians	Magnetic field	4.0 Tesla
radius of ring	41.66666 meters	beam circulation time	0.87327E-06 sec
dilated muon life time	0.10397E-02 sec	turn by turn decay constant	0.8399E-03

TABLE 1. Parameters of an idealized muon storage ring

(W threshold), 175 GeV (top threshold) as well as half the neutral Higgs mass, which could be as low as 55 GeV in some SUSY scenarios. We sample all electrons that have energies greater than half the muon energy. Figure 4

$\hat{P}$	PERR	Number of electrons sampled	$\delta\gamma/\gamma oscillations$	$\delta\gamma/\gamma decay$	χ^2 for NDF=1000
-0.90	0.50E-02	41261	0.14568E-05	0.13227E-02	824.
-0.90	0.10E-01	10315	0.22147E-05	0.20124E-02	936.
-0.90	0.20E-01	2579	0.39999E-05	0.36398E-02	1009.
-0.90	0.30E-01	1146	0.58659E-05	0.53457E-02	1030.
-0.74	0.50E-02	41261	0.17418E-05	0.13019E-02	843.
-0.74	0.10E-01	10315	0.26183E-05	0.19591E-02	954.
-0.74	0.20E-01	2579	0.46981E-05	0.35229E-02	1021.
-0.74	0.30E-01	1146	0.68765E-05	0.51672E-02	1039.
-0.50	0.50E-02	41261	0.25903E-05	0.12813E-02	888.
-0.50	0.10E-01	10315	0.38407E-05	0.19029E-02	973.
-0.50	0.20E-01	2579	0.68338E-05	0.33972E-02	1026.
-0.50	0.30E-01	1146	0.99744E-05	0.49749E-02	1041.
-0.26	0.50E-02	41261	0.51242E-05	0.12688E-02	898.
-0.26	0.10E-01	10315	0.75317E-05	0.18791E-02	1004.
-0.26	0.20E-01	2579	0.13324E-04	0.33447E-02	1053.
-0.26	0.30E-01	1146	0.19380E-04	0.48950E-02	1061.

TABLE 2. Results of fits for $\delta\gamma/\gamma$ as a function of polarization $\hat{P}$ and noise PERR. Also shown is the χ^2 of the fit for 1000 turns.

shows the variation of $\delta\gamma/\gamma$ as a function of muon beam energies that straddle these values. It can be seen that $\delta\gamma/\gamma$ first decreases as one gets close to the resonance and then blows up on the spin resonance. As one approaches the spin resonance, the oscillations slow down. It is nevertheless possible to fit the slowed down oscillations by a rapidly oscillating theoretical function to high accuracy on either side of the resonance. At the resonance, the oscillations die completely, which results in a large value of $\delta\gamma/\gamma$. It may be possible to use this blow-up in $\delta\gamma/\gamma$ to find the spin resonance accurately and (paradoxically) determine γ at resonance accurately. This would depend on the width of the spin resonance, an analysis of which would take us beyond the scope of this paper.

Variation of $\delta\gamma/\gamma$ as a function of beam energy spread

We now calculate the variation of polarization as a function of turn number for an ensemble of muons with initial value of polarization $\hat{P}$ = -0.26 and values of momentum spread $\delta p/p$ varying from 0.02E-2 to 0.00125E-2. This variation is plotted in figure 5. For the larger values of momentum spread, there is a significant degradation of polarization as a function of turn number, due to differential spin precession of the individual beam particles. We note that when the beam energy is at 175 GeV, the spin tune is significantly higher

Machine	Spin tune ν_0	Quadrupoles	RMS Kl_Q meters^{-1}	σ_y meters	$\delta\nu$	$\sigma_{\delta\nu}$
46 GeV LEP	100.47	≈ 600	0.032	0.5E-3	5.7E-6 $\equiv$ 3KeV	6.1E-5 $\equiv$ 30KeV
50 GeV MC	0.5517	70	0.274	0.5E-3	-0.26E-8 $\equiv$ -0.24KeV	1.66E-8 $\equiv$ 1.46KeV

TABLE 3. Predictions for spin tune shift $\delta\nu$ and spread in spin tune shift $\sigma_{\delta\nu}$ caused by quadrupoles for LEP compared to the 50 GeV Muon Collider (MC) ring

and the depolarization is more rapid. Despite this depolarization, there is enough information from the first few hundred turns to extract the excellent value of $\delta\gamma/\gamma$ for 175 GeV beam energy as shown in figure 4.

Figure 6 shows the variation of the fractional energy resolution, $\delta\gamma/\gamma$ as a function of fractional beam energy spread for a muon beam with $\hat{P} = -0.26$, with 41261 electrons sampled. There is little dependence of $\delta\gamma/\gamma$ on the momentum spread. This is due to the fact that the momentum spread is determined from the spin tune and not from the spin oscillation amplitude and the fact that the depolarization is not significant for the first few hundred turns for any of the beam momentum spreads considered here.

Optimization of the electron energy cut

We now vary the cut on electron energy and study the dependence on $\delta\gamma/\gamma$ on the cut. Figure 7 shows the variation of $\delta\gamma/\gamma$ with the cut on individual electron energies for $\hat{P} = -0.26$ for 41261 and 1146 electrons sampled. The fractional error on the average energy of electrons is much smaller than the fractional error on the total energy of electrons. It is possible to measure the average electron energy by counting the number of electrons going into the calorimeter with a scintillator array. However, the precession information is contained increasingly in the number of electrons rather than their average energy as we increase the electron energy cut. Figure 7 shows the variation of $\delta\gamma/\gamma$ calculated from average as well as total electron energy as a function of the electron energy cut. For smaller values of the electron energy cut, the average method produces superior errors than the total energy method. However, with 40,000 electrons or more sampled a total energy method with a cut of 25 GeV or higher seems optimal. It should however be pointed out that the average energy method does not require a model for the rate of decay of muon intensity in the machine, which in practice could be a complicated function of turn number. As such the systematics associated with this would not be present in the average energy method.

EFFECTS DUE TO DEPARTURES FROM THE IDEAL CASE

So far we have considered a planar collider ring with uniform vertical magnetic field and no electric fields. The actual collider ring will depart from the ideal in three respects; a)It will have RF electric fields to keep the muons bunched, b) it will have radial horizontal magnetic fields experienced by partcles in an off-center trajectory at quadrupoles and at vertical correction dipoles, and c) it will have longitudinal magnetic fields due to solenoidal magnets in the interaction region(s). We now consider the effect due to each of these departures from the ideal.

Electric fields

Equation 2 implies that there is no spin precession due to longitudinal electric fields ($\vec{\beta} \times \vec{E} = 0$). RF electric fields are longitudinal, so there will be no precession due to the RF electric fields. At present there are no plans to install electrostatic separators to separate the beams. If and when this happens, one should consider the effect due to the transverse electric fields thus introduced.

Effect of radial magnetic fields

Particles which are off-axis at quadrupoles will experience radial as well as vertical magnetic fields. Even though the net integral of these off-axis fields around the ring is zero, the spin rotation along a horizontal axis followed by spin rotation about a vertical axis (caused by a bend dipole) followed by a reverse rotation in the horizontal direction still produces a net effect since the rotations about the horizontal and vertical axes do not commute. The effects have been analyzed by Assmann and Koutchouk [5] who show that this results in both a net spin tune shift $<\delta\nu>$ as well as a spread in tune $\sigma_{\delta\nu}$.

$$<\delta\nu> = \frac{cot\pi\nu_0}{8\pi}\nu_0^2 \left(n_Q(Kl_Q)^2\sigma_y^2 + n_{CV}\sigma_{\theta CV}^2\right) \quad (14)$$

where $\nu_0 \equiv a\gamma$ is the spin tune of the collider ring, n_Q are the number of quadrupoles with integrated gradient Kl_Q, σ_y is the misalignment spread of the closed orbit at the quadrupoles, n_{CV} is the number of vertical correction dipoles and $\sigma_{\theta CV}$ is the rms beand angle in the vertical correctors. The spread in tune is given by,

$$\sigma_{\delta\nu} = \frac{<\delta\nu>}{cos\pi\nu_0} \quad (15)$$

Table 3 shows the values for $<\delta\nu>$ and $\sigma_{\delta\nu}$ obtained by Assman and Koutchuk [5] for LEP. We compare this with to the current design for the

50 GeV muon collider ring [6]. Including the low beta section, there are 70 quadrupoles with an RMS value of $Kl_Q = 0.27$ m^{-1}. The effects due to correction dipoles may be neglected in both the LEP and the muon collider cases. We assume a beam misalignment of 5mm at the quadrupoles, which is the same value used in the LEP calculation. This is probably being conservative. The tune shift for LEP corresponds to a shift in beam energy calibration of 3.0 KeV. The tune spread for LEP corresponds to a spread in beam energy calibration of 30 KeV. For the muon collider, the tune shift corresponds to a shift in beam energy calibration of -0.24 KeV and a spread of 1.46 KeV, both of which are negligible. The reason for the smallness of this effect for the muon collider is twofold. Since the circumference of the muon collider is smaller than LEP, there are fewer quadrupoles. Secondly, the muon is two hundred times more massive than the electron and has has a spin tune $a\gamma$ that is smaller by the same factor. The spin tune shift depends on the the square of the spin tune. It should be noted that the above formulae are not valid for a fractional spin tune of 0.5.

Solenoidal magnetic fields

The experimental region will in all likelihood contain a solenoidal magnet. This solenoidal field, if uncorrected, will rotate the spin vector of the muons about the beam direction by a constant amount θ_s per turn, which can be derived using equation 2.

$$\theta_s = -\frac{e}{\gamma m_\mu}(1+a)B_s = -(1+a)\frac{B_s l}{B\rho} \quad (16)$$

where B_s is the field due to the solenoid of length l, B is the dipole bending field of the ring of radius ρ. For a solenoid of 1.5 Tesla and length 6 meters, $\theta_s = 3.09$ degrees for the planar storage ring parameters of table 1. It can be shown analytically [8] that this produces a spin tune shift $\delta\nu$ given by

$$\nu + \delta\nu = \frac{1}{\pi}arccos\left(cos(\pi\nu)cos(\frac{\theta_s}{2})\right) \quad (17)$$

yielding a spin tune shift $\delta\nu$ = -1.901E-5, or a fractional spin tune shift of $\delta\nu/\nu$ = -3.45E-5. For a 50 GeV muon beam, this is a shift in energy calibration of -1.72 MeV. In LEP, a similar solenoid will have a much smaller fractional tune shift [8], since the tune is 200 times larger for electrons. It is important to correct the effect due to the solenoids, since this is cumulative turn by turn. At LEP this is done by a series of vertical orbit correctors [9] followed by normal lattice followed by vertical orbit correctors of reverse polarity, which has the effect of rotating the spin by half the amount produced by the solenoid. A similar set of corrections is inserted after the solenoid to complete the correction. This method depends on a non-zero value of $g-2$ and as such will be 200 times less effective for muons than for electrons, for any given magnet strength. The most effective method to correct for the solenoid is to surround it on either side by compensating solenoids of minimal radius large enough to allow the beam to go through.

COSY studies

We have studied the effects due to non-linearities in the aperture using COSY [7], a beam optics program based on differential algebra techniques. Figure 8 shows the polarization as a function of turn number for three different cases of emittance. We have tracked 1000 muons from the interaction point with an initial polarization of 0.25. The three different cases considered are

- emittance = 297πmm-mr and $\delta p/p = 0.0025$E-2
- emittance=85π and $\delta p/p = 0.02$E-2 and
- emittance=40π and $\delta p/p = 0.02$E-2.

We have also studied other cases of $\delta p/p$. The general conclusion is that the main depolarization effect seems to be the non-linearities sampled by the larger emittance and not the beam momentum spread, as can be evidenced by the fact the depolarization effects are smaller when one goes from an emittance of 297π to 85π despite the fact the momentum spread is worse for the latter case. However, we have shown that depolarization effects of the type exhibited here can still be tolerated provided that there is initial polarization of the order of 0.25 which can be maintained for a few hundred turns. Cosy gives the spin tune as a function of the position, angle and energy variables of the beam as

$$\nu = 0.5517 + 0.5915\kappa - 64.61x^2 - 0.1017x'^2 - 69.81y^2 - 0.1088y'^2 - 8.341x\kappa - 0.392 \tag{18}$$

to second order in the variables $\kappa \equiv$ change in kinetic energy/kinetic energy of a 50 GeV/c momentum muon, x, y are deviations from the closed orbit in centimeters and x' and y' are defined as p_x/p and p_y/p where p_x and p_y are the momentum components of a muon of beam momentum $p \equiv 50$ GeV/c. It may be possible to run the muon collider ring with much less sextupole correction for $\delta p/p$=0.0025E-2 case. This would have the effect of reducing the rate of depolarization as a function of time. Studies are under way to see if this is so.

CONCLUSIONS

We have demonstrated that it is feasible to measure the energy of a 50 GeV muon collider to a few parts per million using the $g - 2$ spin precession technique, provided it is feasible to maintain a muon polarization of the order of $\hat{P}=0.25$ in the ring for a thousand turns. In order to explore the Higgs resonance, it is necessary to measure the bunch by bunch variation in energy to a few parts per million. We have demonstrated that the $g-2$ technique is capable of doing so. It is still possible to tolerate a spin tune shift in the overall energy scale of a few percent, which will act only as a systematic error on the Higgs mass and width.

We would also like to note in passing that polarization information from a calorimeter of the type proposed here can be used in conjunction with a neutrino detector placed along the line of the neutrinos produced in association with the electrons to estimate the variation in the energy spectrum of the muon neutrinos and electron antineutrinos in the beam. Such information can be a valuable tool in untangling various possible neutrino oscillation scenarios.

The authors would like to thank Martin Berz, Weishi Wan and Carol Johnstone for help with COSY calculations and would like to acknowledge useful conversations with Alain Blondel and Robert Rossmanith.

REFERENCES

1. V.Bargmann, L. Michel and V.L. Telegdi, Phys. Rev. Lett. 2, 10 (1958) 435.
2. Particle Data Group, R.M.Barnett et al., Physical Review D54,1 (1996).
3. G.Barr, T.K. Gaisser and T. Stanev, Phys. ReV. D 39(1989) 3532.
4. MINUIT is a CERNLIB program written by F. James.
5. R.Assmann and J.P.Koutchouk, "Spin tune shifts due to optics imperfections", Cern SL/94-13. See also L. Arnaudon et al, "Accurate determination of the LEP beam energy by resonant depolarization ", Z.Phys.C66: 45-62,1995.
6. Carol Johnstone, Private Communication.
7. COSY INFINITY Version 7 User's guide and reference manual, M.Berz, Michigan State University Preprint MSUCL-977.
8. J.P.Kouthcouk, "Spin tune shift due to solenoids", CERN SL-note/93-26 (AP) (1993).
9. R.Rossmanith, LEP Note 525 (1985). A. Blondel, LEP note 629, (1990).

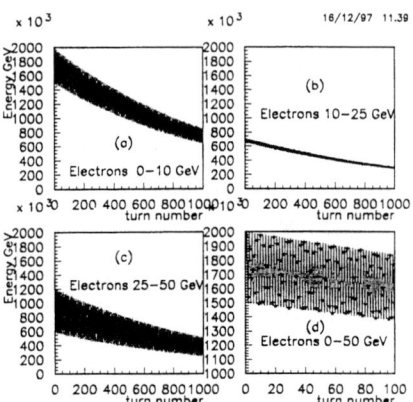

FIGURE 1. (a)Total energy observed as a function of turn number for $\hat{P} = -1.0$ with individual electron energies in the range 0-10 GeV for 100,000 muon decays. (b) Electron energies in the range 10-25 GeV (c) 25-50 GeV (d) All electrons included. Superimposed is a functional form defined by equation 8

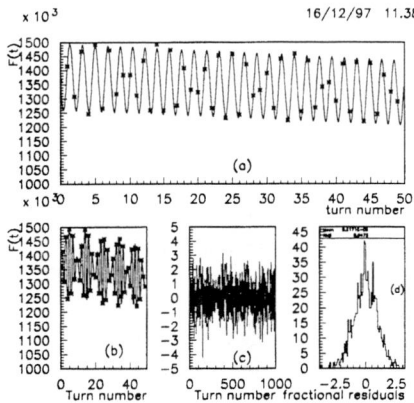

FIGURE 2. (a)Energy detected in the calorimeter during the first 50 turns in a 50 GeV muon storage ring (points). An average value of $\hat{P}=-0.26$ is assumed and a fractional fluctuation of 0.5E-2 per point. The curve is the result of a MINUIT fit to the functional form in equation 10. (b) The same fit, with the function being plotted only at integer turn values. A beat is evident. (c) Pulls as a function of turn number (d)Histogram of pulls.

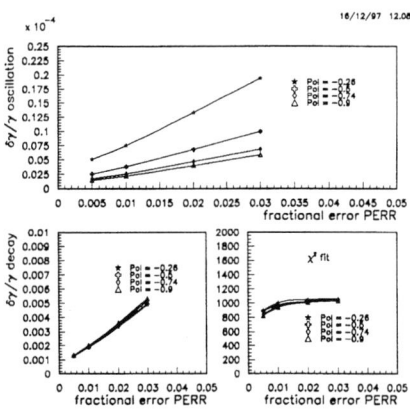

FIGURE 3. (a)Fractional error in $\delta\gamma/\gamma$ obtained from the oscillations as a function of polarization $\hat{P}$ and the fractional error in the measurements PERR (b) Fractional error in $\delta\gamma/\gamma$ obtained from the decay term as a function of polarization $\hat{P}$ and the fractional error in the measurements PERR (c) The total χ^2 of the fits for 1000 degrees of freedom

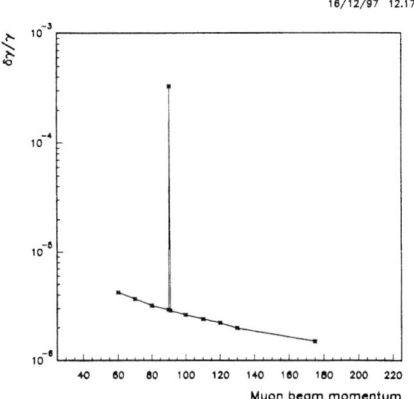

FIGURE 4. Fractional error in $\delta\gamma/\gamma$ obtained from the oscillations as a function of muon beam momentum

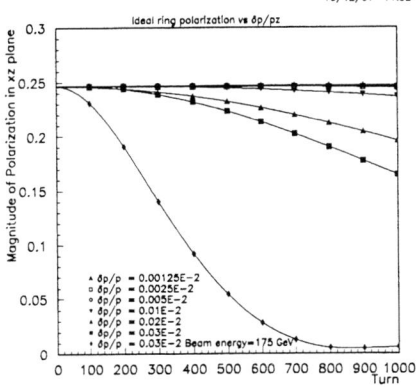

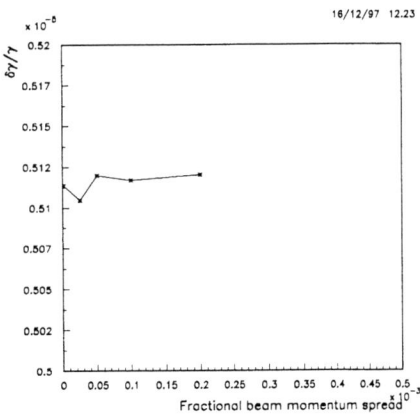

FIGURE 5. Variation of polarization as a function of turn number for 50 GeV muons with initial $\hat{P} = -0.26$ and various values of $\delta p/p$ in an ideal collider ring. The bottom curve is for 175 GeV muons and shows a more rapid depolarization due to the higher spin tune.

FIGURE 6. $\delta\gamma/\gamma$ versus fractional beam energy spread for 50 GeV muons with PERR=.5E-2 and $\hat{P} = -0.26$

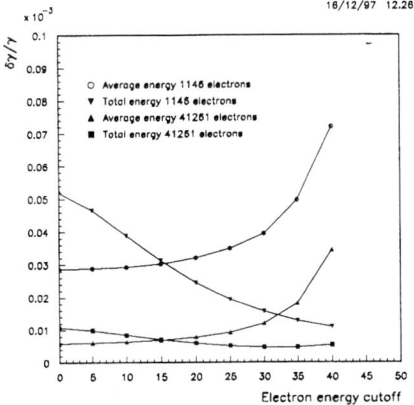

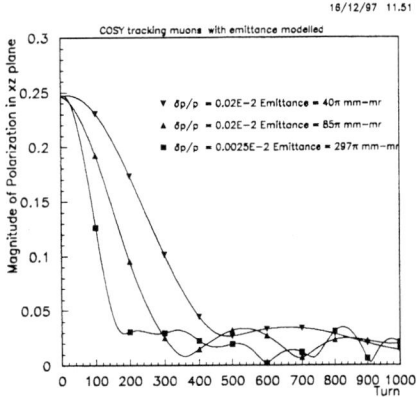

FIGURE 7. The variation of $\delta\gamma/\gamma$ as a function of the electron energy cut for 41261 and 1146 electrons $\hat{P} = -0.26$. We fit the total energy in the calorimeter as well as the average energy per electron

FIGURE 8. Polarization as a function of turn for 1000 muons in COSY. Three separate cases of beam emittance and beam energy spread are considered.

Scraping Beam Halo in $\mu^+\mu^-$ Colliders

A. Drozhdin, N. Mokhov, C. Johnstone and W. Wan

Fermi National Accelerator Laboratory, Batavia, IL 60510[1]

A. Garren

Lawrence Berkeley National Laboratory, Berkeley, CA 94720

Abstract. Beam halo scraping schemes have been explored in the 50×50 GeV and 2×2 TeV $\mu^+\mu^-$ colliders using both absorbers and electrostatic deflectors. Utility sections have been specially designed into the rings for scraping. Results of realistic STRUCT-MARS Monte-Carlo simulations show that for the low-energy machine a scheme with a 5 m long steel absorber suppresses losses in the interaction region by three orders of magnitude. The same scraping efficiency at 2 TeV is achieved only by complete extraction of beam halo from the machine. The effect of beam-induced power dissipation in the collider superconducting magnets and detector backgrounds is shown both for the first few turns after injection and for the rest of the cycle.

INTRODUCTION

High background rates in the detectors are one of the most serious problems on the road towards a high-luminosity $\mu^+\mu^-$ collider [1,2]. It was shown at an early stage [3] that detector backgrounds originating from beam halo can exceed those from decays in the vicinity of the interaction point (IP). Only with a dedicated beam cleaning system far enough from the IP can one mitigate this problem [4]. Muons injected with large momentum errors or betatron oscillations will be lost within the first few turns. After that, with active scraping, the beam halo generated through beam-gas scattering, resonances and beam-beam interactions at the IP reaches equilibrium and beam losses remain constant throughout the rest of the cycle. Two beam cleaning schemes are studied in this paper: beam halo extraction with an electrostatic deflector and standard collimation (see Fig. 1,2). The resulting effect on the superconducting (SC) magnets and detector backgrounds is described in detail for a 50×50 GeV $\mu^+\mu^-$ collider, plus some results and conclusions for the 2×2 TeV case (see for more details [4]).

---

[1] Work supported by the Universities Research Association, Inc., under contract DE-AC02-76CH00300 with the U. S. Department of Energy.

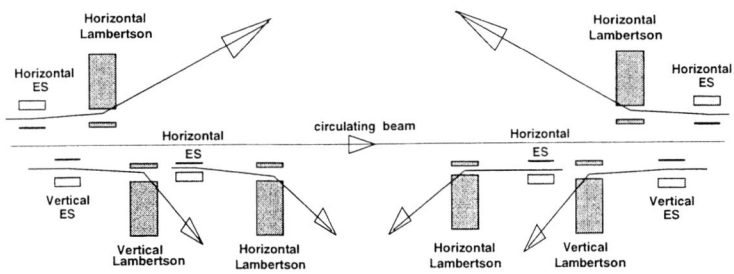

FIGURE 1. Schematic view of a $\mu^+\mu^-$ collider beam halo extraction.

BEAM HALO EXTRACTION

A 3-m long electrostatic deflector (Fig. 1) separates muons with amplitudes larger than 3σ and deflects them into a 3-m long Lambertson magnet, which extracts these downwards through a deflection of 17 mrad. A vertical septum magnet is used in the vertical scraping section instead of the Lambertson to keep the direction of extracted beam down. The shaving process lasts for the first few turns. To achieve practical distances and design apertures for the separator/Lambertson combinations, β-functions must reach a kilometer in the 2-TeV case, but only 100 m at 50 GeV. The complete system consists of a vertical scraping section and two horizontal ones for positive and negative momentum scraping (the design is symmetric about the center, so scraping is identical for both μ^+ and μ^-). Always, the halo is extracted down into the ground downstream of the utility section (US).

Three possible layouts were investigated. The first consisted of two horizontal electrostatic deflectors (not shown in Fig. 1) separated by 180° in phase (the second deflector is in the shadow of the first) and using the same Lambertson magnet for extraction. The horizontal deflectors are followed by a vertical one which uses a septum magnet. After vertical scraping, a second horizontal scraping system is inserted, but on the opposite side of the US. The first horizontal deflection scrapes off-momentum muons with momenta greater than the central momentum. The sec-

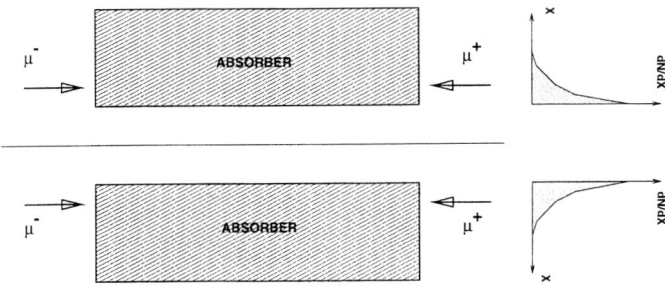

FIGURE 2. Scraping muon beam halo with a 5-m steel absorber.

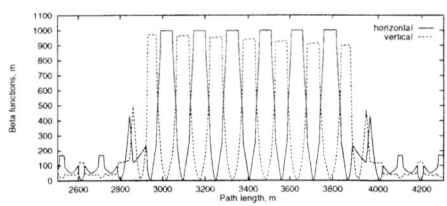 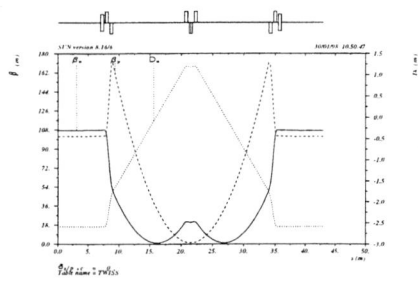

FIGURE 3. Lattice of the $\mu^+\mu^-$collider utility section: (a) 2×2 TeV with halo extraction; (b) 50×50 GeV with halo collimation.

ond scrapes muons with lower-than-average momentum. The entire scraping layout has reflection symmetry about the center to make scraping identical for μ^+ and μ^-. This scheme is designed and found to be optimal for a 2×2 TeV $\mu^+\mu^-$collider [4]. Using only one horizontal electrostatic deflector with the Lambertson magnet (Fig. 1) instead of two, gave a calculated efficiency which was several times lower. A final combination consisted of electrostatic deflectors on both sides of the ring in a single high-β region (Fig. 3(a)). Its efficiency was somewhere between the first two layouts. Its advantage, however, is that it is much more compact, occupying only three large high-β regions. Therefore, it is best suited to the compact 50×50 GeV $\mu^+\mu^-$collider.

Realistic Monte-Carlo simulation of beam halo effects is done in three stages. Primary muon interactions with electrostatic deflector wires (or collimator in the other approach to scraping which is described in a later section) are simulated using the MARS13(97) code [5]. Multi-turn tracking of muons scattered out of the deflector (or collimator) in the collider lattice and analysis of their loss in the collider elements is done using the STRUCT code [6]. For the third stage, a full hadronic and electromagnetic shower simulation in collider and detector elements is performed after returning to the MARS13(97) code. A 8.5σ aperture is assumed in the arcs (85π and 50π mm-mr normalized rms emittance at 50-GeV and 2-TeV,

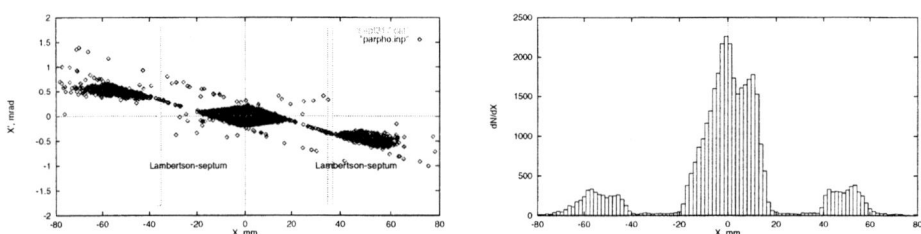

FIGURE 4. 50 GeV/c muon halo distributions in horizontal plane at the Lambertson magnet entrance in the symmetric scheme.

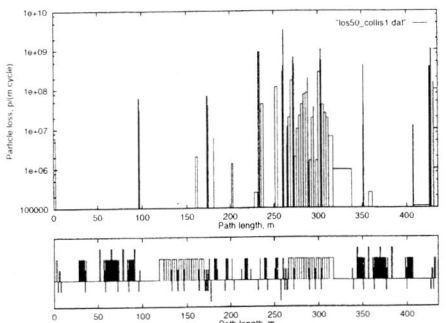
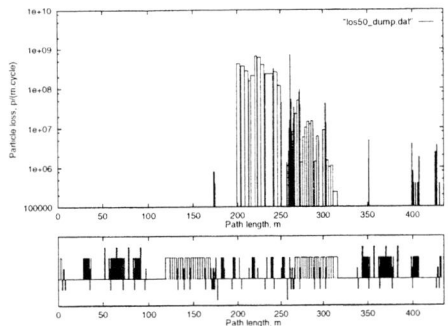

FIGURE 5. 50 GeV muon beam loss distributions at collisions with halo extraction (left) and halo collimation with the internal absorber (right). 1% of the beam intensity is intercepted.

respectively) and only 5σ in the interaction region (IR). The aperture is enlarged to $>8.5\sigma$ in the scraping section. To protect the IR magnets against irradiation, the aperture of tungsten insertions between the magnets is 4σ as in [7,8]. Only the halo muons with betatron amplitudes of 2.5 to 4σ and with $\sigma_{\Delta p/p}=0.0004$ are used in these simulations. Deflector septa (wires) are placed at 12.65 mm from the closed orbit to shave both halo and large-amplitude muons and also muons with positive and negative momentum deviations. In the distance from one high-β region to the next, halo muons are sufficiently separated from the circulating beam (Fig. 4) to be cleanly extracted by a Lambertson magnet. Extracting large-amplitude and off-momentum muons decreases dramatically beam loss in the IR. Calculations show that 83% of halo is extracted from the collider over the first few turns. About 30% of beam halo pass through the electrostatic deflector wires. These muons loose on average 0.6% of their energy and are lost at the limiting apertures along the collider, mostly in the first 70 m after the US (see Fig. 5). About 4% of halo muons just get an angular (amplitude) kick without noticeable momentum loss and are lost in the IR resulting in detector background. Assuming the interception of 1% of the circulating beam in the beam cleaning process, 8×10^8 muons are lost in the final focus quadrupoles (just a few meters from the IP) over the first few turns after injection. After that, the scraping system becomes very efficient as beam halos are regenerated by beam-gas and beam-beam scattering, ground motion and resonances. The step size (particle betatron amplitude rise during one turn) at this process is of the order of a few μm. Because of that, disturbed muons will interact first with the electrostatic deflector wires. According to the simulations, 60% of regenerated halo is extracted from the collider, with only 4.6% of the scraped muons passing through the material of the low-β quadrupoles.

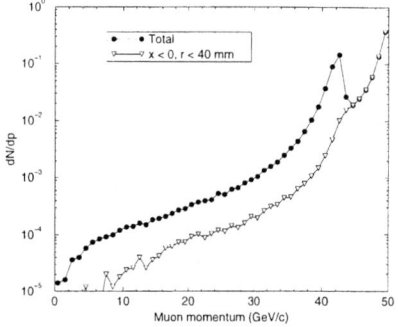 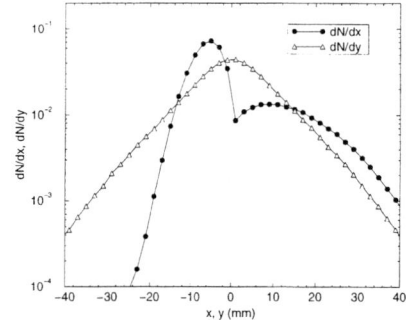

FIGURE 6. Muon leakage from a 5-m steel half-absorber for 50 GeV/c muon beam: (a) Momentum spectrum; (b) Space distributions.

BEAM HALO COLLIMATION

An alternative scheme is to collimate the halo using a solid absorber (Fig. 2). Our studies [4] showed that no absorber, ordinary or magnetized, will suffice for beam cleaning at 2 TeV; in fact the disturbed muons are often lost in the IR. At 50 GeV, on the other hand, collimating muon halos with a 5-m long steel absorber (Fig. 2) in a simple compact US (Fig. 3(b)) does an excellent job. As Fig. 6 shows, muons loose a significant fraction of their energy in such an absorber (8% on average) and have broad angular and spatial distributions. Therefore, almost all of these muons are lost in the first 50-100 m downstream of the absorber as shown in Fig. 5, with only 0.07% of the scraped muons reaching the low-β quadrupoles in the IR. This is 60 times better than with the halo extraction scheme. At the same time, the peak beam loss in SC magnets downstream of the US is six times higher compared

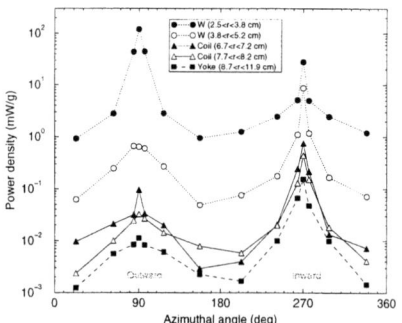 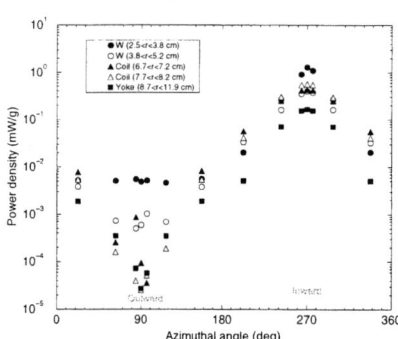

FIGURE 7. Azimuthal distributions of power density in the ring SC dipoles for 50 GeV/c muon beam: (a) Beam decays; (b) Scraping with the absorber.

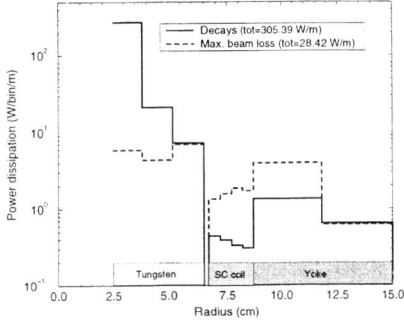 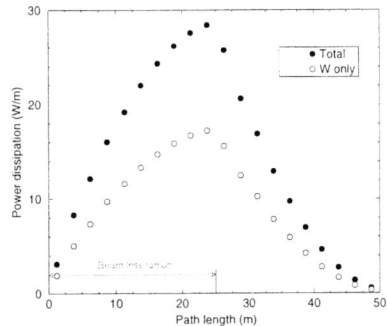

FIGURE 8. Power dissipation in the ring SC dipoles for 50 GeV/c muon beam: (a) Radial distribution for beam decays and beam halo; (b) Longitudinal distribution for beam halo.

to the halo extraction (Fig. 5). Without halo scraping, a full 1% of the beam is lost in the IR, i.e., the collimation system reduces beam loss in the IR by almost a factor of 1500. One percent of the steady-state beam loss on the collimators results in a total of 1.4×10^7 muons lost in the low-β quadrupoles during the cycle. Halo collimation has a further advantage in that the lattice required (Fig. 3(b)) is shorter and simpler compared to Fig. 3(a). It could, in fact, be placed in the matching sections on either side of the IP leaving the US for injection and extraction and reducing the overall accelerator circumference.

EFFECT ON SC MAGNETS AND DETECTOR

As shown in [1,7,8], to protect the SC coils of a 2×2 TeV $\mu^+\mu^-$ collider from the excessive heat load due to unavoidable muon decays, a tungsten absorber (liner) of up to 6 cm thick is required inside the SC magnets. Our new calculations for a 50 GeV machine with 3.3×10^{12} μ/bunch at 15 Hz, show that such a liner should be 3 to 4 cm thick. The power density distribution in the SC magnet components is strongly non-uniform azimuthally (Fig. 7), so an alternative design would be with cold or warm iron and SC coils completely separated on the mid-plane [1]. If one takes the liner approach, then a 3-4 cm thick tungsten absorber protects the SC magnets at 50 GeV even where the beam loss peaks, which is just downstream of the US. Opposite to the decay-induced heat load, the power density from scraping peaks on the one, inner, side of the magnet aperture relative to the ring center (Fig. 7), and more power is dissipated deeper in the magnet body (Fig. 8(a)). As Fig. 8(b) shows, at 50 GeV halo muons are absorbed in the lattice elements within about 30 m after the beam loss region. With a collimator-based scraping system, this occurs within a 100 m region downstream the US. Another words, with such a system on the opposite side of the ring from the IR, halo-induced detector backgrounds are not an issue in the 50×50 GeV $\mu^+\mu^-$ collider.

REFERENCES

1. $\mu^+\mu^-$ Collider: A Feasibility Study, The $\mu^+\mu^-$Collider Collaboration, BNL–52503; Fermilab–Conf–96/092; LBNL–38946, July 1996.
2. Mokhov, N. V., Nucl. Phys. B, **51A**, 210-218 (1996).
3. Foster, G. W. and Mokhov, N. V., in *AIP Conference Proceedings* **352**, *Sausalito, November 1994*, pp. 178–190; also Fermilab–Conf–95/037 (1995).
4. Drozhdin, A. I., Johnstone, C. C., and Mokhov, N. V., '2×2 TeV $\mu^+\mu^-$ Collider Beam Collimation System', Workshop on Muon Colliders, Orcas Island, WA, May 1997.
5. Mokhov, N. V., Fermilab–FN–628 (1995).
6. Baishev, I. S., Drozhdin, A. I., and Mokhov, N. V., 'STRUCT Program Reference Manual', SSC-MAN-0034 (1994).
7. Johnstone, C. C. and Mokhov, N. V., *Snowmass 96 Proceedings*; Fermilab–Conf–96/366 (1996).
8. Johnstone, C. C. and Mokhov, N. V., presented at the IEEE Particle Accelerator Conference, Vancouver, May 1997.

Towards Ultimate Luminosity Polarized Muon Collider (problems and prospects)

A.Skrinsky

Budker Institute of Nuclear Physics
Novosibirsk, Russia

Abstract. The aim of this paper is to present the "ultimate options" for crucial stages of muon collider, as the subject for comparison of different "practical options".

INTRODUCTION

In all this article considerations I shall have in mind the general scheme of muon collider complex presented in Fig.1, considered initially in /1,2/. The updated literature on the subject is presented in many publications of Muon Collider Collaboration /3/.

Let us first remind the prime importance of the final stage of ionization cooling for reaching the ultimate muon collider luminosity (for given number of muons).

If at this stage, the normalized transversal emittance $\varepsilon_{neqtran}$ and longitudinal emittance $\varepsilon_{neqlong}$ were reached, and collider magnetic field H_{coll} provides for muons N_{life} effective turns before they decay, the ultimate luminosity would be:

$$L_{\mu\mu \, max} = \frac{N_\mu^2}{4\pi} \cdot \frac{\gamma_{coll}^2}{\varepsilon_{neqtran} \cdot \varepsilon_{neqlong}} \cdot \frac{\Delta E_{max}}{E} \cdot N_{life} \cdot f_0 \quad .$$

The luminosity of "optimal" collider ($E_\mu = 2$ TeV + 2 TeV, $N_\mu = 1*10^{12}$, $H_{coll} = 10$ T, $f_0 = 15$ s^{-1}, $\frac{\Delta E_{max}}{E} = 0.3\%$, fraction of the sum of cooling decrements transferred to the longitudinal direction $\kappa_{long} = 1/3$, space charge problems at cooling stage are neglected) with the equilibrium emittances reachable as ultimate limit in cooling process (see below), as a function of cooling kinetic energy, is shown at Fig. 2.

But if calculate the beta-value at collision, assumed as always here to be equal to the muon bunch length, we would get 5 microns!

CP441, *Physics Potential and Development of mu-mu Colliders:* Fourth International Conference
edited by David B. Cline
© 1998 The American Institute of Physics 1-56396-723-5/98/$15.00

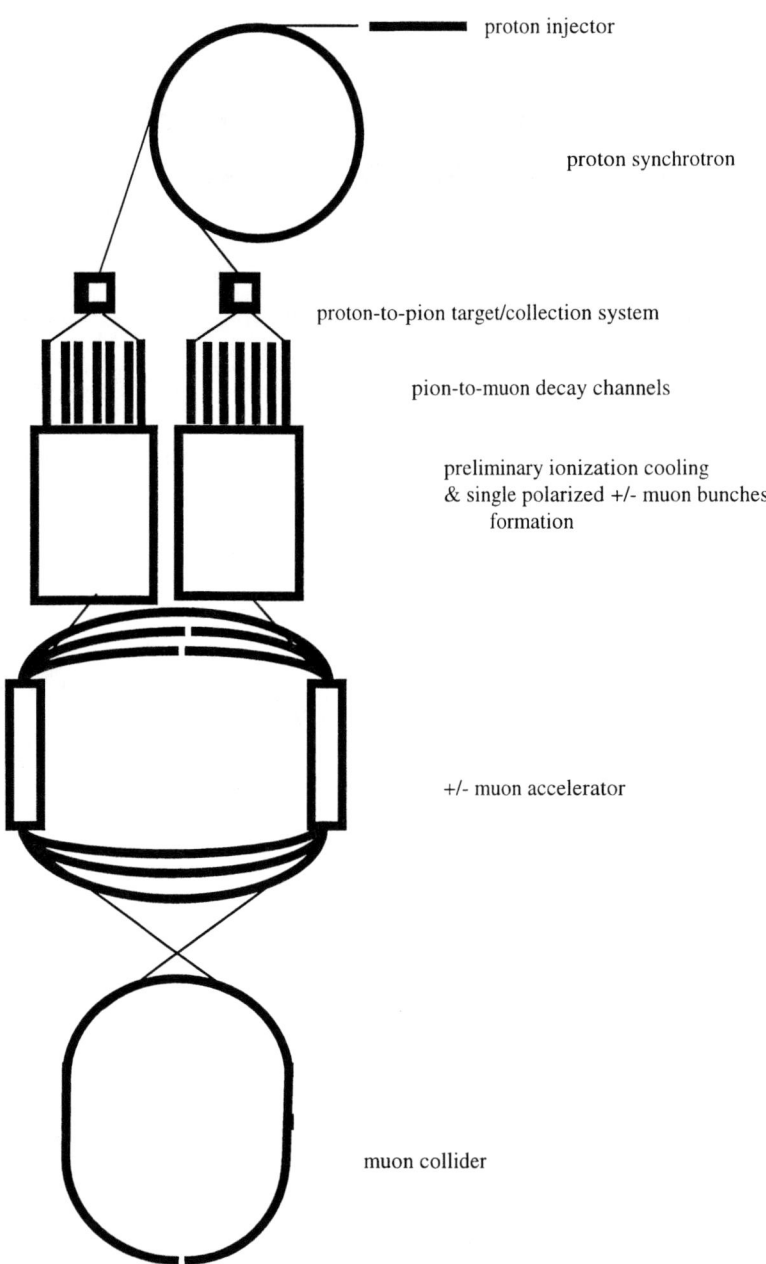

Figure 1. Schematic layout of muon collider complex.

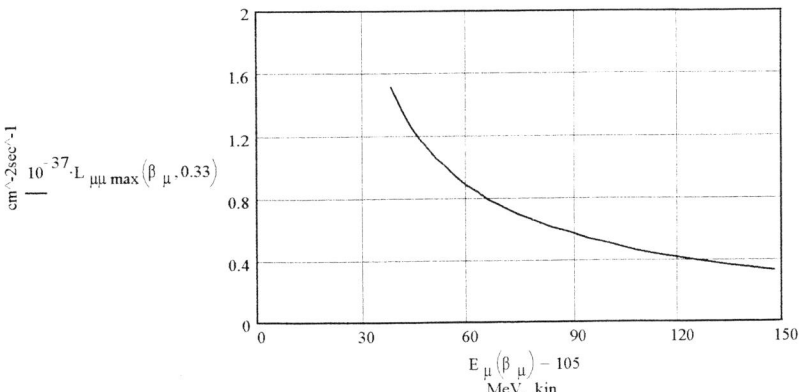

Figure 2. "Maximum" collider luminosity (without "emittance gymnastics" and with no limitation for final focusing and energy spread).

Such bunch length and beta-function look impractical. If we limit these values as $\sigma_{longcoll}$, the following formula for "<u>practical</u>" maximum luminosity should be valid:

$$L_{\mu\mu} = \frac{N_\mu^2}{4\pi} \cdot \frac{\gamma_{coll}^{\frac{3}{2}}}{\varepsilon_{neq6}^{\frac{1}{2}} \sigma_{longcoll}^{\frac{1}{2}}} \cdot \left(\frac{\Delta E_{max}}{E}\right)^{\frac{1}{2}} N_{life} f_0 .$$

For reasonable limit of $\sigma_{longcoll} = \beta_{coll} = 300$ microns, the luminosity dependence on cooling energy is shown in Fig. 3.

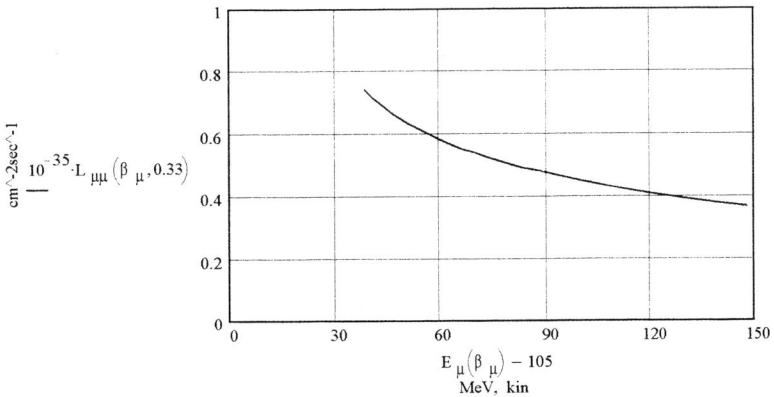

Figure 3. "Practically achievable" collider luminosity.

Hence, "under practical circumstances", the equilibrium normalized 6-dimensional emittance ε_{neq6} enters the maximum luminosity directly. And the aim of final ionization cooling is to reach minimum of

$$\varepsilon_{neq6} = (\varepsilon_{neqtran})^2 \varepsilon_{neqlong} \, .$$

In a special section we discuss the way to rearrange the structure of this emittance properly - to reach the highest luminosity, necessary monochromaticity and/or polarization degree.

FINAL COOLING STAGE

Following /4,5/, the 6-dimensional decrement (in cm^{-1}) for muons of velocity $\beta_\mu c$ passing the ionization substance, can be written down as

$$\delta_\Sigma = \frac{2 P_{fr}}{M_\mu c^3} \cdot \frac{\sqrt{1-\beta_\mu^2}\left(1+\beta_\mu^2\right)}{2\beta_\mu^3} + \frac{1}{M_\mu c^3} \cdot \frac{dP_{fr}}{d\beta_\mu} \cdot \frac{\left(1-\beta_\mu^2\right)^{\frac{3}{2}}}{\beta_\mu^2},$$

where P_{fr} is the power of ionization losses:

$$P_{fr} = 4\pi \cdot m_e c^3 r_e^2 N_e \beta_\mu^{-1} \cdot \left(\ln \frac{2 m_e c^2 \beta_\mu^2 \gamma_\mu^2}{I\left(1+2\gamma_\mu \frac{m_e}{M_\mu}\right)} - \beta_\mu^2 \right),$$

N_e - electron density of the substance, I - the effective ionization potential of the substance atoms.

The "cooling length" for 6-emittance in lithium is presented at Fig. 4:

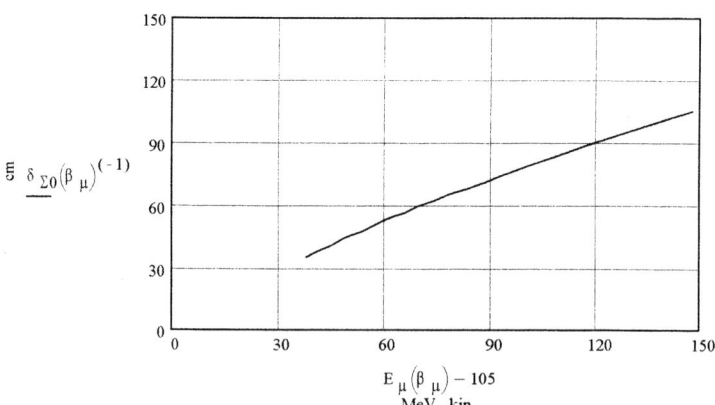

Figure 4. Cooling length for 6-emittance.

Transversal cooling shrinks the muon angular spread down to equilibrium one, acquired due to multiple scattering at one transversal emittance cooling length (κ_{long} - fraction of full emittance transferred to the longitudinal degree of freedom):

$$\vartheta^2_{x,y} = 4\pi \cdot r^2_\mu N_e (Z+1) L_c \frac{1}{\gamma^2_\mu \beta^4_\mu} \cdot \left(\frac{1-\kappa_{long}}{2} \cdot \delta_\Sigma \right)^{-1}.$$

The resulting equilibrium transversal emittance (normalized) depends on the focusing (on transversal beta-function value β_{tran}) inside the cooling matter:

$$\varepsilon_{neqtran} = 4\pi \cdot r^2_\mu N_e (Z+1) L_c \frac{1}{\gamma_\mu \beta^3_\mu} \cdot \left(\frac{1-\kappa_{long}}{2} \cdot \delta_\Sigma \right)^{-1} \cdot \beta_{tran}.$$

For the longitudinal, direction situation is more complex. "Naturally", when the longitudinal damping is caused only by direct dependence of energy losses on muon energy, at kinetic energies above 200 MeV the longitudinal cooling length is 5 times longer than the transversal one. This "natural" length of longitudinal cooling is presented at Fig. 5.

And for most attractive cooling energy around 100 MeV and lower, the ionization cooling transforms even into anti-damping. As a result, in this energy range the longitudinal emittance (the energy spread) grows exponentially reasonably fast, with the increment length shown at Fig. 6.

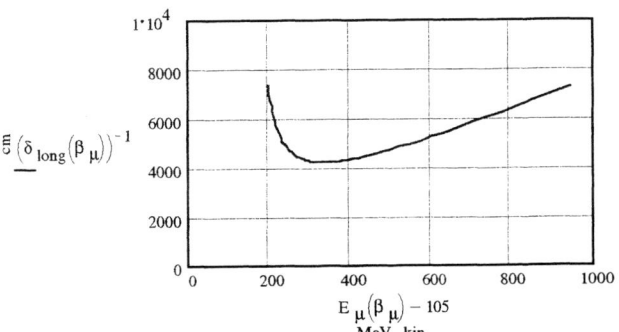

Figure 5. "Pure" longitudinal cooling length at high energies.

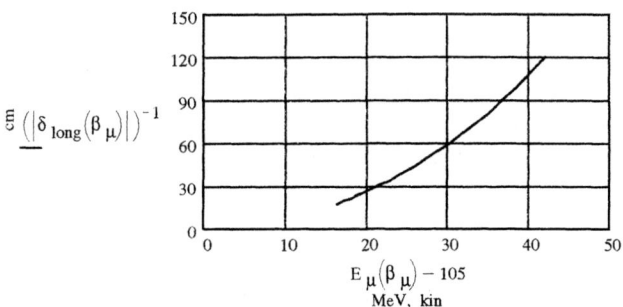

Figure 6. "Pure" longitudinal anti-damping length at low energies.

Hence, it is obligatory to <u>redistribute</u> the decrements sum in favour of longitudinal degree of freedom. If we arrange the transfer of κ_{long} fraction of δ_Σ to this degree, the equilibrium relative energy spread shall be

$$\Delta_E^2 = \frac{2\pi \cdot r_\mu^2 N_e \left(2 - \beta_\mu^2\right)}{\kappa_{long} \delta_\Sigma}.$$

The dependence of this spread on cooling energy for $\kappa_{long} = 1/3$ is presented at Fig. 7.

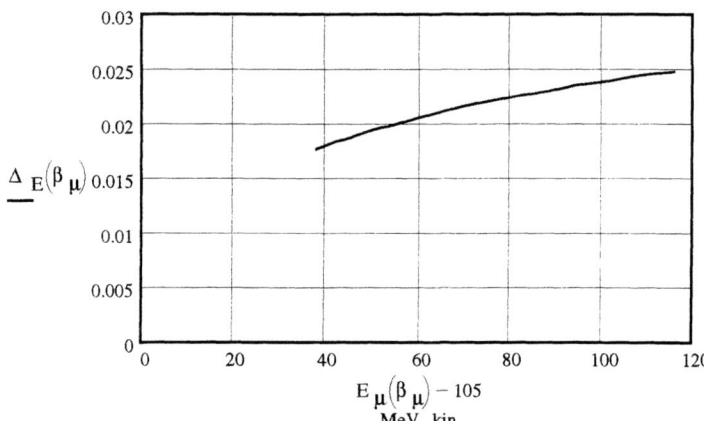

Figure 7. Equilibrium relative energy spread.

The resulting equilibrium longitudinal emittance is proportional to the effective longitudinal beta-value β_{long} (and this value should be arranged as small as possible):

$$\varepsilon_{neqlong} = \Delta_E^2 \cdot \beta_\mu \gamma_\mu \cdot \beta_{long} .$$

The 6-dimensional equilibrium emittance ε_{neq6} shall be

$$\varepsilon_{neq6} = \varepsilon_{neqtran}^2 \varepsilon_{neqlong} .$$

Let us underline: in order to reach minimal 6-emittance one needs to arrange proper cooling of all the three degrees of freedom <u>simultaneously</u>.

In my opinion, the option of final cooling which is non-self-contradictory and gives the ultimately low 6-emittance, is the following /5/, presented at Fig. 8.

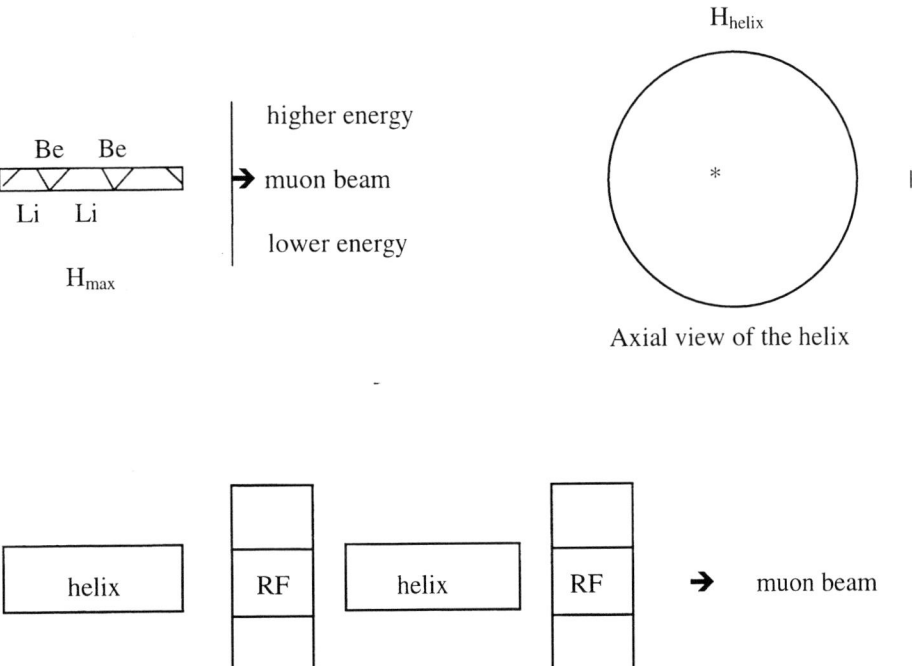

Figure 8. The schematics for simultaneous transversal and longitudinal cooling.

The lithium cylinder of small radius a_{Li} carries high pulsed current (of the same direction as muon current, thus focusing them); the reverse current goes backward along the insulated conductive cover. The high current liquid lithium lenses technology was developed at Novosibirsk INP for antiproton collection systems /6,7/. Effective transversal beta-value in such a cylinder for muons of full energy E_μ travelling along its axis is

$$\beta_{Li} = \sqrt{\frac{\beta_\mu E_\mu a_{Li}}{eH_{max}}},$$

where H_{max} is magnetic field on the lithium surface. The cylinder radius is determined by the requirement: its acceptance should be high enough to prevent muon losses. At the final stage, it should be few times bigger than equilibrium transversal emittance of muon beam under cooling. In this case, the optimal fraction of summary decrement transfer to longitudinal direction equals to 0.25, giving the smallest equilibrium normalized 6-emittance (the resulting luminosity dependence is presented at Fig.9).

The necessary transfer of summary cooling decrement to the longitudinal degree of freedom is arranged by converting the cylinder into helix (the local helical curvature of radius about 10 cm is produced by corresponding external magnetic field H_{helix}), and by positive electron density gradient. The latter aim can be reached by combining in the cooling cylinder of lithium and beryllium - more beryllium at the outer side of the helix - with teeth-like boundary between lithium and beryllium. The helical parts of the final cooling device, of alternating direction of curvature, are separated by RF accelerating sections - to compensate the average ionization energy losses; in these RF sections, a proper and strong enough focusing should be arranged to match emittances in the neighbouring cooling sections.

For such a device (in case H_{max}=10 Tesla; $2a_{Li}$ is of few mm, providing transverse acceptance 3 times bigger than the equilibrium one; H_{helix}=5 Tesla; one quarter of summary decrement is transferred to longitudinal direction; the RF wave length is 10 cm, RF average amplitude per cm is 3 times higher than average ionization losses) the equilibrium normalized emittances, as a function of cooling kinetic energy, are presented at Figures 10 - 12.

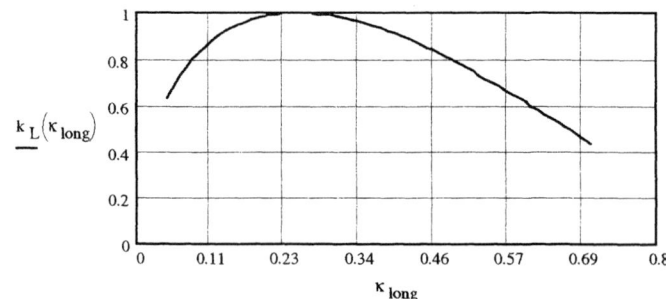

Figure 9. The loss factor in luminosity, in dependence on fraction of 6-decrement transferred to longitudinal direction.

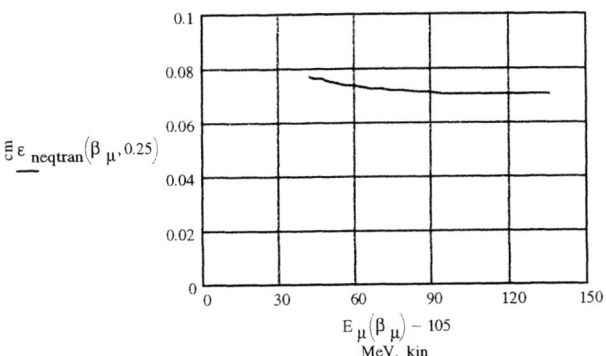

Figure 10. Equilibrium normalized transversal emittance.

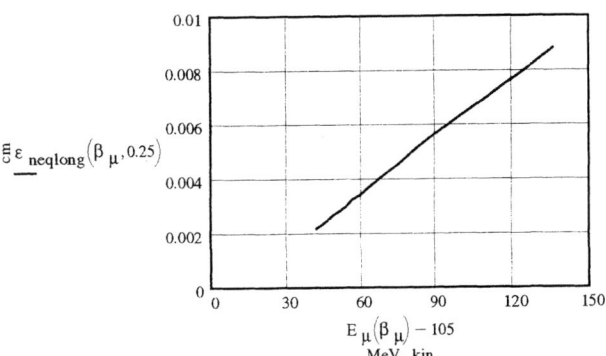

Figure 11. Equilibrium normalized longitudinal emittance.

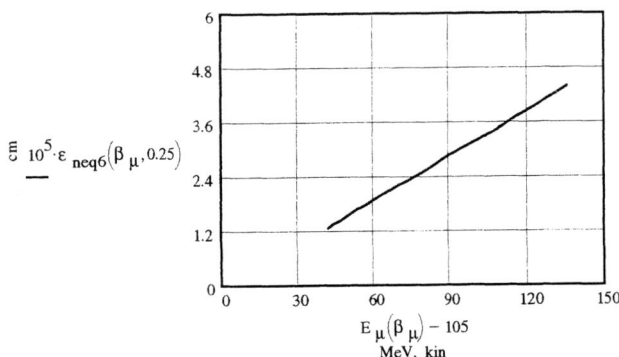

Figure 12. Equilibrium normalized 6-emittance.

In all the above considerations, we did neglect all the space charge effects. At the planned muon bunch intensities $1*10^{12}$ and higher, these effects can limit the achievable 6-emittance and should be a subject for careful design.

Space charge limitation (transversal) at cooling stage, when self-defocusing cancels external focusing, can be evaluated as:

$$N_{b\mu cool} = \frac{2}{r_\mu} \cdot \beta_\mu \gamma_\mu^2 \cdot \frac{\varepsilon_{neqtran} \sigma_{long}}{\beta_{loc}},$$

where β_{loc} is the local beta-function.

The use of lithium as a cooling substance can ease these limitations significantly: high conductivity of lithium screens the muon bunch electromagnetic field.

EMITTANCE GYMNASTICS

As we discussed earlier, to "extract" full usefulness of low 6-emittance reached at the final cooling stage, we need to rearrange this emittance according to the muon bunch length in the collider acceptable, taking into account muon-muon monochromaticity desired (or the acceptable energy spread in the collider) and the necessity to keep polarization degree at an acceptable level. For this emittances gymnastic purpose, we need to use combination of dispersive and septum elements, RF accelerating/decelerating structures, delay lines (but not ionization components, which damage 6-density!). Such a transformation is worth to arrange at any convenient energy. The schematic layout of such a device for maximising collider luminosity is presented at Figure 13. The "monochromatic" collider option (as in the case of the "low energy Higgs Factory") could require rearrangement in opposite direction.

In-coming muon bunch: "wide, but short"

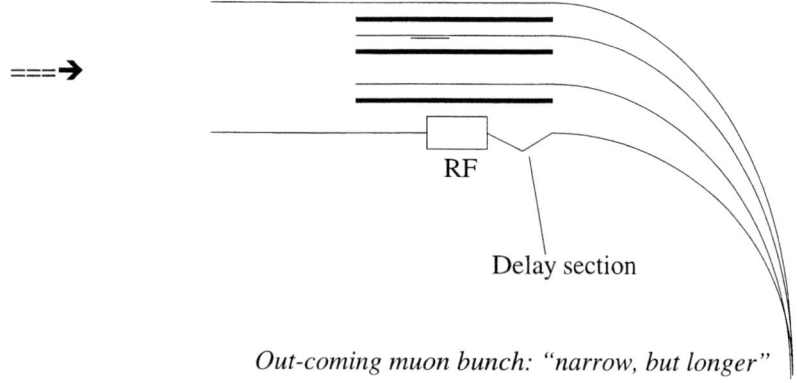

Out-coming muon bunch: "narrow, but longer"

Figure 13. The transformation of the bunch with big transversal emittance into the bunch with smaller one but with proportionally higher momentum spread and longitudinal emittance (an option).

POLARIZED MUONS

High degree of polarization is very important to extract full physics information out of muon collider experiments. Hence, first of all, it is worth to find the way to produce highly polarized intense muon beams /5,8/. We assume, positive and negative pions are collected being generated by different proton bunches.

The sketch of a possible option for protons-to-pions multi-channel conversion system, followed by multi-channel pions-to-muons decay channels, is presented at Figure 14. Maybe, it is reasonable to arrange a sectioned target (using additional channels). This could be especially useful at high proton energy - around 100 GeV.

In each pion collecting straight channel, with the use for the initial matching focusing of one-dimensional "thin surface-current-carrying lenses" doublets, it is necessary to direct pions of wide spectrum in many independent channels. In each channel, in θ-direction beam transversal emittance is big, but in ϕ-direction is quite small. These beams can easily be transported straight away of the target space, and the following channel gymnastics will happen in reasonably free room.

The next step is to arrange in each channel the energy dispersion in this smaller emittance direction, and then to direct each of the +/- 5% momentum spread pion beams in additional separate strong focusing decay channels.

Such narrow momentum spread pion beams (with very small emittance in one direction), upon passing about 2 decay lengths (proportional to the pion energy in each channel, around $15\beta_\pi\gamma_\mu$ meters), generate muon beams of momentum spread about +/- 30% (Fig.15), with strong correlation of muon spin direction and its momentum.

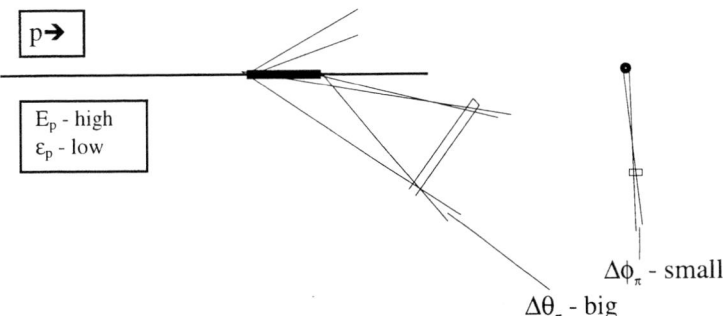

Figure 14. Schematics for multi-channel proton-to-pion conversion system.

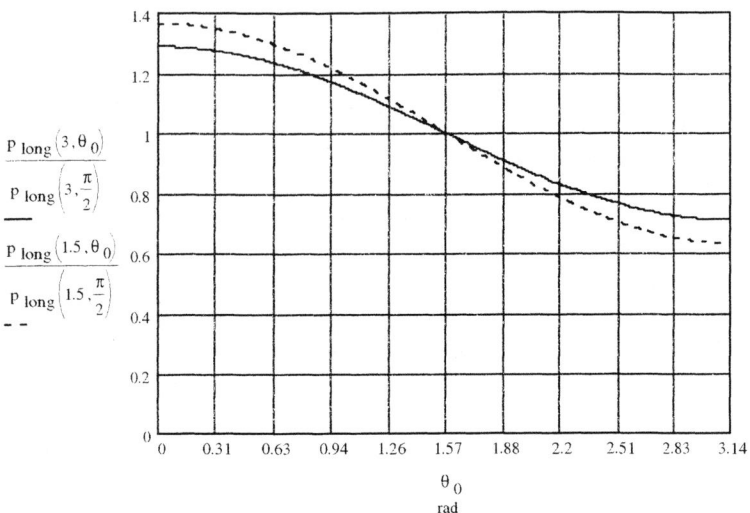

Figure 15. The dependence of relative muon momentum in lab system on polar angle of decay θ in CoM system is shown, for pion kinetic energies 300 MeV (solid line) and 60 MeV (dotted line).

Consequently, we, for every particular muon beam, cut away the middle 30% of muon spectrum and direct the upper and the lower parts with opposite helicities in to 2 separate sub-channels. At the next phase, we shift the energy in each muon channel by RF acceleration/deceleration to the energy optimal for ionization cooling (below 100 MeV kinetic). And then, upon preliminary cooling, we combine all the "upper sub-channels" muons in one longitudinally polarized bunch, and all the "lower sub-channels" muons in another bunch of opposite helicity, each with 70% polarization degree. If it would happen useful, this procedure could be arranged in few stages. Then, all 4 bunches (μ^+ and μ^-) will be cooled down to lowest 6-emittance.

Afterwards, we can reverse helicity of the "lower" bunches at the later stage upon acceleration to 45 GeV (by additional non-accelerating full turn) and then combine two bunches in one (one μ^+ bunch + one μ^- bunch) with 6-emittance 2 times higher than reachable at the final cooling.

The helicity reversing of muons happens because of their anomalous magnetic moment. Positive spin-to-velocity relative rotation is very slow at low energy (e.g., at cooling stage), thus not damaging initial muon beam polarization degree; but it becomes faster proportionally to the muon energy and at 45 GeV each full turn of muon trajectory results in reversing of muon helicity. Let us keep in mind that all the muon spin motion proceeds in the median plane of the collider.

Helicities of colliding bunches are modulated at relative frequency

$$v_{spin} = \frac{\mu_{anom}}{\mu_0}\gamma_\mu = \frac{E_{GeV}}{90} \ .$$

Due to this modulation, at integer spin resonances helicity remains in collision process the same all the time. At half-integer resonances the helicity reverses at consequent turns. At intermediate energies the spin-at-collision modulation proceeds with non-integer fraction of ν_{spin}.

Relative helicities of muon bunches at the interaction region (from ++/-- to +-/-+) can be controlled by choosing a proper injection path (e.g., by additional non-accelerating turn of one beam at, say, 45 GeV).

At high energy, when $\nu_{spin} \gg 1$, non-complete coherence of spin rotation becomes important, and this effect can lead to the loss of polarization degree due to beam energy spread. The loss becomes to be significant if the spin frequency difference in the beam reverses the relative spin orientation at a half of synchrotron oscillation period. The effective polarization degree loss factor ζ_{eff} (compared to the initial degree) can be expressed as

$$\zeta_{eff} = 1 - \frac{1}{2} \cdot (2\pi \cdot \frac{\nu_{spin}}{\nu_{synch}} \cdot \Delta_{Ecoll})^2,$$

where Δ_{Ecoll} is the muon beam energy spread in collider, ν_{synch} - relative synchrotron frequency (this evaluation is meaningful if ζ_{eff} is not very far from 1; otherwise, the polarization degree goes to zero).

MUON COLLIDER OPTIONS

Let us present now the potential luminosity $L_{\mu\mu}$ and polarization factor ζ_{eff} for muon collider at different energies (the actual polarization degree at collision stage would be close to the product of ζ_{eff} and polarization degree at muon production).

For all the options, we assume the "ultimately" low beta-value at the interaction $\beta_0 = 3$ mm, equal to the muon bunch length $\sigma_{longcoll}$. The average magnetic field in the collider H_{coll} is assumed always to be equal to 10 Tesla, except for "Higgs Factory" for which it is assumed to be lower - 7 Tesla, since at this small collider the bigger fraction should be used for interaction regions, etc. The relative betatron frequency in the collider ν_R, RF wave length λ_{RF} and total RF voltage V_{RF} (in MV) are chosen in correspondence with relative energy spread in the collider $\frac{\Delta E_{max}}{E}$.

TABLE: Collider options for $5*10^{12}$ muons per bunch (luminosity per detector).

$2E_{coll}$	N_μ 10^{12}	R_{coll} m	$\frac{\Delta E_{max}}{E}$	ν_r	V_{0coll} MV	ζ_{eff}	$L_{\mu\mu}$ cm^{-2}s^{-1}	
100GeV	5	25	$3*10^{-5}$	20	0.1	1	$7*10^{32}$	Higgs factory
1TeV	5	180	$3*10^{-4}$	30	100	0.8	$2*10^{35}$	
4TeV	5	720	$2.5*10^{-4}$	30	1000	0.8	$7*10^{35}$	
10TeV	5	1800	$1*10^{-4}$	30	1000	0.8	$1.5*10^{36}$	
10 TeV	5	1800	$1*10^{-3}$	300	1000	0	$6*10^{36}$	Ultim.collider

The average muon current at $H_{coll} = 10$ T and $f_0 = 15$ Hz for $N_\mu = 5*10^{12}$ per bunch is the same for all the energies - around 40 mA. For energies of 10 TeV (total) and higher, the muon current becomes almost constant at this level in spite of decay, and the muon generation/acceleration repetition rate should go down.

Beam-Beam Tune Shift Compensation

For higher energies, the beam-beam tune shift ξ (per interaction region) grows higher and higher, and for the "ultimate collider" it reaches more than 0.5. Such a tune shift, in spite of short lifetime of muons, cannot be tolerated without damage for luminosity. Possible approach for preventing the damage is to compensate the shift with the liquid lithium jet of radius σ_{long}, crossing the interaction region /.../. For proper compensation of coherent fields of the colliding muon bunches, the electron density n_e in the "neutralizing" substance (which for lithium is, probably, of one third of full electron density N_{eLi}) should be high enough, to provide

$$\omega_{pl}^{-1} \ll \frac{r_{coll}}{c},$$

where r_{coll} - transverse size of muon bunch at the collision, and plasma frequency is

$$\omega_{pl} = c\sqrt{4\pi r_e \cdot \frac{N_{eLi}}{3}}.$$

In this case, the attenuation factor k_{atten} for beam-beam tune shift is

$$k_{atten} = \frac{2}{\pi \cdot r_e \cdot \dfrac{N_{eLi}}{3} \cdot r_{coll}^2}.$$

For "ultimate collider" the factor is few percent. Let us notice that the use of plasma instead of lithium jet looks practically excluded in the case - the electron density needed is much too high.

The important observation is, the beam-beam compensation approach under consideration can be applicable for (short living) muon beams only. For other types of cyclic colliders, the interaction cross sections are too high and destroy colliding beams too fast. For linear colliders, the beam density at the interaction region is too high - higher than electron density in condensed lithium!

Remarks on Background

The additional background resulting due to muon-lithium interaction needs careful consideration; a hope is all the interaction products fly in very forward direction, and could be lead out of detector region.

Figure 16. Schematics of dipoles (left) and skew-quads (right) for "open median plane" collider.

The background problem is one of the most important for muon collider experiments: all the muons decay inside the collider - 10^{14} muons per sec! Decay electrons go to the inner radius, but the synchrotron radiation photons emitted by them go to the outer direction - and both remain strictly in median plane of the collider. Hence, it is worth to try to remove all the superconducting coils (and other "thick" components) out of median plane completely - to let all this high energy species to leave collider and detector almost without interaction. In principle, there is such an option both for dipoles and for quadrupoles (Fig.16). For latter, it means transition to the use of skew-quads exclusively in the whole collider optics. This approach is in coherence with equality of transversal emittances as result of ionization cooling.

Of course, all the muons outside of the useful for luminosity beam emittance should be cut off carefully prior to injection into collider.

Other Collisions

Let us underline, the muon collider provides almost automatically the same full luminosity for μ^- - proton collisions. Additional care and arrangements are needed to get longitudinally polarized protons at collision. To reach this, we need to install proper proton spin rotators on both sides of the interaction region(s) - to transform vertically polarized protons in the arc to longitudinally polarized at collision, and to restore transversal polarization prior to entering the next arc. The influence of these spin rotators on muon polarization is negligibly small.

To arrange (if interesting) the same sign collisions, we need double-ring collider; the other components of the complex remain almost unaltered.

ACKNOWLEDGEMENTS

It is my great pleasure to thank V.Parkhomchuk, G.Silvestrov, T.Vsevolozhskaya, K.Lotov for long lasting and fruitful collaboration, and R.Palmer, D.Cline, J.Gallardo, Z.Parsa for interesting and useful discussions.

REFERENCES

1. Skrinsky A.N., "Colliding Beams Program in Novosibirsk," presented at International Seminar on Prospects in High Energy Physics; Morges, CERN, 1971.

2. Skrinsky A.N., "Accelerator and Instrumentation Prospects of Elementary Particle Physics," in *Proceedings of the XX International Conference on High Energy Physics, Madison, 1980*, New York, 1981, v.2, p.1056-1093; and in *Uspekhi Fiz. Nauk*, Moscow, 1982, v.138, 1, pp.3-43.

3. Muon Collider Collaboration, "Muon Collider Design", in *Proceedings of the Symposium on Physics Potential and Development of $\mu^+\mu^-$ Colliders*, San Francisco (1995); Nuclear Physics B, Proceedings Supplement, v. 51 A, November 1996, pp. 61-84.

3a. Palmer R.B. and Gallardo J.C., "Muon-Muon and Other High Energy Colliders", *report of BNL-64148*, 1997.

4. Parkhomchuk V.V. and Skrinsky A.N., "Methods of Cooling of Charged Particles Beams," in *Physics of Elementary Particles and Atomic Nuclei*, Moscow, 1981, v. 12, pp. 557-613; and *Soviet Journal on Particles and Nuclei*, v. 12(3), May-June 1981, pp. 223-232.

5. Skrinsky A.N., "Ionization Cooling and Muon Collider", in *Proceedings of 9th ICFA Beam Dynamics Workshop: Beam Dynamics and Technology Issues for Muon-Muon Colliders*, Montauk, NY (1995) and *Nuclear Instruments and Methods*, A 391, (1997) pp. 188-195.

6. Silvestrov G.I., "Problems of Intense Secondary Particle Beams Production", in *Proceedings of the XIII International Conference on High Energy Accelerators*, Novosibirsk (1986), v. 2, p. 258-263.

7. Silvestrov G.I., "Lithium Lenses for Muon Colliders", in Proceedings of 9th ICFA Beam Dynamics Workshop: Beam Dynamics and Technology Issues for $\mu^+\mu^-$ Colliders, Montauk, NY (1995).

8. Skrinsky A.N., "Polarized muon beams for muon collider", in *Proceedings of the Symposium on Physics Potential and Development of $\mu^+\mu^-$ Colliders*, San Francisco (1995); Nuclear Physics B, Proceedings Supplement, v. 51 A, November 1996, pp. 201-203.

Luminosity Monitoring at μp and $\mu\mu$ Colliders

Ilya F. Ginzburg
Institute of Mathematics. 630090. Novosibirsk. Russia.
E-mail: ginzburg@math.nsc.ru

Abstract. We propose procedures for luminosity monitoring at μp and $\mu^+\mu^-$ colliders.

INTRODUCTION

There are two problems related to the luminosity monitoring. First, it is necessary to determine luminosity to measure cross sections of interesting processes with high accuracy. Second, it is desirable to measure luminosity fast (even with low accuracy) for quick improvement of collision. The solutions of these problems may or may not coincide.

To solve the first problem, it is useful to have some specific process with well known cross section — *monitoring process* which measurement will give us the absolute normalization of luminosity. The necessary accuracy here should not be better than the statistical inaccuracy for the typical processes under interest. It is desirable that the cross section of the monitoring process will be of the same order of value or larger than those of processes under interest. In the discussion, we have in mind that the construction of discussed colliders will be justified if their luminosity will be so high to measure the cross sections like $(\alpha/E)^2$ with reasonable accuracy.

The problems in the luminosity monitoring are very different for both colliders discussed. This difference is connected with the difference in the scale of typical cross sections there.

I THE μP COLLIDER

We start with the rough estimate of cross section of main process here $\mu p \to \mu + hadrons$. In this estimate we consider the cross section σ_0 of subprocess $\gamma p \to hadrons$ to be energy independent, $\sigma_0 \sim 100$ μb. Then, the total cross section is

$$\sim \frac{\alpha}{\pi}\sigma_0 \ln\frac{2E^2}{2m_p m_\pi} \ln\frac{2E^2 m_\rho}{m_\mu^2 m_\pi},$$

about 40 μb for E=50 GeV and 100 μb for E=2 TeV.

A The Proposal and Basic Process

The way for luminosity monitoring that we propose here is similar to that proposed for the large proton colliders, Tevatron and LHC [1] and earlier for ISR [2][1].

We propose to record the electrons and positrons from the process $\mu p \to \mu p e^+ e^-$ in anticoincidences with other particles. Some two–stage procedure for realization of this idea is essential part of proposal.

The total cross section of the process $\mu p \to \mu p e^+ e^-$ is [3]

$$\sigma = \frac{28}{27\pi} \frac{\alpha^4}{m_e^2} \left[(\ell - 2.12)^3 + 2.2(\ell - 2.12) + 0.4 \right], \quad \ell = \ln \frac{s}{m_\mu m_p}. \tag{1}$$

It is $\sim$ 1.2 mb for $\sqrt{s} = 100$ GeV and $\sim$ 6.6 mb for $\sqrt{s} = 4$ TeV. These values are higher than those of process studied.

The main contribution to this process is calculable within QED with accuracy better $< (m_e/m_\pi)^2 \sim 0.1\%$ without any phenomenological parameters. The real accuracy of monitoring is limited by the accuracy of signature. The precise equations for the process in tree approximation (i.e. with accuracy better 0.5%) are written in ref. [2]. For this main part of cross section the protons remain within the beam, they cannot be detected. The qualitative description of this process is well known (see e.g. refs. [2], [1].)

The main part of cross section is concentrated within the momentum region (k_z and ϵ are total longitudinal momentum of produced e^+e^- pair and its energy)

$$m_e \frac{E}{m_\mu} \gtrsim k_{z\,e} \gtrsim -m_e \frac{E}{m_p}. \tag{2}$$

Here, the produced pairs distribute almost uniformly in the rapidity scale. In more detail, *the distribution over the total energy of pair* is

$$d\sigma = \begin{cases} \frac{14\alpha^4}{9\pi m_e^2} \frac{dk_z}{\epsilon} \left[\ell^2 \ln^2 \frac{\epsilon}{m} \right] & \text{at} \quad m_e \frac{E}{m_\mu} \gtrsim k_{z\,e} \gtrsim -m_e \frac{E}{m_p}; \\ \sim \frac{\alpha^4}{m_e^2} \frac{d\epsilon}{\epsilon} \left(\frac{4E^2 m_e^2}{m_\mu m_p \epsilon^2} \right)^2 & \text{at} \quad k_z \gtrsim m_e \frac{E}{m_\mu}, \quad k_z \lesssim -m_e \frac{E}{m_p}. \end{cases} \tag{3}$$

Here k_z is the longitudinal momentum of the pair and ϵ is its energy, $|k_z| \approx \epsilon$.

Typically, the electron and positron energies in each pair differ greatly from each other. Therefore, the total energy of pair is near the energy of one positron or electron. It is the reason why *the distribution over the energy of one electron ϵ_1, emitted along the motion of initial μ^+*, has the form of eq. (3) with the replacement

---

[1] We use below equations from this paper without additional referring.

of total energy and longitudinal momentum of pair to the energy and longitudinal momentum of electron $\epsilon \to \epsilon_1$, $k_z \to k_{1z}$.

Let $\vec{k}_{i\perp}$ ($i = 1, 2$) be the transverse components of the momenta of electron or positron. *The distribution over the total transverse momentum of the pair* $\vec{k}_\perp = \vec{k}_{1\perp} + \vec{k}_{2\perp}$ is peaked near zero:

$$d\sigma \sim \frac{\alpha^4}{m_e^2} \frac{dk_\perp^2}{k_\perp^2} \cdot \begin{cases} \frac{28}{9\pi} [\ln(\gamma^2 k_\perp^2/m_e^2)]^2 & \text{at} \quad m_e/\gamma < |k_\perp| < m_e; \\ \frac{m_e^2}{k_\perp^2} & \text{at} \quad |k_\perp| > m_e. \end{cases} \quad (4)$$

The other important *distribution* is that *over acoplanarity angle* 2ψ, which is the angle between $\vec{k}_{1\perp}$ and $-\vec{k}_{2\perp}$:

$$d\sigma \propto \frac{\ln(\gamma^2 \sin^2 \psi)}{\sqrt{\sin^2 \psi + 1/\gamma^2}} d\psi. \quad (5)$$

The scale of effective sizes for the process m_e^{-1} is much higher than that for the hadron processes, form factors etc. ($< m_\pi^{-1}$.) It is the reason why the main contribution to this process is calculable within QED with accuracy better $(m_e/m_\pi)^2 \sim 0.1\%$ without any phenomenological parameters.

B Proposed Procedure

Two difficulties seem to forbid the use of this process for luminosity monitoring.

- The energies of produced electrons and positrons are small in comparison with that of collided muons and protons, their escape angles are small too. It is necessary to have the special devices for recording these soft particles under very small angles.
- It is almost impossible to see e^+ and e^- of interest from a large number of background pions produced in the main process.

There are two stages in overcoming these difficulties:

First, one should use the time when collider works with smaller luminosity than in the final operations, so that the probability of μp interaction per bunch crossing $\mathcal{P}$=0.01-0.1. In this period one should:

- **to record e^+e^- pairs in the events without additional particles beyond beams**. Last veto deletes physical background almost entirely. **Good timing resolution** can be used to reduce beam – gas backgrounds, etc. To avoid electrons from μ decay (if it is necessary) one can consider only pairs that move to the side of protons.

Result: precise normalization of luminosity (in this stage.)

- to observe the hadronic processes with simple and definite signature and large enough cross section (*process C*), for example, to observe within the angular interval 30–50 mrad something like:
1. The protons with energy $E_p > E/2$.
2. The pion flux with $p > 0.01E$.
3. The total energy flow.
4. The muons scattered in this angle.

The result: precise absolute value of the cross section for the process C.

Second, (at normal luminosity) **the process C is used for the permanent luminosity monitoring.**

C What to Measure

Since the probability of process per bunch crossing is $\mathcal{P} \ll 1$ at the first stage, one can consider e^+e^- pair production and usual hadron processes separately. To track the pairs one can use Cerenkov counters — for the electrons and positrons with the energy 20–50 MeV (characteristic angle 10-20 mrad) and the specific Roman pots at the suitable distance from the IP for the electrons and positrons with energy 50–500 MeV (characteristic angle 1-10 mrad.)[2] (I remind — the energies of electron and positron within the pair usually are different. So, perhaps, both detectors should be used for one pair.) The magnetic field within detector is also essential in this problem.

In our process protons remain within beam, they are invisible. The sharp distributions over total transverse momentum and acoplanarity angle can be used for the additional testing of origin of events.

The production of e^+e^- pairs by the residual gas, at the body of detector etc. can be removed by cuts on timing and vertex reconstruction. Our old estimates of residual gas e^+e^- pairs for ISR gave small background.

II THE $\mu^+\mu^-$ COLLIDER

For the $\mu^+\mu^-$ collider, at the reasonable energies almost all processes with good signature have cross sections $\sim (\alpha/E)^2$. In addition, their cross sections depend strongly on the muon longitudinal polarization (with mean degree $\lambda_\pm$.) Here the independent measurements of luminosity and polarization are necessary.

The best process for the luminosity monitoring here seem to be the small angle Bhabha scattering. For the muon scattering angle θ the cross section of this process is $\sim \theta^{-2}$ times larger than the typical cross sections of interest. Moreover, the polarization corrections to the cross sections is given by

---

[2] These quantities are typical for the pairs emitted along proton beam at its energy about 1 TeV

$$\sim \lambda_+\lambda_-\theta^2. \tag{6}$$

These corrections can be neglected at θ_i(30-50) mrad.

It is useful to note that the accuracy of tree calculations here is much better than those for the e^+e^- collider. Indeed, the radiative corrections here are damped due to high mass of muon. In particular, the Initial State Radiation is almost absent here.

The measurements at $\theta \sim 1$ rad (with much lower cross section) can be used here for the independent testing of muon polarization degree (value of quantity $\lambda_+\lambda_-$.)

For the additional testing of luminosity and solving of the second basic problem noted in abstract, one can try to use the process $\mu^+\mu^- \to \mu^+\mu^- e^+e^-$ with (polarization independent) cross section about 2.2 mb for $E = 50$ GeV and about 9.5 mb for $E = 2$ TeV.

I am grateful to N. Mokhov and V. Serbo for discussions. I am thankful to A.M. Sessler, S. Chattopadhyay and M. Zolotorev for warm hospitality in LBNL before and during Conference and to Organizing Committee for the invitation. This work is supported by grants INTAS - 93-1180 ext and RFBR - 96-02-19114.

REFERENCES

1. I.F. Ginzburg. NIMR, in print.
2. V.M. Budnev, I.F. Ginzburg, G.V. Meledin, V.G. Serbo. Phys. Lett. **39B** (1972) 526; Nucl.Phys. **B63** (1973) 519.
3. G. Racah. Nuovo Cim. **14** (1937) 93.

Phase Space Exchange in Thick Wedge Absorbers for Ionization Cooling

David Neuffer

Fermilab, PO Box 500, Batavia IL 60510

Abstract

The problem of phase space exchange in wedge absorbers with ionization cooling is discussed. The wedge absorber exchanges transverse and longitudinal phase space by introducing a position-dependent energy loss. In this paper we note that the wedges used with ionization cooling are relatively thick, so that single wedges cause relatively large changes in beam phase space. Calculation methods adapted to such "thick wedge" cases are presented, and beam phase-space transformations through such wedges are discussed.

INTRODUCTION

Recent innovations have intensified research on the possibility of a high-intensity high-luminosity $\mu^+\mu^-$ collider.[1,2] In this collider concept ionization cooling is used to increase the phase space density of the beam by a factor of ~10^6. In ionization cooling the beam passes through an absorber, in which it loses momentum parallel to its motion, followed by reacceleration in which it regains only longitudinal momentum. The transverse components of the motion are effectively reduced and therefore cooled in each step, but the procedure is inefficient in cooling longitudinal phase space. However, phase-space exchange between longitudinal and transverse phase space is possible by passing the beam through a wedge absorber (in which absorber thickness or density depends on transverse position), and the degree of exchange is controllable by changing the dispersion at the absorber. The scenario for phase space cooling alternates energy-loss transverse cooling with transverse-longitudinal exchanges, in order to obtain cooling in all 6-D phase space dimensions.[3]

In previous discussions of ionization cooling,[4, 5, 6] this phase-space exchange has been included in the differential equations of cooling, following a similar exchange occurring in synchrotron radiation damping.[7] That treatment implies a small exchange per integration step. In ionization cooling relatively large exchanges can occur in a single wedge, and a formalism more appropriate to these large exchanges in a single step is desirable, particularly in the development of explicit designs. In this paper, we separate the cooling and heating aspects of the ionization from the phase-space exchange, and consider the exchange with dispersion as a linear transformation. Based on the betatron function formalism[8] and following a related application by J. Peterson [9], we present simplified formulae for the amount of phase space exchange and the related transformations of transverse and longitudinal phase-space parameters (emittances, betatron functions, dispersion, momentum width, etc.). We apply these to particular examples, and deduce optimized conditions and cases for phase-space exchange.

FORMALISM

Figure 1 shows a stylized view of the passage of beam with dispersion through an absorber. The key parameters are the beam properties entering and exiting the wedge, as well as the properties of the wedge itself. The beam properties entering the wedge are the beam energy and momentum E_0, p_0, the beam momentum width $\Delta p_0 = \Delta E_0/(v/c)$ and the beam transverse emittance ε_0. We use the relative momentum width defined by $\delta_0 = \Delta p_0/p_0$. The beam transport properties are given by the betatron function β_0, and the dispersion η_0. To simplify discussion, the beam is focussed to a betatron and dispersion waist at the wedge: β_0', $\eta_0' = 0$. (This is an optimal choice and permits us to avoid any changes in β', η' in the wedge.) The beam properties after the wedge are represented by the same symbols with the subscript 1: E_1, p_1, δ_1, ε_1, β_1, η_1. The wedge is represented by its relative effect on the momentum offsets δ of particles within the bunch at position x:

$$\frac{\Delta p}{p} = \delta \to \delta - \frac{(dp/ds)\tan\theta}{p} x = \delta - \delta'x$$

dp/ds is the momentum loss rate in the material (dp/ds = β^{-1}dE/ds), which is dependent on the material and the particle energy. For beryllium with beam at the minimimum-ionizing energy, dp/ds is ~3.0 MeV/c/cm. x tanθ is the wedge thickness at transverse position x (relative to the central orbit at x=0), and we have introduced the symbol δ' = dp/ds tanθ /p to indicate the change of δ with x. At typical values for ionizing cooling wedges (δ = 0.01, p =300 MeV/c, tanθ up to 1, x = 1cm), changes in δ are relatively large, so a thick wedge treatment is desirable.

Note that changes in the transverse beam only occur in the coordinate with wedge thickness variation (horizontal for horizontal bends and wedges). The perpendicular beam projection is unchanged. Horizontal and vertical effects can be balanced by including both horizontal and vertical bend and wedge sections.

In practical applications, we would like to constrain the wedge geometry so that the beam entrance and exit angles are not too steep; a maximum value of θ of ~45° would seem reasonable (tan θ = 1). The angles could be increased if a symmetric wedge of angle ϕ is used (see figure 2); the two wedges are equivalent if tanθ =2 tan(ϕ/2). The symmetric wedge is also practically constrained to ϕ < 45°, but this extends the range of tan θ to ~2.

In first order, we ignore the central beam energy loss, as well as multiple scattering and energy straggling in the absorber, and consider only the effects of δ' and η_0. In that case,

δ' and η_0 can be represented as linear transformations in the x-δ phase space projections and the transformations are phase-area preserving. Thus the dispersion can be represented by the matrix:

$$\mathbf{M}_\eta = \begin{bmatrix} 1 & \eta_0 \\ 0 & 1 \end{bmatrix}, \text{ since } x \Rightarrow x + \eta_0 \delta$$

and the wedge can be represented by the matrix: $\mathbf{M}_\delta = \begin{bmatrix} 1 & 0 \\ -\delta' & 1 \end{bmatrix}$, so that the dispersion + wedge becomes: $\mathbf{M}_{\eta\delta} = \begin{bmatrix} 1 & \eta_0 \\ -\delta' & 1 - \delta'\eta_0 \end{bmatrix}$.

Under the general assumption of smoothly-populated beam distributions, we can represent the x-δ beam distribution as a phase space ellipse:

$$g_0 x^2 + b_0 \delta^2 = \sigma_0 \delta_0$$

where σ_0 is the initial zero-η beam size (related to transverse emittance and betatron function by $\sigma_0 = (\varepsilon_0 \beta_0)^{1/2}$). Also $g_0 = \delta_0/\sigma_0$, and $b_0 = \sigma_0/\delta_0$. After the dispersion plus wedge, the beam is within another equal-area phase-space ellipse, given by:

$$g_1 x^2 + 2a_1 x\delta + b_1 \delta^2 = \sigma_0 \delta_0$$

where the revised ellipse parameters are found from the old ones and the transfer matrix by standard betatron function transport techniques:[8]

$$b_1 = b_0 + (\eta_0)^2 g_0$$
$$a_1 = \delta' b_0 - \eta_0 (1 - \delta'\eta_0) g_0$$
$$g_1 = \delta'^2 b_0 + (1 - \delta'\eta_0)^2 g_0$$

From these revised parameters, the rotated beam parameters can be deduced.

The energy width is changed to:

$$\delta_1 = \sqrt{g_1 \sigma_0 \delta_0} = \delta_0 \left[(1 - \eta_0 \delta')^2 + \frac{\delta'^2 \sigma^2}{\delta_0^2} \right]^{1/2}.$$

In the transformations the bunch length is unchanged. The longitudinal emittance, the area of the beam in longitudinal phase-space (energy-width × bunch-length), has therefore changed simply by the ratio of energy-widths, which means that the longitudinal emittance has changed by the factor δ_1/δ_0.

From emittance conservation, the transverse emittance has changed by the inverse of this factor:

$$\varepsilon_1 = \varepsilon_0 \left[(1 - \eta_0 \delta')^2 + \frac{\delta'^2 \sigma^2}{\delta_0^2} \right]^{-1/2}.$$

This same factor can also be obtained by noting that the transverse beam size can be written as $\sigma_1 = (b_1 \sigma_0 \delta_0)^{1/2}$, calculating b_1 and noting that transverse velocities are unchanged. The emittance, which is simply the product of size and velocity, can then be obtained from $\varepsilon_1 / \varepsilon_0 = \sigma_1 / \sigma_0$.

The x-δ phase space ellipse equation can be rewritten into the form:

$$g_1 \left(x + \frac{a_1}{g_1} \delta \right)^2 + \frac{\delta^2}{g_1} = \sigma_0 \delta_0,$$

where we have used $b_1 g_1 = 1 + a_1^2$. From this form we can see that the new value of the dispersion is: $\eta_1 = -a_1 / g_1$, which can be written as

$$\eta_1 = -\frac{a_1}{g_1} = \frac{\eta_0 (1 - \eta_0 \delta') - \delta' \frac{\sigma^2}{\delta_0^2}}{(1 - \eta_0 \delta')^2 + \delta'^2 \frac{\sigma^2}{\delta_0^2}}$$

The rms uncorrelated beam size (dispersion-removed) has also changed to:

$$\sigma_1^2 = \langle (x - \eta_1 \delta)^2 \rangle = \sigma_0 \delta_0 / g_1,$$

which means: $\sigma_1 = \sigma_0 \left[(1 - \eta_0 \delta')^2 + \frac{\delta'^2 \sigma_0^2}{\delta_0^2} \right]^{-1/2}$. Combining this with our expression for the transverse emittance, and using $\sigma_1^2 = \varepsilon_1 \beta_1$, we find that the transverse betatron function has also been changed in the same proportion:

$$\beta_1 = \beta_0 \left[(1 - \eta_0 \delta')^2 + \frac{\delta'^2 \sigma^2}{\delta_0^2} \right]^{-1/2}.$$

Note that the change in betatron functions (β_1, η_1) implies that the following optics should be correspondingly rematched.

PARTICULAR CASES

We have, in the above equations, completely re-characterized the phase-rotated beam. We can apply the above results to particular cases, some of which are of great practical importance, and which can simplify the cooling process.

A. Thin Wedge

In the limit where $\delta' \to 0$ (which means small tilt angle θ and/or large p and/or small dp/ds), the emittance change per wedge is small. In that limit:

$$\varepsilon_1 \to \varepsilon_0(1+\eta_0\delta')$$
$$\delta_1 \to \delta_0(1-\eta_0\delta')$$
$$\eta_1 \to \eta_0(1+\eta_0\delta') - \delta'\frac{\sigma^2}{\delta_0^2}$$
$$\beta_1 \to \beta_0(1+\eta_0\delta')$$

This is the same result previously derived in reference 4 for the case of continuous small changes. Note that a dispersion develops from a wedge absorber even in the absence of initial dispersion, and a second-order emittance exchange develops from integration of that.

B. Maximum Exchange

A maximal exchange effect occurs when $\delta' = 1/\eta_0$, and many terms in the equations are cancelled out. The relationship between initial and resulting parameters are greatly simplified:

$$\delta_1 = \delta_0 \frac{\sigma_0}{\eta_0\delta_0}, \varepsilon_1 = \varepsilon_0 \frac{\eta_0\delta_0}{\sigma_0}, \beta_1 = \beta_0 \frac{\eta_0\delta_0}{\sigma_0}, \eta_1 = -\eta_0.$$

In this case the dispersion beam size $\eta\delta$ and the emittance beam size σ are simply exchanged.

In this "maximal exchange" condition, the emittances change simply as the ratio of the momentum beam size $\eta_0\delta_0$ to the initial (emittance) beam size σ_0, obtaining a large exchange when that ratio is large. We also note that dispersion actually changes sign in the absorber. The equations also show a large inverse exchange for small η. However this inverse exchange may require a large δ', and δ' is limited by our geometrical constraint of $\tan \theta < \sim 1$. (The case of large δ' with $\eta=0$ is discussed below.)

C. Dispersion-matched exchange

One can similarly choose δ' and η_0 to obtain dispersion-cancelling; that is, a solution in which the resulting dispersion η_1 is zero. This can be obtained when $a_1 \to 0$, which implies that

$\delta' = \dfrac{1}{\eta_0\left[1 + \dfrac{\sigma_0^2}{\eta_0^2 \delta_0^2}\right]}$, a similar but somewhat reduced case from maximal exchange.

In this case, somewhat simplified emittance exchange fomulae are obtained:

$$\delta_1 = \dfrac{\delta_0}{\sqrt{1 + \dfrac{\eta_0^2 \delta_0^2}{\sigma_0^2}}}, \varepsilon_1 = \varepsilon_0 \sqrt{1 + \dfrac{\eta_0^2 \delta_0^2}{\sigma_0^2}}, \beta_1 = \beta_0 \sqrt{1 + \dfrac{\eta_0^2 \delta_0^2}{\sigma_0^2}}, \eta_1 = 0.$$

This case is particularly useful, because the cancelling of η implies that dispersion matching after the wedge is not necessary. This greatly simplifies the optics. As in the previous case the degree of exchange possible depends on the ratio of the momentum beam size $\eta_0 \delta_0$ to initial (emittance) beam size σ_0, obtaining a large exchange when that ratio is large.

D. Wedge only.

As previously discussed, passing the beam through a wedge absorber without an initial dispersion introduces a dispersion in the beam. In a "thick" absorber that dispersion generates an emittance exchange, and this increases the longitudinal emittance and energy spread while decreasing the transverse emittance. The equations for emittance change and final dispersion are :

$$\delta_1 = \delta_0 \sqrt{1 + \dfrac{\delta'^2 \sigma^2}{\delta_0^2}}, \varepsilon_1 = \dfrac{\varepsilon_0}{\sqrt{1 + \dfrac{\delta'^2 \sigma^2}{\delta_0^2}}}, \beta_1 = \dfrac{\beta_0}{\sqrt{1 + \dfrac{\delta'^2 \sigma^2}{\delta_0^2}}}, \eta_1 = -\dfrac{\delta' \sigma^2}{1 + \delta'^2 \sigma^2}.$$

This is also a very useful case, since it does not require dispersion matching into the wedge.

INTEGRATION WITH COOLING AND HEATING

In the linearized model, the mean beam energy is assumed to remain constant and no scattering is included. However the wedges are sufficiently thick to have significant energy loss and with that some emittance cooling, multiple scattering, and energy straggling. These effects should be included in a full discussion.

The differential equation for transverse cooling is:

$$\dfrac{d\varepsilon_N}{ds} = -\dfrac{1}{\beta^2 E}\dfrac{dE}{ds}\varepsilon_N + \dfrac{\beta_\perp E_s^2}{2\beta^3 m_\mu c^2 L_R E}$$

where the first term is the frictional cooling effect and the second is the multiple scattering heating term. Here L_R is the material radiation length, $\beta_\perp$ is the betatron function, and E_s is the characteristic scattering energy (~14 MeV). In a first approximation we can estimate the effect by simply multiplying by the thickness of the wedge Δz at the beam center. The required thickness is set by the requirement that the wedge cover at least $\pm 2\sigma$ in momentum and position, and the thickness goes to zero at $x = -\Delta z/\tan\theta$. This means $\Delta z > 2\eta\delta_0 \tan\theta$, and $\Delta z > 2\sigma_0 \tan\theta$ are required. Therefore:

$$\Delta\varepsilon_N\big|_{material} = -\frac{1}{\beta^2 E}\frac{dE}{ds}\varepsilon_N \Delta z + \frac{\beta_\perp E_s^2}{2\beta^3 m_\mu c^2 L_R E}\Delta z$$

and a choice of Δz as the geometric sum of $2\eta\delta_0 \tan\theta$ and $2\sigma_0 \tan\theta$ is reasonable.

Similarly the longitudinal cooling and heating equation is:

$$\frac{d\sigma_E^2}{ds} = -2\frac{\partial \frac{dE}{ds}}{\partial E}\sigma_E^2 + 4\pi\left(r_e m_e c^2\right)^2 n_e \gamma^2\left(1-\frac{\beta^2}{2}\right)$$

the first term is the frictional cooling (or heating) term, which is naturally small, except at low-energies ($E_\mu < 100$ MeV) where it heats the beam. The second term is the energy-straggling term, where n_e is the material electron density. From the above discussion, we estimate for moderate to high μ energies that

$$\Delta\delta^2\big|_{material} \cong +4\pi\left(r_e m_e c^2\right)^2 \frac{n_e}{\beta^4 (m_\mu c^2)^2}\left(1-\frac{\beta^2}{2}\right)\Delta z$$

These effects are included in our discussions of examples. In optimal examples these effects should be relatively small.

NUMERICAL EXAMPLES AND SIMULATION RESULTS

In developing a possible scenario for the assembly of high-intensity bunches for a possible μ^+-μ^- collider, a beam cooling and compression scenario is being developed.[3] In that scenario a beam with a very large emittance, including large energy spread, is cooled by many passages (20—100) through material absorbers followed by rf reacceleration sections. That cooling sequence includes many wedge absorbers for exchange of transverse and longitudinal emittance. The cooling sequence begins with wedge absorbers to reduce the energy spread, followed by ~20 steps of transverse energy cooling intermixed with wedge exchangers. In the final steps, a relatively small longitudinal emittance with small energy-spread is wedge-exchanged to obtain a greatly reduced transverse emittance. In the full-sequence the transverse normalized emittances ($\varepsilon_N = \varepsilon_\perp \beta\gamma$) are reduced by ~ two orders of magnitude (from $\varepsilon_N \cong 0.01$m-rad to ~0.00005m-rad). The longitudinal emittance is reduced by ~ one order of magnitude, although it has been reduced by more than two orders of magnitude before the final emittance exchanges.

In this section we describe several characteristic emittance exchange steps. These examples will illustrate the cooling parameters as well as demonstrate the use of the present exchange model in designing a matched and optimized cooling system.

The first example we consider corresponds to beam conditions near the beginning of the system, where the energy spread and emittances are both quite large. In this case we choose μ-beam at an initial kinetic energy of 300 MeV (momentum of 391.7 MeV), an initial rms momentum spread δ_1 of 7.4%, and rms normalized emittance of 0.015m-rad (geometric emittance of $\varepsilon_\perp$ = 0.004). The beam is focussed onto a beryllium wedge absorber (dE/ds = 3 MeV/cm) with β_1 =0.34m (σ=3.7cm) at a dispersion of 1m, so the ratio of momentum to emittance beam size is 2. The example is matched to obtain zero dispersion after the wedge (case C), which implies $\tan\theta$ = 1. The wedge is therefore designed to reduce the energy spread by $\sqrt{5}$ while increasing transverse emittance by the same factor.

This first example has also been simulated using SIMUCOOL.[10,11] Even though this is a wedge case with very large energy spread and large emittance, results in good agreement with the linear model were obtained. The wedge must be thick enough to accommodate the full momentum spread as well as the full beam size, which means that cooling and scattering (rms heating terms) are nonnegligible. With a wedge with a 17 cm thickness at the beam center; the mean energy decreases ~50 MeV in energy to 340 MeV/c. Without a reoptimization, the dispersion was reduced from 1 m to <0.05 m and an exchange of a factor of ~2 has been obtained (rather than $\sqrt{5}$ = 2.236), accompanied by ~10% emittance cooling.

The second case we consider corresponds to conditions near the middle of the cooling sequence. The beam momentum is 200 MeV/c and a 0.035m Be wedge at η = 0.5m and β = 0.15m reduces the momentum spread from 4% to 2.5% while emittance is increased by a factor of 1.6. This case is also matched to zero dispersion exiting the wedge, and the exit of the wedge would be an appropriate place to put an extended absorber for transverse cooling (possibly a Li or Be focussing rod for extended cooling). SIMUCOOL results are in good agreement with this general process.

The third case we consider is one from near the end of the cooling sequence, where the longitudinal phase space is increased in order to reduce the transverse emittance (see Fig. 3), in order to obtain minimal final emittances for the μ^+-μ^- collider. This wedge is arranged so as to increase the energy spread, and that condition is obtained by choosing $\tan\theta$ < 0, or alternatively a negative dispersion. The beam energy is 25 MeV with a normalized emittance of 61 mm-mrad and initial δ = 0.0081, and the beam is focussed to small β^* (0.014m) at small dispersion (0.0105m) with a 0.0017 m thick wedge. We expect a decrease of emittance by a factor of 1.6 with a corresponding increase of $\delta p/p$. SIMUCOOL results are also in reasonable agreement with this simplified model.[11]

DISCUSSION

As shown in the above examples, typical wedges used for phase-space exchange are relatively large, and a large exchange formalism is useful. Transformations through a wedge significantly change emittances, betatron functions and dispersion. Matching of these changes is posssible in the present formalism. Particular examples, such as dispersion-suppressing wedges, will be useful in generating compact cooling scenarios.

REFERENCES

1. D. Neuffer and R. Palmer, Proc. European Particle Accelerator Conference EPAC94, London, p.52, 1994.
2. J. Gallardo, ed. Proc. 9th Advanced ICFA Beam Dynamics Workshop, "Beam Dynamics and Technology Issues for μ^+-μ^- Colliders", Montauk NY 1995, AIP Conf. Proc. 372 (1996).
3. R. B. Palmer et al, ibid, AIP Conf. Proc. 372, 3 (1996).
4. D. Neuffer, Particle Accelerators, 14, 75 (1983).
5. D. Neuffer, Nucl Inst. and Meth. A350, 27 (1994).
6. A. N. Skrinsky and V. V. Parkhomchuk, Sov. J. Nucl. Phys. 12, 3 (1981).
7. M. Sands, SLAC-121 (1970), also in Proc. of the Int. School of Physics, B. Touschek, ed. Varenna 1969, Academic Press NY (1971).
8. E. Courant and H. L. Snyder, Annals of Physics 3,1 (1957).
9. J. M. Peterson, Proc. 1983 Particle Accelerator Conference, IEEE Trans. NS-30, 2403 (1983).
10. A. Van Ginneken, Nucl. Inst. and Meth. A 362, 213 (1995).
11. A. Van Ginneken and D. Neuffer, studies in progress (1996).

Table 1: Parameter list for a 4 TeV μ^+-μ^- Collider

Parameter	Symbol	Value
Energy per beam	E_μ	2 TeV
Luminosity	$L = f_0 n_s n_b N_\mu^2 / 4\pi\sigma^2$	10^{35} cm^{-2}s^{-1}
Source Parameters		
Proton energy	E_p	30 GeV
Protons/pulse	N_p	$2 \times 3 \times 10^{13}$
Pulse rate	f_0	15 Hz
μ-production acceptance	μ/p	.2
μ-survival allowance	N_μ/N_{source}	.33
Collider Parameters		
Number of μ /bunch	$N_{\mu\pm}$	2×10^{12}
Number of bunches	n_B	1
Storage turns	$2n_s$	2000
Normalized emittance	ε_N	3×10^{-5} m-rad
μ-beam emittance	$\varepsilon_t = \varepsilon_N/\gamma$	1.5×10^{-9} m-rad
Interaction focus	β_0	0.3 cm
Beam size at interaction	$\sigma = (\varepsilon_t \beta_0)^{1/2}$	2.1 µm

Figure 1: Overview of the beam transformation in passing through a wedge absorber. The upper portion shows a stylized view of a beam passing through a dispersive transport into a wedge absorber; the lower portion shows the projection of the 6-D beam phase space ellipse into x-δ phase space, and its changes passing through the system. Dispersion imposes an x-δ correlation (ellipse tilt), and the wedge reduces the beam energy δ(x), with energy loss a function of x: $\Delta\delta = x \, dp/ds \, \tan\theta/p$. Note that the x-δ ellipse area remains the same (in the limit where average energy loss is zero).

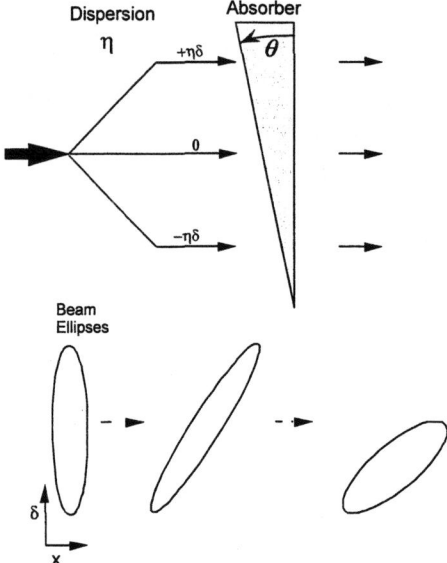

Figure 2: equivalent wedge absorbers, tilted and symmetric cases.

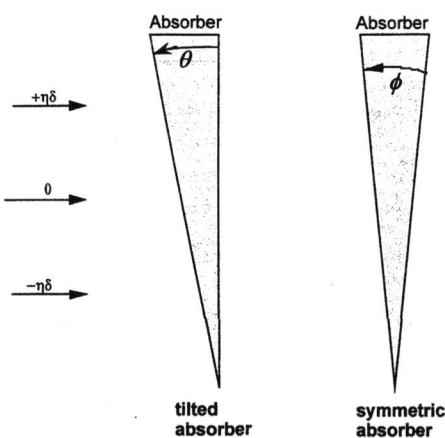

$$\tan(\theta) = 2\tan(\phi/2)$$

Figure 3. Transformation of phase space ellipses where the wedge is designed to increase energy spread; the wedge is oriented so that lower energy particles go through the thicker end of the wedge.

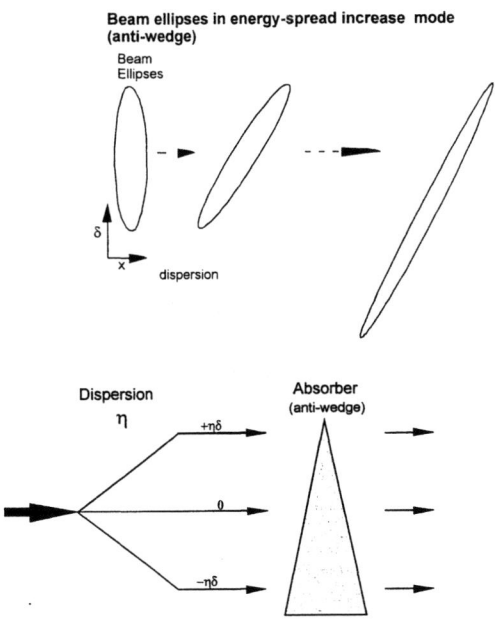

Table 2: Examples of Phase Space Exchange

Exchange Parameter	Example 1	Example 2	Example 3
E_{beam}	300	120	25
P_{beam}	392	199.4	76.9
$\delta_0 = \delta P/P$ (initial)	0.074	0.04	0.0081
η_0 initial dispersion	1.0m	0.5	-0.105
σ_0 - initial beam size	0.037m	0.017	0.001
β_0 - initial betatron function	0.34m	0.15	0.013
ε_0 - initial transverse emittance (unnormalized)	0.004 m-rad	0.00212	84×10^{-6}
Wedge material	Be	Be	LiH
dp/ds (MeV/c/cm)	3.0	3.4	5.70
tan θ	1.0	0.65	0.6
$\delta' = $ dp/ds tan θ/p	0.80	1.107	0.044
Exchange factor	0.44	0.67	1.60
thickness (2σ)	0.17	0.035	0.0017

Muon Dynamics in a Toroidal Sector Magnet

Juan C. Gallardo[1], Richard C. Fernow and Robert B. Palmer

Physics Department, Brookhaven National Laboratory, Upton, NY 11973, USA

Abstract. We present a Hamiltonian formulation of muon dynamics in toroidal sector solenoids (bent solenoid).

I INTRODUCTION

The present scenario for the cooling channel in a high brightness muon collider [1] calls for a quasi-continuous solenoidal focusing channel. The beam line consists of a periodic array of hydrogen absorbers immersed in a solenoid with alternating focusing field and rf linacs at the zero field points.

The simple $\frac{dE}{dx}$ energy loss in conjunction with multiple scattering and energy straggling leads to a decrease of the normalized transverse emittance. Reduction of the longitudinal emittance could be achieved by wedges of material located in dispersive regions; at least in principle, this scenario seems appropriate to obtain effective 6-D phase space reduction. [2]

A conventional chicane is a dispersion element but its use presents a serious challenge, as it is very difficult to integrate it with the solenoidal channel. The matching into the periodic solenoidal system imposes constraints on the Twiss parameters of the beam which seems not easily achievable. A possible alternative is the use of curved solenoids in conjunction with wedge absorbers as suggested by one of the authors. [3]

Solenoids and toroidal sectors have a natural place in muon collider design given the large emittance of the beam and consequently, the large transverse momentum of the initial pion beam or the decay muon beam. Bent solenoids as shown in Fig.1 were studied for use at the front end of the machine, as part of the capture channel [4] and more recently as part of a diagnostic setup to measure the position and momentum of muons. [5]

---

[1] Email:gallardo@bnl.gov

II TOROIDAL SECTOR SOLENOID

If we restrict ourselves for the moment to a horizontal bending plane, the magnetic field inside of the solenoid and near the axis has a gradient (field lines are denser at smaller radius) described approximately by $\vec{B}(x,y,s) \approx B_s \vec{e}_s$ with

$$B_s(x,y,s) \approx \frac{B_s(0,0,s)}{(1+hx)} \tag{1}$$

where s is the coordinate along the particle trajectory and $h = \frac{1}{R_o}$ is the curvature at the position s, with R_o the radius of curvature. As a consequence of the curvature of the trajectory and the corresponding magnetic gradient, the center of the particle guide orbit, averaged over the Larmor period, drifts in a direction perpendicular to the plane of bending [6]. The combined drift velocity can be written as,

$$\frac{d\vec{r}}{dt} = v_\| \frac{\vec{B}}{B} + \frac{m_\mu}{2q_\mu}(2v_\|^2 + v_\perp^2) \frac{(\vec{R}_o \times \vec{B})}{(1+hx)R_o^2 B^2}. \tag{2}$$

and the magnitude of the transverse drift velocity is

$$v_{\text{drift}}^T = \frac{m_\mu}{2q_\mu}(2v_\|^2 + v_\perp^2) \frac{h}{(1+hx)B_s} \tag{3}$$

Clearly a y-position versus energy ($v_\|$) correlation will develop as the muon beam travels along the toroidal sector solenoid.

From Eq.2 above we notice that if we include an additional vertical field, a dipole with a curvature equal to that of the bent solenoid for the reference energy, i.e.

$$\vec{B}_D \approx -\frac{|B|}{v_\|} v_{\text{drift}}^T \vec{e}_y \tag{4}$$

then Eq.2 reduces to,

$$\frac{d\vec{r}}{dt} = v_\| \frac{B_s}{|B|} \vec{e}_s + (v_\| - v_\|^o) \frac{B_D}{|B|} \vec{e}_y \tag{5}$$

and consequently, particles with the chosen energy will not drift vertically and will remain on the axis of the bent solenoid. Those particles with larger energy will drift upward (positive y-direction) and those with lower energy downward (negative y-direction), achieving the needed dispersion. The magnitude of the dispersion is given by

$$D_y = 2\pi \frac{p_o}{q} \frac{B_D}{B_s^2} = 2\pi h \left(\frac{p_o}{qB_s}\right)^2 \tag{6}$$

where p_o is the chosen momentum corresponding to zero dispersion.

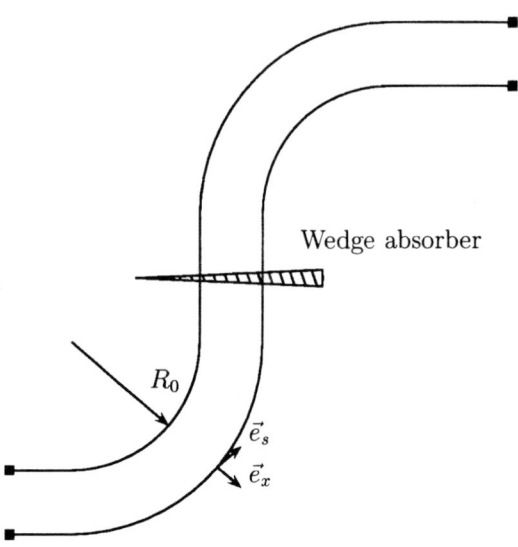

FIGURE 1. Schematic of emittance exchange system with a bent solenoid

III DYNAMICS IN A TOROIDAL SOLENOID. HAMILTONIAN FORMULATION

From general principles of classical mechanics and following the usual approximations in accelerator physics [7] the normalized Hamiltonian reads,

$$H(x, p_x, y, p_y, z, \delta; s) \approx -\frac{q}{p_0}(1 + hx)A_s(s) - (1 + \delta)hx$$
$$+ \frac{(1+hx)}{2(1+\delta)}\left\{(p_x - \frac{q}{p_0}A_x)^2 + (p_y - \frac{q}{p_0}A_y)^2\right\} \quad (7)$$

where the path length s is the independent variable, p_x, p_y are normalized momenta with respect to p_0, the initial reference total momentum $p_0 = \sqrt{p_x^2 + p_y^2 + p_s^2}$; $z = s - \beta_0 ct$, $\delta = \frac{(p-p_0)}{p_0}$ and $\vec{A} = (A_x, A_y, A_s)$ is the vector potential. The vector potential satisfies the gauge condition $\nabla \cdot \vec{A} = 0$.

In the accelerator frame of reference, i.e. the Frenet-Serret coordinate system defined by the metric,

$$d\sigma^2 = dx^2 + dy^2 + (1 + hx)^2 ds^2 \quad (8)$$

the equations for the coordinate unit vectors are

$$\frac{d\vec{e_x}}{ds} = h(s)\vec{e_s} \quad , \quad \frac{d\vec{e_y}}{ds} = 0 \quad , \quad \frac{d\vec{e_s}}{ds} = -h(s)\vec{e_x} \quad (9)$$

The magnetic and electric fields are obtained from

$$\vec{B} = \nabla \times \vec{A} \quad \text{and} \quad \vec{E} = c\beta_0 \frac{\partial \vec{A}}{\partial z} \tag{10}$$

with

$$\nabla \times \vec{A} = \frac{1}{(1+hx)} \{\partial_y(1+hx)A_s - \partial_s A_y\} \vec{e}_x$$
$$+ \frac{1}{(1+hx)} \{\partial_s A_x - \partial_x(1+hx)A_s\} \vec{e}_y$$
$$+ \{\partial_x A_y - \partial_y A_x\} \vec{e}_s \tag{11}$$

and

$$\nabla \cdot \vec{A} = \frac{1}{(1+hx)} \left[\frac{\partial}{\partial x}(A_x(1+hx)) + \frac{\partial}{\partial y}(A_y(1+hx)) + \frac{\partial}{\partial s} A_s \right] \tag{12}$$

The lowest order approximation for a toroidal solenoidal field is given by Eq.1. The corresponding vector potential in the next order is [8],

$$\vec{A} = -\frac{1}{2} B_o \frac{y}{(1+hx)} \vec{e}_x + \frac{B_o}{2h} \ln(1+hx) \vec{e}_y \tag{13}$$

and the corresponding magnetic fields are:

$$B_x = -\frac{1}{(1+h(s)x)} \frac{1}{2h(s)} \left\{ \frac{dB_o}{ds} \ln(1+h(s)x) \right.$$
$$\left. - \frac{B_o}{h(s)} h'(s) \ln(1+h(s)x) + h'(s) \frac{B_o x}{(1+h(s)x)} \right\} \tag{14a}$$

$$B_y = -\frac{1}{2(1+h(s)x)^2} \left\{ \frac{dB_o}{ds} y - h'(s) \frac{B_o(s)xy}{1+h(s)x} \right\} \tag{14b}$$

$$B_s = \frac{B_o}{1+h(s)x} \tag{14c}$$

One possible second order approximation for the vector potential of a dipole is

$$A_x = -\frac{B_D}{2h(s)} \frac{h''(s) \, x(y^2 - \frac{1}{3}x^2)}{(1+h(s)x)} \tag{15a}$$

$$A_y = \frac{B_D}{2h(s)} \frac{h'(s) \, xy}{(1+h(s)x)} \tag{15b}$$

$$A_s = -\frac{B_D}{2h(s)} \left\{ 1 + h(s) \, x - h'(s) \, (y^2 - x^2) \right\} \tag{15c}$$

Substituting the total vector potential into the Hamiltonian, and dropping some constants we can write,

$$H^{\text{dip}}_{\text{tor.sol.}}(x, p_x, y, p_y, z, \delta; s) \approx \tfrac{1}{2}(hx)^2 - \delta h x$$
$$+ \tfrac{(1+hx)}{2(1+\delta)}\left\{(p_x - \tfrac{q}{p_o}A_x)^2 + (p_y - \tfrac{q}{p_o}A_y)^2\right\} \quad (16)$$

We have written a simple Fortran program to solve the equations of motion from the above Hamiltonian ; its result for a few representative cases of interest are shown in Fig.2.

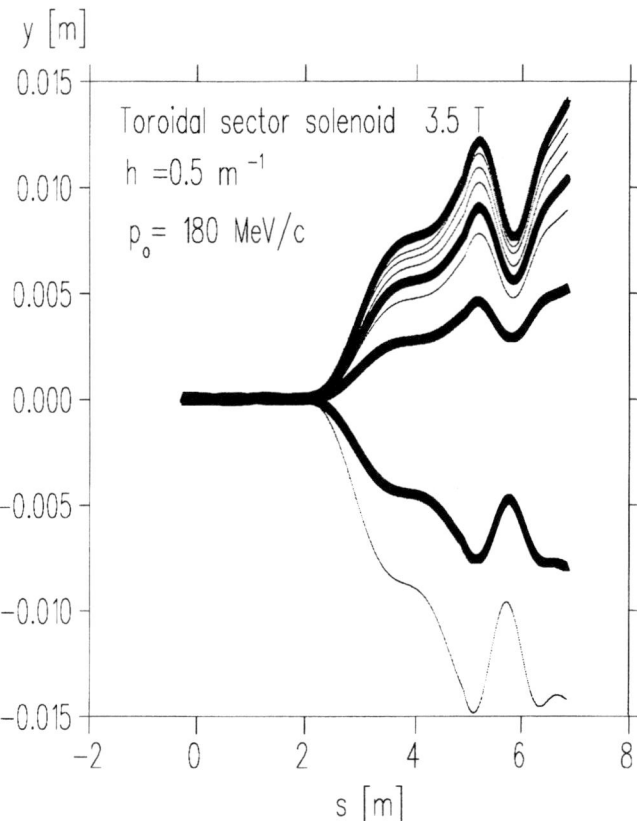

FIGURE 2. Example of dispersion created by a toroidal sector solenoid plus a vertical dipole. We show several tracks with energies (±1%) larger and smaller than the reference energy p_o.

A second order expansion of the bent solenoid magnetic field given in Eqs. 14 has been used together with a second order expansion of the dipole magnetic field in the cooling simulation program ICOOL [9]. Fig.3a shows an example of the dispersion D_y in a bent solenoid obtained in ICOOL as a function of the dipole strength B_D. It is apparent that the dependence of Eq. 6 on B_D is well satisfied. Likewise, Fig 3b shows simulation results for the dispersion as a function of B_s^{-2}. Again we see that the mentioned equation gives a good representation of the results.

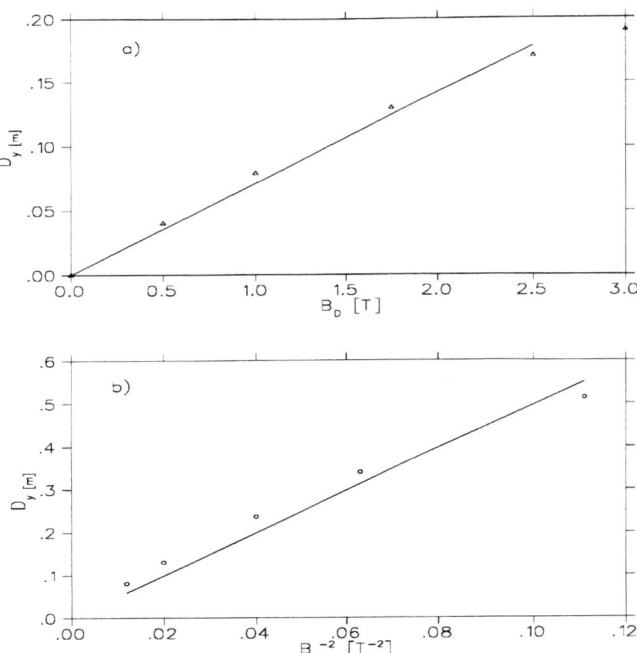

FIGURE 3. a) Dispersion, D_y vs. dipole magnetic field, B_D; b) Dispersion, D_y vs. solenoid magnetic field B_s^{-2}.

IV ACKNOWLEDGEMENTS

This work was supported by the US DoE under Contract No. DE-AC02-76CH00016.

REFERENCES

1. R. B. Palmer, presentation at FNAL Muon Collider Collaboration Meeting, July 1995, unpublished; R. B. Palmer, et al., *Muon Colliders*, 9th Advanced ICFA Beam Dynamics Workshop, Ed. Juan C. Gallardo, AIP Press, AIP Conference Proceedings 372, (1996).
2. R. B. Palmer, *Cooling Theory*, in preparation.
3. R. B. Palmer, private communication.
4. $\mu^+\mu^-$ **Collider: A Feasibility Study**, New Directions for High-Energy Physics. Proceedings of the 1996 DPF/DPB Summer Study on High-Energy Physics Snowmass'96, Chapter 4; see also the Muon Collider Collaboration WEB page http://www.cap.bnl.gov/mumu/

5. C. Lu, K. T. McDonald and E. J. Prebys, *A Detector Scenario for the Muon Cooling Experiment*, Princeton/$\mu\mu$/97-8, July 1997.
6. F. Chen, *Introduction to Plasma Physics*, Chap.2, Plenum Press (1979).
7. C. Wang and A. Chao, *Notes on Lie algebraic analysis of achromats*, SLAC/AP-100, Jan. 1995; E. D. Courant and H. S. Snyder, *Theory of the alternating-gradient synchrotron*, Ann. of Phys. 3,1 (1958); R. Ruth, *Single Particle Dynamics in Circular Accelerators*, Physics of Particle Accelerators, AIP Conference Proc. 153, Ed. M. Month and M. Dienes, Vol.1, pag. 150, (1987); C. Wang and A. Chao, *Transfer matrices of superimposed magnets and RF cavity*, SLAC/AP-106, Nov. 1996; H. Wiedemann, *Particle Accelerator Physics II*, pag. 51, Springer (1995).
8. See C. Wang and A. Chao [7] and A. Morozov and Solov'ev, *Motion of charged particles in e.m. fields*, Review of Plasma Physics, vol. II, Ed. M. A. Leontovich, Consultants Bureau, Division of Plenum Publishing Company, New York (1966).
9. R. Fernow, ICOOL, fortran program to simulate muon ionization cooling.

Ionization Cooling and Muon Dynamics

Zohreh Parsa

Brookhaven National Laboratory [1]
Physics Department, 901A
Upton, New York 11973-5000, USA.
e-mail: parsa@bnl.gov

Abstract. Muon colliders potential to provide a probe for fundamental particle physics is very interesting. To obtain the needed collider luminosity, the phase-space volume must be greatly reduced within the muon life time. The Ionization cooling is the preferred method used to compress the phase space and reduce the emittance to obtain high luminosity muon beams. We note that, the ionization losses results not only in damping, but also heating. We discuss methods used including moments methods, Focker-Plank Equation, and Multi Particle Codes. In addition we show how a simple analysis permits us to estimate the most part of the optimal system parameters, such as optimal damping rates, length of the system and energy.

INTRODUCTION

To obtaine very intense muon beams needed for a high-luminosity muon collider a high intensity proton source is bunch compressed and focused on a heavy metal target. The pions generated are captured and transferred to a decay channel. Since the muons are created through pion decay into a diffuse phase space, some form of cooling is necessary, to compress the phase space and reduce the emittance. Ionization is the preferred method, where the muons passing through a material medium lose momentum and energy through ionization interactions in transverse and longitudinal directions. The normalized emittance is reduced due to transvers energy losses. Although Ionization cooling studies dates back to 1970 (see e.g. [1–5]), experimentally it is a new technique that has not yet been demonstrated. We discuss the formalism including equations for damping rates of particles, optimization criterion, etc. We show how to estimate the optimal damping, energy and length of the system and illustrate the use of our formulations, with a numerical example.

---

[1] Supported by US Department of Energy Contract No. DE-AC02-76CH00016.

FORMALISM

At present time principles of the ionization cooling are well known and established. However, some details of the theory are not completely clear. A milestone of the theory are equations for damping rates of an individual particle. If longitudinal ionization cooling is absent transverse damping rates are defined by the following equation:

$$\lambda_x(z) = \lambda_y(z) = \frac{\frac{dE^{ion}}{dz}(z)}{mc^2\beta^2(z)\gamma(z)} = \lambda_\perp(z) \tag{1}$$

Here x, y are transverse variables and z is a longitudinal variable.
$\frac{dE^{ion}}{dz}(z)$ = ionization energy losses; m is the muon mass; β, γ are the relativistic muon parameters. Where $\lambda_\perp$ and $\lambda_\parallel$ are natural transverse and longitudinal damping rates (decrements) respectively.

Longitudinal ionization cooling may be introduced artificially, if: 1) we use section with dependence $\frac{dE^{ion}}{dz}(x, z)$; 2) this section is designed in such a way as to have dispersion function Ψ_x inside. Then

$$\lambda_\parallel(z) = \lambda_\perp(z)\frac{\Psi_x(z)}{L(z)} \tag{2}$$

Here

$$L(z) = \frac{\left(\frac{dE}{dz}\right)^{ion}(0, z)}{\frac{d}{dx}\left(\frac{dE}{dz}\right)^{ion}(0, z)} \tag{3}$$

However, introduction of $\lambda_\parallel(z)$ changes a corresponding damping rate $\lambda_x(z)$ [2,9]

$$\lambda_x(z) = \lambda_\perp(z) - \lambda_\parallel(z) \tag{4}$$

In general case, there is a fundamental theorem [2], that

$$\sum_i \lambda_i(z) = 2\lambda_\perp(z) \tag{5}$$

Thus, we see, that we can only redistribute the damping rates.

However, each section with ionization losses give not only damping, but a heating as well: transverse heating appears due to multiple Coulomb scattering and longitudinal one is due to so named "straggling" of the ionization losses (we note that, this straggling is produced by fast "knock-on" ionization electrons).

From theoretical point of view, a situation with ionization cooling completely corresponds to a situation with radiation cooling whose theory is well developed.

There is some standard "hierarchy" of methods for analyzing such systems: i) Moments methods, ii) Focker-Planck equation, iii) Multi-particles codes.

i) Method of moments (MM)

Using equation of motion for separate particles or kinetic equation, it is easy to derive equations for the second order moments ($\langle x^2 \rangle, \langle xx' \rangle, \langle x'^2 \rangle$ and etc.). (see, for example, [11]). Analysis of these equations permits us to find equilibrium emittances and to investigate evolution of emittances with time for linear external fields. It is necessary to be very careful when using this method (MM) e.g., 1) In transverse plane a strong focusing usually is used, and therefore "smoothing" of the motion (change $w_\perp^2(z) \rightarrow w_0^2$) must be done very accurately. 2) Often some authors have solved only the equations for $\langle x'^2 \rangle$ (which, really, is proportional to beam temperature) *without* solving the equations for the last moments. However, to calculate evolution on time accurately we must solve the system of coupled equations for the moments [6,7,11].

ii) Focker-Planck (FP) equation

To find a form of a distribution function and to estimate beam losses due to diffusion it is possible to use the FP equation. The one-dimensional FP equation has the following form:

$$D\frac{\partial}{\partial I}\left(I\frac{\partial f}{\partial I}\right) + \alpha\frac{\partial}{\partial I}(fI) = \frac{\partial f}{\partial z} \qquad (6)$$

Where f = the distribution function ($f = f(I,z)$); I = action; z = longitudinal variable; D = the diffusion coefficient; and α = the damping coefficient. The FP equation must be solved using the following boundary condition:

$$f_m(I_m, z) = 0 \qquad (7)$$

where I_m is the aperture in I-variable. If $I_m = \infty$, Eq. (5) has a stationary solution:

$$f_0 = \frac{1}{I_o}\exp\left(-\frac{I}{I_0}\right) \qquad (8)$$

Where $I_0 = \frac{D}{\alpha}$.

Let us underline, that if $I_m \gg I_0$, we can estimate particle losses using a simple formula:

$$\frac{1}{N}\frac{dN}{dz} \simeq D\frac{I_m \partial f_0}{\partial I}(I = I_m) = \frac{DI_m}{I_0^2}\exp\left(-\frac{I_m}{I_0}\right) =$$
$$\alpha\frac{I_m}{I_0}\exp\left(-\frac{I_m}{I_0}\right) \qquad (9)$$

If $\frac{I_m}{I_0} \simeq$ constant,

$$\frac{\Delta N}{N} \sim \alpha L \frac{I_m}{I_0}\exp\left(-\frac{I_m}{I_0}\right) \qquad (10)$$

For example, if $\frac{I_m}{I_0} = 9$ and $L = \frac{5}{\alpha}$ (that corresponds to damping of an initial emittance in 150 times), three-dimensional diffusion losses $\frac{\Delta N}{N} \sim 0.017$

If condition $I_m \gg I_0$ is not performed, we can find a beam evolution by numerical solution of FP equation.

<u>iii) Multi-particle codes.</u>

We see that method of moments (MM) and Focker-Planck (FP) equation gives only limited information. A whole information about the beam, including large-angle scattering, interactions with residual gas, increase of emittances due to non-linearity of the external field and so on can be found only by multi-particle codes. Map methods are not convenient for our problem due to presence of large noises. However, we note, that it is a correct procedure that can be used to validate accuracy of the final results of numerical calculations.

OPTIMIZATION AND CALCULATION OF PARAMETERS

Problem of choice of optimal parameters for the muon cooling system is open, and its solution is far from trivial due to infinite numbers of possible options. The process of optimization can usually be divided into five stages: 1) choice of optimization criterion; 2) calculation of "primary" parameters by use of simple analytical models; 3) choice of design of focusing and accelerating systems; 4) search of optimum for chosen focusing and accelerating systems by use of more sophisticated algorithms; 5) validation of chosen scheme by use of multi-particles codes, etc. Here we discuss the first two points (others are not of scope of this paper).

• <u>Optimization:</u> Luminosity of collider L is defined by the following expression:

$$L \sim \frac{N^2 f}{g_x g_y} = \frac{N^2 f}{\epsilon_\perp^f \cdot \beta_\perp^f} \tag{11}$$

Where N = a number of muons per bunch, f = mean repetition frequency of collisions, $\epsilon_\perp^f$ = emittance at collision point and $\beta_\perp^f$ = β-function at collision point. Usually $\beta_\perp^f$ is limited by condition:

$$\beta_\perp^f \geq \sigma_z^f \tag{12}$$

where σ_z^f is a longitudinal bunch size. Let us assume, that: 1) $\frac{\Delta p_f}{p}$ is known (monochromatic experiments); 2) we can redistribute emittances inside a given six-dimensional phase volume. Then, taking into account losses in the cooling system, we can rewrite Eq. (11) in the following form:

$$L \sim \frac{N_0^2 \exp\left(-\frac{2}{cT_0} \int_0^z \frac{dz}{\gamma(z)}\right) D^2}{\sqrt{V_6^N \cdot \epsilon_\parallel^f}} \cdot \left(\frac{\Delta p}{p}\right)_\parallel^f \tag{13}$$

Here "N_0" is a number of particles at an entrance of the cooling system, "exp" describes muon decay, "D" describes muon losses in cooling section, and "V_6^N" is an invariant six-dimensional phase volume of muon beam.

Thus we can introduce "merit factor" which describes a quality of muon cooling system. We obtain

$$R = \frac{D^2 \exp\left[-\frac{2}{cT_0} \int_0^z \frac{dz}{\gamma(z)}\right]}{\sqrt{V_6^N}} \tag{14}$$

Note that, the dependence on V_6^N may be stronger. With account of all the circumstances, we can write

$$R \sim (V_6^N)^\alpha \tag{15}$$

with α in interval (0.5; 2/3).

• Calculation of primary parameters

Let us neglect muon losses ($D = 1$, $\exp\left[-\frac{2}{cT_0} \int_0^z \frac{dz}{\gamma(z)}\right] = 1$). Then to maximize R we must minimize V_6^N. Taking into account standard formula for equilibrium phase volumes, we have:

$$V_6^N = \frac{W_\perp^2 \cdot W_\|}{\alpha_\| \left(\alpha_\perp^0 - \frac{\alpha_\|}{2}\right)^2} \tag{16}$$

This equation has a solution relative to $\alpha_\|$:

$$\alpha_\perp^{\min} = \alpha_\|^{\min} = \frac{2\alpha_\perp^0}{3}; \tag{17}$$

$$(\sqrt{V_6^N})_{\min} = \frac{27}{8} \frac{W_\perp^2 \cdot W_\|}{(\alpha_\perp^0)^3} \tag{18}$$

However, a situation will change if we take into account the muon losses.

As an example, let us consider a simplest case: 1) $D = 1$; 2) $\lambda_\| \ll \Omega$; 3) $\gamma = \gamma_0$. In this case (for $\lambda_\perp = \lambda_\|$)

$$R \simeq \frac{\exp[-2L/cT_0 \gamma_0]}{(V_6^N)_{\min}^{\frac{1}{2}} \left[1 + \frac{\epsilon_\perp^0 - \epsilon_{eq}}{\epsilon_{eq}} \exp(-\lambda L)\right]^{-3/2}} \tag{19}$$

It is easy to see that in this case there is an optimal system length L, which is defined by the following equation:

$$L_0 = \frac{1}{\lambda} \ell n \left(\frac{\frac{AB}{\lambda}}{\frac{3}{2}B - \frac{A}{\lambda}}\right)^{-1} \tag{20}$$

Where $A = \frac{2}{cT_0 \gamma_0}$ and $B = \frac{\epsilon_1^0 - \epsilon_{eq}}{\epsilon_{eq}}$.

Numerical Example: We consider a numerical example with $\gamma_0 = 1.5$; $\lambda = 5 \cdot 10^{-2} m^{-1}$; $B = 10$. Then we obtain:

$$\tau_0^{max} = \lambda L = 3.6$$
$$R_{max} \simeq \frac{0.6}{\sqrt{(\sqrt{V_6^N})_{min}}} \tag{21}$$

DISCUSSION

We discussed the formulation and methods used to study the muon dynamics. We did not try to present all the questions in detail. We illustrated and would like to underline, that simple analysis permits us to estimate the most part of the optimal system parameters, such as optimal damping rates, length of the system, energy, etc.

REFERENCES

1. A. Skrinsky, Colliding Beams Program in Novosibirsk, International Seminar on Prospects in High Energy Physics; Morges, 1971.
2. V. Parchochuk and A. Skrinsky, Methods of cooling of charged particle beams, *Phys. of Elementary Part. and Atomic Nucl.*, **12** 557-613 (1981).
3. Physics Potential and Development of $\mu^+\mu^-$ Colliders Workshop - AIP Conf. Proceedings, 352, AIP Press, Woodbury, NY.
4. The 9th Advanced ICFA Beams Dynamics Workshop - AIP Conf. Proceedings, 372.
5. $\mu^+\mu^-$ Collider. A Feasibility Study, BNL-52503.
6. T. Vserolozhskaya, Kinetics of ionization cooling of muons, AIP CP 372, p. 159, (1996).
7. Z. Parsa and P. Zenkevich, Kinetics of Muon Longitudinal cooling in *Proceedings of the 1996 Beam Stability and Nonlinear Dynamics Symposia* AIP CP 405, p.165-172.
8. Z. Parsa and P. Zenkevich, unpublished, ITP-96.
9. D. Neuffer, Muon cooling and applications, *Montreux Conference on Beam Cooling, Proc.*, p. 49, 1993.
10. Z. Parsa and P. Zenkevich, Application of Moments Method to Dynamics of Muon Cooling System in *Beam Stability and Nonlinear Dynamics* AIP CP 405, p.183-188, 1997.
11. Z. Parsa, Emittance and Moments in Accelerators and Beamlines Parts (I) and (II), Reports and BNL-44618, (10/1989); W. Lysenko, Z. Parsa, in *Beam Stability and Nonlinear Dynamics* AIP CP 405, p. 211-222, 1997.

New μ± Cooling for μ± Colliders and Possible Realization at JHF/KEK

K. Nagamine

Meson Science Laboratory, Institute of Materials Structure Science
High Energy Accelerator Research Organization(KEK)
1-1 Oho, Tsukuba-shi
Ibaraki-ken 305, JAPAN
and
Muon Science Laboratory, The Institute of Physical and Chemical Research (RIKEN)
Hirosawa, Wako-shi, Saitama-ken 351-01, JAPAN

Abstract. Based upon the proposed concept of the front-end design of the μ± colliders combined with the concept of ultra-slow μ+ generation via laser resonant ionization of thermal muonium, a new scheme of "zero-energy cooling" is proposed. Results of some design calculations for the facility to be realized at JHF/KEK are presented. Also, a design of a pilot-station as well as R & D program at either RIKEN/RAL or at KEK/MSL are described.

INTRODUCTION

In the widely-approved conceptual design of the μ± colliders (1,2,3), there are the important components such as proton accelerator, pion collector, π→μ decay section, rf rotation, muon cooling, muon accelerating ring and muon colliding ring. Among these components, the muon cooling is now known to be the most difficult and important component which should be subject to the various types of R & D works.

At the same time, an achievement of the muon cooling is important not only for the muon source for further acceleration like the one to be used for the μ± colliders but also for the one to be used in various advanced muon science studies (4).

According to the presently accepted design of μ± colliders, the muon cooling is expected to be mainly carried out by the ionization cooling by using more than 10 stages of the configurations comprising of Li absorber, wedge absorber and linear accelerator. There, one-drectional accelertion under homogenious energy-loss gives us a transverse-directional cooling. Such ionization cooling is expected to produce μ± beam of 3 x 10^{12} μ/bunch, 20 MeV and a normalized emittance of ε_N = 4 x 10^{-5} rad · m.

Contrary to the ionization cooling method the present author has been proposing the concept of the "Zero-energy" cooling where the muons are once completely stopped inside the materials followed by a re-emission of the slow muons from the surface of the materials (5).

ZERO-ENERGY COOLING

Let us summarize the excellent features of the zero-energy cooling method described in an earlier paper (5) and in the later part of the present paper.

1. Once the method becomes realized, the transverse emittance becomes extremely small. For example, the thermal μ^+ obtained by a laser-ionization of the 0.2 eV thermal Mu produceed from hot tungsten surface has the transverse momentum (P_t) of 6×10^{-3} MeV/c in the length (d) of 1 cm, so that the normalized emittance $\varepsilon_N(1\ \text{TeV})$ of the 1 TeV μ^+ is $(P_t/P_L) d = 4 \times 10^{-11}$ rad · m. As described later, the slow μ^- from the ($t^3\text{He}\mu^-$) decay method has P_t of 1 MeV/c, so that $\varepsilon_N(1\ \text{TeV})$ is 6×10^{-8} rad · m.

2. Once the zero-energy cooling is realized, the produced muon is ready for further accelerations.

3. Even without any accelerations, the produced muon is almost promptly able to be used for advanced muon science experiments; e.g. surface science studies, etc.

NEW μ^+ COOLING; RECENT DEVELOPMENT

Since the late 80's, in low energy muon science community, there have been several proposals for the method of ultra-slow μ^+ generation. Some distinguished examples are as follows; 1) laser resonant ionization of thermal muonium (Mu) in vacuum produced from hot metal surface (6) 2) degrading energetic μ^+ in rare gas solid (Ar, Kr, Xe) which has a finite band-gap (7); 3) phase space compression of energetic μ^+ by deceleration and acceleration by electric fields (8); 4) ionization cooling or frictional cooling by one-directional accelerating under homogenizes slowing-down (9); 5) cooling by channeling process of energetic μ^+ through crystal (10).

As described in the previous report (5), only the first method seems to be the most relevant for high-intensity and high-brightness with a low emittance μ^+ beam production method enough to be used for the $\mu^\pm$ colliders in the future. Actually, the earlier two methods are now extensively applied for the μ^+SR spectroscopy on very thin materials and surface magnetism.

As described in Fig.1, there are two processes for a realization of ultra-slow μ^+ source by the thermal-Mu-ionization method: thermal Mu production in vacuum by stopping μ^+ at a rear-side of metallic foil followed by the μ^+ diffusion towards the foil surface and Mu evaporation from the surface; efficient muonium ionization by e.g. laser resonant ionization via 1s →2p→ unbound excitation.

As one of the most powerful step of efficient muonium ionization, we have proposed the method of laser resonant ionization of thermal muonium produced at a hot tungsten placed at the primary proton beam line. In order to realize this ultra-slow μ^+ production idea, a new laboratory space with a dedicated pulsed (50 ns width and 20 Hz repetition rate) proton beam line from the 500 MeV booster synchrotron was constructed at KEK-MSL. Hot tungsten (W) was adopted as the thermal Mu-producing target. All of the target area as well as the following ion optics of slow μ^+ transport was maintained at a pressure below 10^{-8} Torr under the conditions of proton-beam delivery and target heating. The actual target comprises 2 mm thick BN (boron nitride) and 50 µm thick W. The BN, located beside the 2300 K hot W foil, was used for efficient π^+ production with the minimum allowable divergence of the proton beam.

For an efficient ionization method of thermal Mu produced by pulsed primary protons, we adopted a resonant ionization scheme using pulsed lasers (11,12), namely, the single photon resonant transition of 1s→2p (122 nm) followed by the photo-ionization transition 2p→unbound (<366 nm). For this purpose, a laser system was constructed for intense 122 nm VUV light with a 200 GHz frequency width (FWHM), which was matched to the Doppler broadening of the thermal Mu.

There, coherent VUV light was generated by a four-wave sum-difference method developed at Imperial College; the transition to $4P^55P$ [1/2, 0] of Kr levels was induced by two-photon absorption of the 212 nm light, followed by subsequent de-excitation by the 820 nm light, thus producing VUV light at around 122 nm. The other laser light used to ionize the 2p state of the Mu (355 nm) of 30 mJ/ (5 ns) was introduced into the target region through a path bypassing the Kr/Ar chamber.

The basic structure of the ion-extraction optics for the ionized products of μ^+ comprised an SOA immersion lens for 9.2 kV acceleration as well as focusing followed by an electric bend (EB) and a magnetic bend (MB) with axially focusing electric quadrupole lens. This front part of the ion optics was followed by two electric- mirror bending systems (EM1 and EM2) with a focusing electric Q lens.

During the course of the development of the ultra-slow μ^+ production, the following new experimental findings have been obtained.

(i) By measuring yield of the ultra-slow μ^+ as a function of temperature, the activation energy for thermal Mu production from hot W surface was obtained. The similar measurements have been done for the thermal T, D and H, all of which are produced by nuclear reactions, yielding the mass-independent values of activation energies, suggesting a thermionic emission mechanism of the hydrogen-like neutrals from the hot W surface.

(ii) The ultra-slow μ^+ yield, at this stage of the project development, becomes 10 μ^+/s · µA (500 MeV proton)· 10µJ/ pulse (112 nm laser power).

(iii) Polarization of the μ^+ produced by a spin-unresolved 1s→ 2p→ unbound laser ionization was found to be 50 %.

In the previous report for a possible application of the zero-energy cooling method (5), we proposed a rather straight extension of the original method; by adopting high intensity proton beam and multi-layer of BN and hot W complex irradiated by multiple laser light beams, we have claimed that one can obtain a 10^{10}/s level of ultra-slow μ^+.

Encouraged by the development of the design work on the front-end part of the $\mu^\pm$ colliders, we are now proposing a different scheme as summarized in Fig. 2. There, after placing a large-acceptance pion capture solenoid which is similar to the front-end structure of the present design of $\mu^\pm$ colliders, a long decay solenoid with a RF rotation cooling capability will be installed which is connecting to the original zero-energy μ^+ cooling scheme. The overall schematic view can be seen in Fig. 3.

Details of the design calculation for the future accelerator project of Japan Hadron Facility (JHF) as well as for the existing facilities will be described in the later part of this manuscript.

NEW μ^- COOLING; RECENT DEVELOPMENT

In the similar manner as the zero-energy cooling method for μ^+, we can expect to have a zero-energy cooling of the μ^- which is obtained by a re-emission of ultra-slow μ^- after a complete stopping of high energy μ^- inside the specially arranged materials. The inherent difficulty due to the μ^- loss by muonic atom formation is not easy to be overcome. Compared to the conventional cooling methods like an ionization/frictional cooling or a cyclotron trap, a realization of the zero-energy cooling seems to be not easy at all.

Recently, three proposals have been made for the zero-energy cooling method for the μ^-: (a) a slow μ^- production via muon catalyzed fusion (μCF) (15); (b) a slow (dμ) production with a multi-photon ionization for 2.0 keV energy difference; (c) slow (^{3}He μ^-)$^+$ production from the decay of (d^{3}He μ) and stripping by an ion collision after some acceleration (16).

Recently, during a series of the μCF experiment on solid and liquid D-T mixture as well as pure T_2 by a RIKEN-KEK-UTMSL-JAERI-ETL-RAL collaboration at RIKEN-RAL muon facility, a remarkable new finding was made which may probably be important for the future development of the zero-energy cooling of the μ^- (13,14). The He^3 produced by a natural t-decay in condensed phases of D-T and T_2 was found to be completely trapped in solid phase, while it was found to be completely releasing in liquid phase. Thus, by using the solid thin layer, a production and maintaing of ^{3}He in D-T or T_2 which was considered to be the most difficult part in the slow (^{3}Heμ^-)$^+$ method (16) is now known to be easily realized. As shown schematically in Fig. 4, the following scheme can be considered. Suppose multi-layer of a thin (a few μm) solid pure T_2 film is formed on a thin cold substrate and wait for some 10 h to accumulate ^{3}He (upto 100 ppm) from the t-decay inside the matrix of solid T_2 thin layer. Then, intense and energetic (10 ~ 100 MeV range) μ^- is introduced, a part of which is stopping inside solid T_2 with ^{3}He impurity. In our experiment (14), it was found that the μ^- in (tμ) transfers to ^{3}He with a rapid rate of $5 \times 10^9 (s^{-1})$ after forming (t^3He μ) molecule. The actual manner of forming the (^{3}Heμ) after the decay of (t^3Heμ) has two possibilities; either a radiative decay of the L=1 molecular state formed at the capture stage to the unbound ^{3}Heμ (a fraction of 80 %), or a non-radioactive decay with energetic (3 keV) particle emission of ^{3}Heμ (20 %). Thus, 20 % of the incoming μ^- is converted to the 3 keV (^{3}Heμ)$^+$. After collecting and accelerating (^{3}Heμ)$^+$ ions upto 1 MeV (corresponding to 30 keV of μ^- energy vs 10 keV μ^- binding energy to ^{3}He) a complete stripping of μ^- is expected after a passage through e.g. a thin graphite foil, producing an intense source of slow μ^-.

INTENSE ULTRA-SLOW μ^+ SOURCE PROJECT AT JHF/KEK

The idea of intense 10 keV μ^+ source with an intensity of 10^{12} μ^+ /s and an emittance of better than 10^{-7} rad· m will be realized at the near-future intense proton accelerator like Japan Hadron Facility (JHF) project at KEK. It will be straightly extended to the front-end structure of the μ^+/μ^- colliders. Here we describe a concept of JHF and its M(Muon)-Arena and somewhat detailed conceptual design works for a possible realization of the intense ultra-slow μ^+ source.

As described in more detail in the recent proposal (17), the JHF is the next intense proton accelerator facility complex to be built at KEK. As seen in Fig. 5a, the accelerator system comprises of 200 MeV linac, 3 GeV x 200 μA rapid cycling synchrotron and 50 GeV x 10 μA synchrotron. Three experimental projects, namely, M-Arena of muon physics and science, N-Arena of neutron science and E-Arena of physics with radioactive beams will be realized at the 3 GeV ring, while K-Arena of Kaon physics etc. will be realized at the 50 GeV ring.

In Fig. 5b, a zeroth facility design of the M-Arena is shown. There, due to a budgetary limitation at proposal-stage, a superconducting decay muon channel for 10 MeV~100 MeV $\mu^\pm$, a 4 MeV surface μ^+ channel and 10 keV ultra-slow μ^+ channel with an injector of a large acceptance surface μ^+ are proposed to be installed around the 3 cm thick carbon target installed at the external beam line of 3 GeV protons. Expected time structure of the pulsed protons from 3 GeV ring will be 4 pulse trains with 100 ns widthes and 200 ns separations. By using a free space between these channel complex and the beam dump, the intense ultra-slow μ^+ channel will be installed as a possible extension program.

A preliminary optics design work has been done with a strong collaboration of K. Ishida (18) and partly with a collaboration of A. Bogacz (19). As summarized in Table 1, by adopting a strong supercoducting solenoid for pion collection and reasonable length of decay-solenoid, one can obtain more than 10^{12} μ^+/s.

As described in Table 1, the momentum spread of the μ^+ obtained at the end of decay solenoid is very large (~40 %). In order to reduce momentum spread of the produced μ^+, a RF-rotation cooling by employing induction linac idea has been considered (19). By adopting 1000 cells with 50 kV voltage with a gradient of 1.5 MV/m, all to be inserted inside the decay solenoid, a model-calculation shows that one can reduduce the momentum spread down to 24 %.

Once the momentum bite after RF rotation cooling becomes small, it becomes possible to stop generated μ^+ inside a multi-layer of hot W in a manner as shown in Fig. 3, so that intense ultra-slow μ^+ beam can be generated by employing the method of laser resonant-ionization of thermal muonium thermionically emitted from the surface of hot W. Simulation calculations are in progress to confirm the characteristic features of produced ultra-slow μ^+ after some optimizations.

POSSIBLE R &D PROJECT AT RIKEN-RAL & KEK-MSL

It is important to conduct a test experiment for the scheme described above by using an existing low intensity proton accelerator facilities like either at RIKEN-RAL or at KEK-MSL.

At RIKEN-RAL, although not considered in detail at all, we may have a possibility to install the large-acceptance pion collector at the new external proton beam which might be constructed in relation to the Second Target Station. Some design calculation has been done. A part of those results are summarized in Table 2. By installing a realistic pion collection solenoid and a decay solenoid with a reasonable length, one can also expect easily more than 10^{12} μ^+/s production.

At the existing KEK-MSL facility, one can expect a similar set-up at the Laboratory-II space. Again, as summarized in Table 2, the expected intensity can be more than 10^{11} μ^+/s.

Both of these R &D projects can straightly be extended to the zero-energy μ^+ cooling instrumentation to produce the world-best ultra-slow μ^+ beam.

CONCLUSION

It is now known to be the most important to develop a new realistic $\mu^\pm$ cooling method for the $\mu^\pm$ colliders. The ionization cooling method is certainly one candidate. However, a realization of the zero-energy cooling is, to the present author's view, at least more than competitive for the μ^+ case. At the same time, the use of high-brightness and intense low energy $\mu^\pm$ beam is inevitable for the future development of the muon sciences, e.g. µSR application for advanced materials characterization, muon catalyzed fusion for energy source, biomedical application of the muon beam.

In any case, a realization of a large solid-angle pion collector is highly requested. Without significant time delay, such device should promptly be installed at existing facilities.

ACKNOWLEDGEMENTS

The author expresses his sincere thanks to Drs K Ishida, Y. Miyake, K. shimomura, A. Bogacz, S.N. Nakamura, Y. Kuno and T. Matsuzaki for their contributions to various parts of the present work.

REFERENCES

1. Proccedings of International Conf. on *Physics Potential & Development of $\mu^+\mu^-$ Colliders* (December, '95, San Fransisco), Nuclear Physics B (Proc. Suppl.) 51A (1996)
2. Reports in Workshop on Physics at the First Muon Collider and at the Front End of the Muon Collider (November, 1997)
3. Neuffer, D.V., Nucl. Instruments a350 (1994) 27.
4. Nagamine, K., *Introductory Muon Sciences*: Cambridge University Press, in print
5. Nagamine, K., Nuclear Physics B (Proc. Suppl.) 51A (1996) 115.
6. Nagamine, K., Miyake, Y., Shimomura, K., et al., Phys. Rev. Lett. 74 (1995) 4814.
7. Morenzoni, E., et al., Phys. Rev. Lett. 72 (1994) 2793.
8. Taqqe, D., Nucl. Instruments A274 (1986) 288.
9. Daniel, H., Muon Catalyzed Fusion 4 (1989) 425.
10. Bogacz, A., Cline, D., private comm.
11. Miyake, Y., et al., Nucl. Instruments B95 (1995) 265.
12. Miyake, Y., et al., Hyperfine Int. 106 (1997) 237.
13. Ishida, K., Nagamine, K., et al., Phys. Rev. Lett., submitted.
14. Matsuzaki, T., Nagamine, K., et al., Phys. Rev. Lett., submitted.
15. Nagamine, K., Proc. Japan Academy 65B(1989) 225.
16. Nagamine, K., Hyperfine Interactions 82 (1993) 539.
17. *Proposal for Japan Hadron Facility*, KEK Report 97-3, JHF-97-1 (May,

1997)
18. Ishida, K and Nagamine, K., contribution to JHF Workshop (March, 1998)
19. Bogacz, A., Ishida, K. and Nagamine, K., *UTMSL Internal Report* (March, 1997)

TABLE 1. Zeroth design of ultra-high intensity muon channel at M-Arena of JHF (18)

	Version 1	Version 2
Proton Source	3 GeV × 200 μA	
μ Production Target	3 em Be	
Pion Collection Solenoid	28T, 7.5 cm Bore (-0.5 m ~ 0.5 m)	14T, 7.5 cm Bore (-0.5 m ~ 0.5 m)
Decay Sction Solenoid	7.5 T, 15 cm Bore (1.5 m ~ 51.5 m)	3.75 T, 15 cm Bore (1.5 m ~ 26.5 m)
N_π (produced)(s^{-1})	2.75×10^{14}	
I_π (captured)(π^-, s^{-1})	2.1×10^{13}	3.0×10^{13}
N_μ (s^{-1})	1.5×10^{13}	3.1×10^{12}
Beam Size (σ_x, σ_y cm)	5.7, 5.7	6.4, 6.4
Divergence (σ_x', σ_y', mr)	190, 190	190, 190
Momemtum ($p_\mu, \Delta p_\mu$, MeV/c)	22.1, 85	138, 49

TABLE 2. Zeroth design of ultra-high intensity muon channel at KEK-MSL and RIKEN-RAL (18)

	KEK-MSL	RIKEN-RAL
Proton Source	0.5 GeV × 5 μA	0.8 GeV × 200 μA
π Production Target	3 cm Carbon	
Pion Collection	14T, 7.5 cm Bore (-0.5 m ~ 0.5 m)	
Decay Sction Solenoid	3.75 T, 15 cm Bore (1.5 m ~ 21.5 m)	
I_p (s^{-1})	3.1×10^{13}	1.25×10^{15}
I_π (captured)(π$^-$, s^{-1})	3.9×10^{10}	2.9×10^{12}
N_μ (s^{-1})	1.5×10^{10}	7.3×10^{11}
Beam Size (σ_x, σ_y cm)	6.1, 6.1	
Divergence (σ_x', σ_y', mr)	190, 190	
Momemtum (p_μ, Δp_μ, MeV/c)	114, 40	

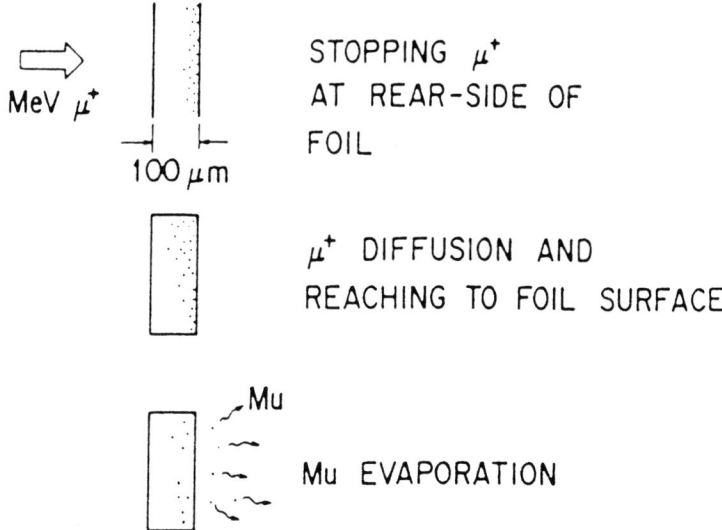

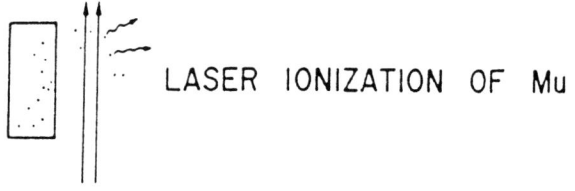

Fig. 1. Concept of ultra-slow μ^+ production by laser ionization of thermal muonium produced from the surface of hot tungsten.

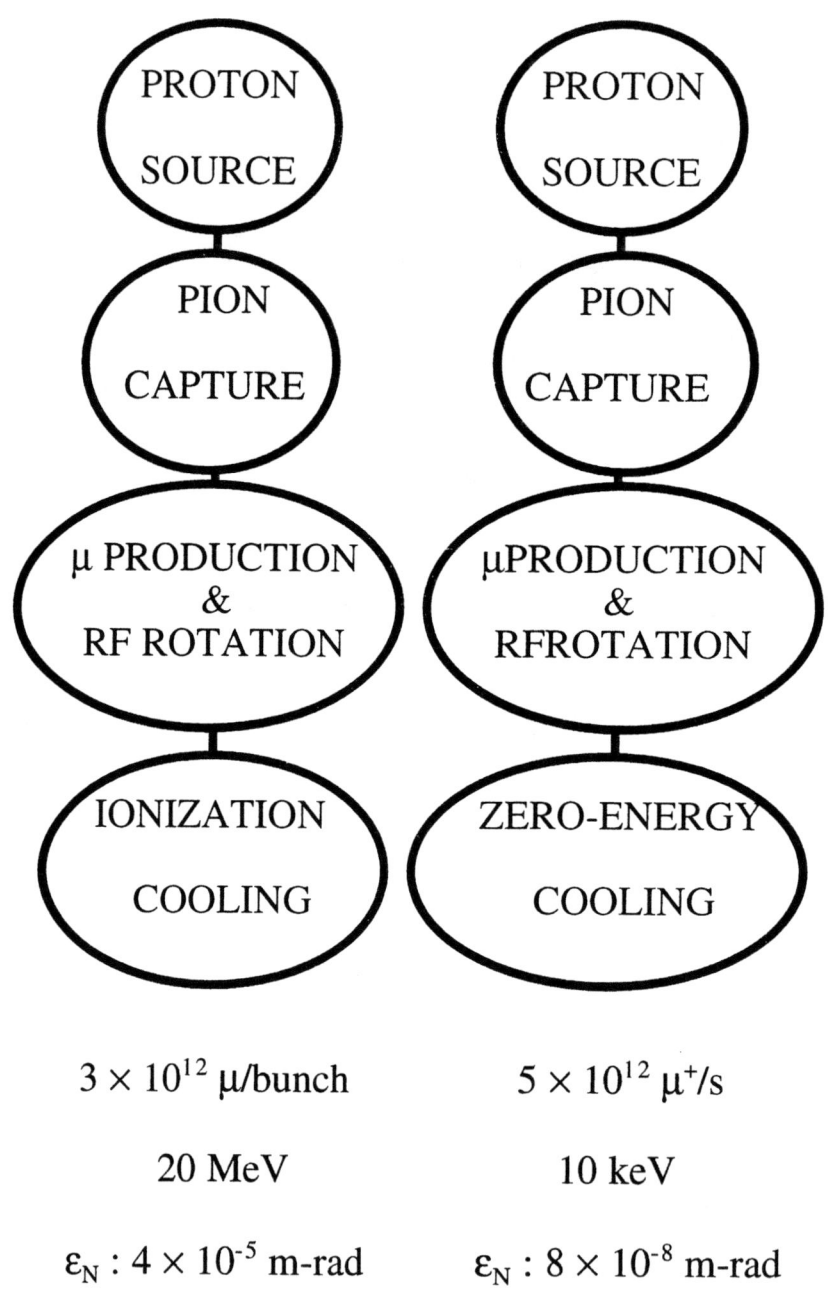

Fig. 2. Comparison between earlier part of the $\mu^+\mu^-$ colliders of the conventional method and that of the zero-energy cooling method.

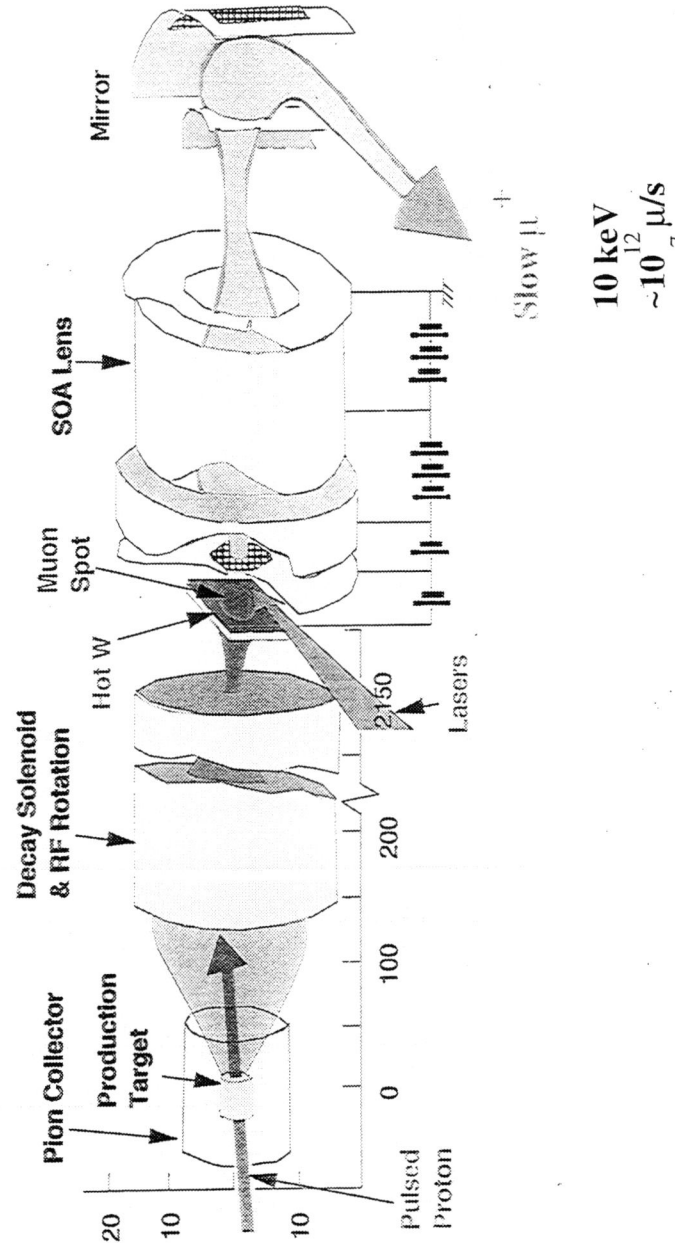

Fig. 3. Intense ultra-slow μ^+ production by the large acceptance pion collector and the decay-section solenoid with RF rotation cooling, connected to the zero-energy cooling device of hot-W and laser muonium ionization.

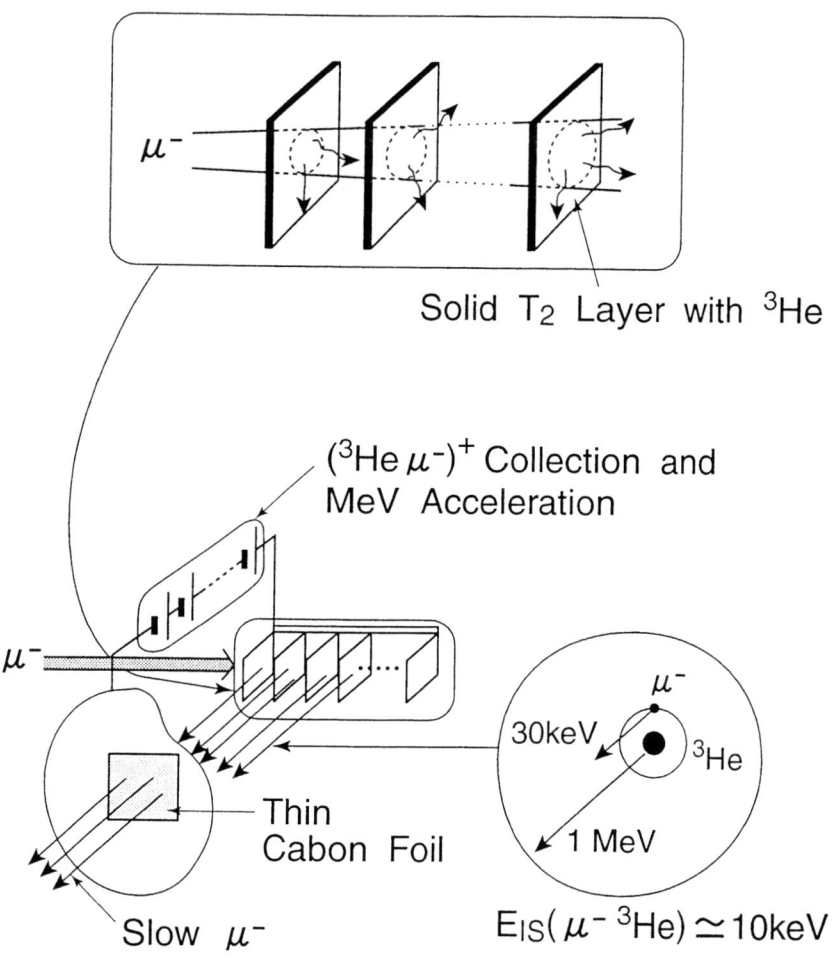

Fig. 4. Proposed large-scale source of slow μ⁻ via a break-up of $(^3He\mu^-)^+$ by using multi-layers of solid T_2 with accumulated ^{3}He of the t-decay product.

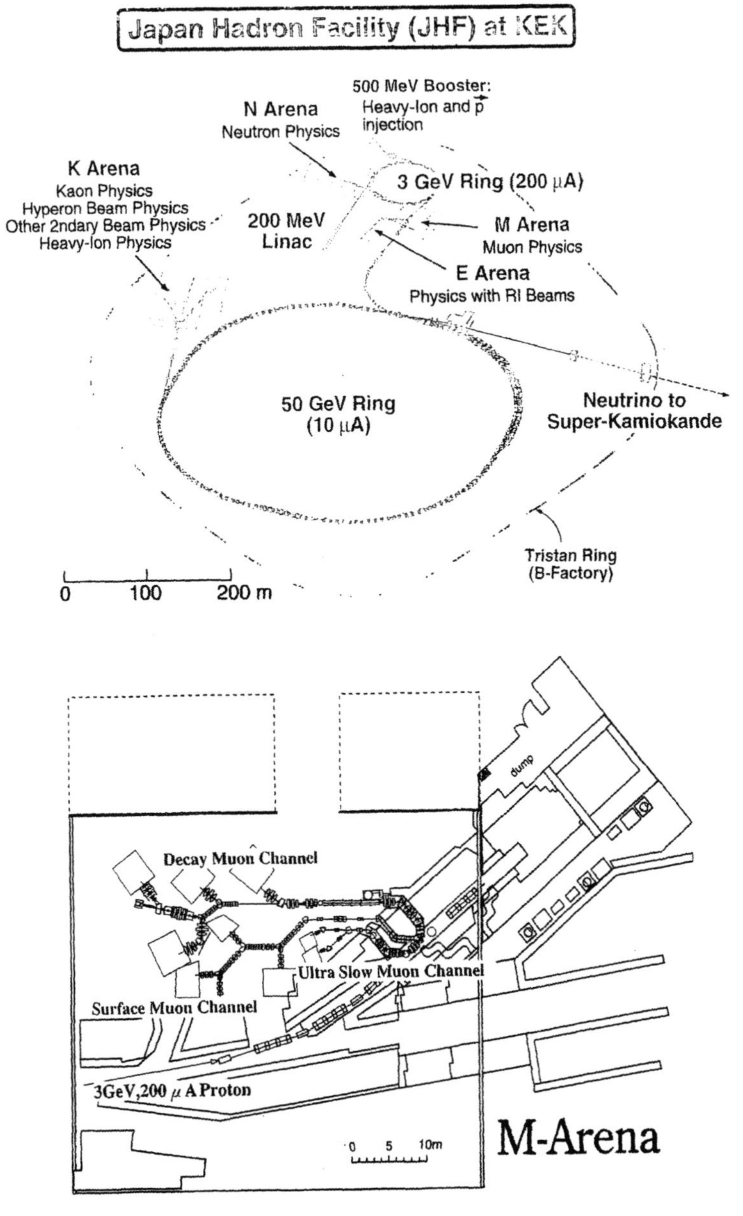

Fig. 5. (a) Overview of the proposed Japan Hadron Project layout; accelerator complex and four experimental projects. (b) zeroth design of the M-Arena experimental hall before an istallation of super-intense muon facility.

Ring Cooler for Muon Collider

V.I. Balbekov and A. Van Ginneken

Fermi National Accelerator Laboratory, Batavia, IL 60510[1]

Abstract. The possibilities of a ring cooler stage in a muon collider are explored. A basic design is examined both with analytic calculations and simulation of the evolution of beam phase space.

INTRODUCTION

This report examines the possibility of using a ring accelerator in the cooling stage of a muon collider. The main merit of such a cooler is its lower projected costs. The expense ratio linear/ring coolers scales roughly as the number of turns needed to achieve effective cooling in the ring, which is about 20 in a typical scenario.

Significant 6D cooling of a bunch is possible mainly by compression of its size since angle and energy spreads are dominated by scattering and straggling in the absorber. A system with decreasing β–function can be used for cooling in a linear scheme. For example, [1] describes a linear cooler using Li lenses with increasing gradient along the path.

The ideal solution for a ring cooler would be a system with a transfer matrix $\lambda \times I$ for each turn where λ is the cooling coefficient. In this case, all variables in 6D phase space are independent and bunch compression without change of angle and energy spread is possible. The following is an attempt to design a system with the appropriate features and to investigate its potential as a stage in a muon collider. The parameters assumed for the injected beam in the calculations are listed in Table 1. (The second column describes the cooled beam, see below). Definitions of normalized transverse and longitudinal emittances adopted here are:

$$\epsilon_x = \sigma_x \sigma_{p_x}/mc, \qquad \epsilon_z = \sigma_T \sigma_E/mc^2$$

SCHEMATIC OF COOLER

A schematic of the cooler is shown in Fig. 1. It includes two bending sections with wedge absorbers and two straight sections which house RF cavities and the

1) Work supported by the Universities Research Association, Inc., under contract DE-AC02-76CH00300 with the U. S. Department of Energy.

CP441, *Physics Potential and Development of mu-mu Colliders:* Fourth International Conference
edited by David B. Cline
 1-56396-723-5/98/$15.00

TABLE 1. Parameters of the Beam

	Injected beam	Cooled beam
Momentum	225 MeV/c	170-225 MeV/c
σ_x	70 mm	9.9 mm
σ_y	70 mm	16 mm
σ_z	1500 mm	69 mm
$\sigma_{x'}$	0.15 rad	0.11 rad
$\sigma_{y'}$	0.15 rad	0.11 rad
$\sigma_{\Delta p/p}$	7.8%	3.3%
Norm. r.m.s. X-emittance	22 mm-rad	2.0 mm-rad
Norm. r.m.s. Y-emittance	22 mm-rad	3.3 mm-rad
Norm. r.m.s. Z-emittance	225 mm	3.7 mm
6D emittance	11×10^4 mm^3	24 mm^3

main absorbers assumed here to be lithium hydride (Fig. 2). Each bending section includes magnets with field index 0.5 for focusing and bending the beam in vertical and horizontal planes. A wedge absorber is placed in the center of the section where the dispersion function is large. Skew quadrupoles are used to control dispersion in the straight sections. Betatron phase advances are 360° to get independent variation of coordinates and angles (which is why turns in two planes are used). Each straight section includes three FODO cells with phase advance per cell of 60° for X and 120° for Y in the leading part of the section, and two 90° FODO cells in the trailing part. This gives betatron transfer matrices $M_{x,y} = \pm\sqrt{\lambda} \times I$ per half-

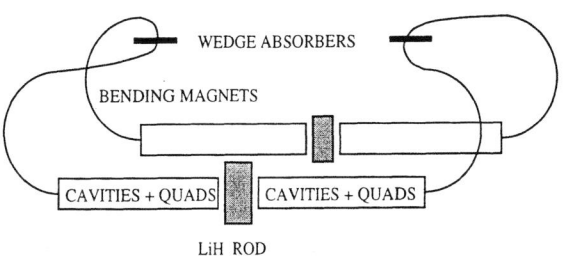

FIGURE 1. Schematic of ring cooler.

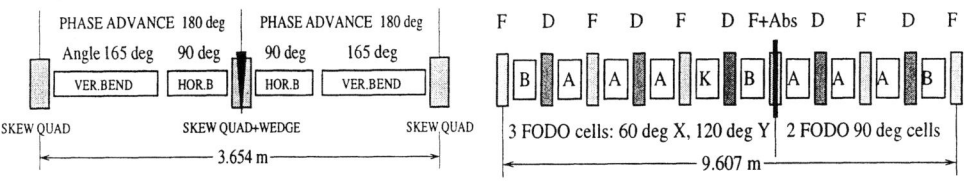

FIGURE 2. Bending and straight sections of ring cooler (F,D-quads, A,B-accelerating cavities and bunchers, K-kicker).

TABLE 2. Parameters of the Cooler

Muon momentum	170 – 225 MeV/c
Circumference	28.4 m
Bending radius	0.411 m
Length of straight sections	9.607 m
Revolution frequency	9.29 MHz
Energy rate in SS (average)	5 MeV/m
Length of main absorber (LiH)	27.6 cm
Angle of the wedge absorber	15.7 deg

turn, and $M_{x,y} = \lambda \times I$ per turn, as desired. Revolution frequency is independent of energy with an appropriate choice of length for the straight section. Bunchers are installed to provide energy-time coupling so as to get a longitudinal transfer matrix $\lambda \times I$. Some parameters of the cooler are listed in Table 2.

COOLING

Cooling is examined both with analytic calculations using transfer matrices—correct up to second moments of the distributions—and Monte Carlo simulations with SIMUCOOL [2]. For now, simulations are limited to what happens to the muons during material traversal. Elsewhere they employ the same transfer matrices as in the analytic approach.

Table 1 gives bunch parameters after cooling. Evolution of r.m.s. sizes and invariant emittances of a cooled bunch as a function of turn–number are presented on Fig. 3 (smooth lines are analytic results, dots – Monte Carlo simulation). The curve labeled N in Fig. 3c shows beam loss during cooling. Approximately 25% of muons are lost because of aperture restriction and almost the same amount by

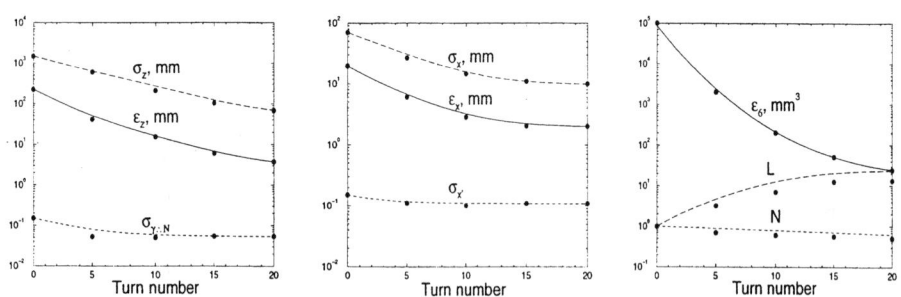

FIGURE 3. Cooling of the bunch: a–longitudinal, b–transverse, c–6D emittance, intensity (N), 'luminosity' (L).

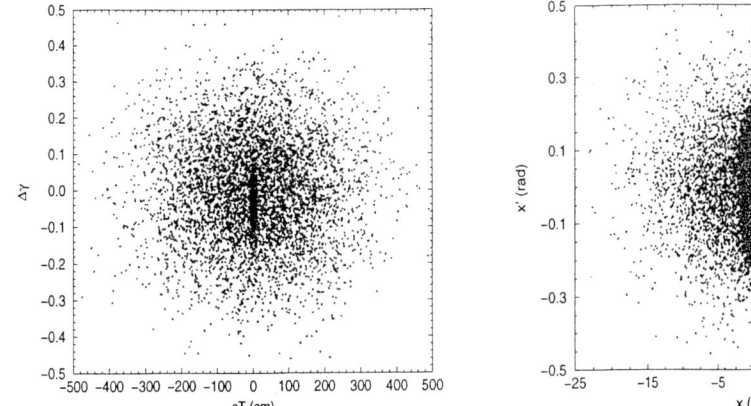

FIGURE 4. Phase space of the bunch before and after cooling: a–longitudinal, b–transverse.

decay (aperture restrictions are not taken into account in the analytic calculations). Luminosity of the collider with optimal β-function depends on beam parameters as $L \propto N^2/\sqrt{\epsilon_6}$. It reaches a maximum after 20 turns which is thus optimal for this cooler. Fig. 4a-b represent initial and final distributions of muons in longitudinal and transverse phase planes (final distributions are the dark central regions).

CONCLUSION

A ring cooler appears capable of satisfactory cooling a muon beam both in transverse and longitudinal directions. The achievable emittance suggests its use as a precooler in a muon collider complex especially for effective bunch shortening which is necessary in any scenario. Full tracking through magnets and cavities will be needed to investigate suppression of both chromatic and nonlinear effects and to study dynamic aperture.

REFERENCES

1. Balbekov, V. I., *Beam Dynamics and Technology Issues for $\mu^+\mu^-$ - Colliders*, New York: AIP Press, 1995, pp. 140-145.
2. Van Ginneken, A., *Nucl. Inst. Meth.* **A362**, p. 213 (1995).

An AGS Experiment to Test Bunching for the Proton Driver of the Muon Collider

J. Norem, C. Ankenbrandt, K-Y. Ng, M. Popovic, Z. Qian

Fermilab, Batavia IL, 60510

L. A. Ahrens, M. Brennan, V. Mane, T. Roser,
D. Trbojevic, W. van Asselt

Brookhaven National Laboratory, Upton NY 11973

Abstract The proton driver for the muon collider must produce short pulses of protons in order to facilitate muon cooling and operation with polarized beams. In order to test methods of producing these bunches we have operated the AGS near transition and studied procedures which involved moving the transition energy γ_t to the beam energy. We were able to produce stable bunches with RMS widths of σ = 2.2 - 2.7 ns for longitudinal bunch areas of ~1.5 V-s, in addition to making measurements of the lowest two orders of the momentum compaction factor.

INTRODUCTION

Muon collider designs (1,2) require that the pions from which muons are produced are generated in a short bunch, both to facilitate subsequent cooling and to improve the separation between polarization states. Calculations done for the Feasibility Study imply that the required initial rms bunch length from the proton driver (3) is $\sigma \sim 1 - 2$ ns, more than a factor of three shorter than the natural bunch length in the Brookhaven AGS (4). Since the proton driver contemplated for the muon collider could have its extraction energy near the natural transition energy, and would probably use a flexible momentum compaction (FMC) (5) lattice with easily adjustable transition energy, we have looked at options which could make use of transition in bunching. This paper summarizes simulations and experiments on the AGS near transition.

Bunching near transition is complicated by the nonlinearity in phase shear which introduces nonlinear motion in synchrotron space, as well as the small value of $|\eta|$ which reduces the synchrotron frequency, making these nonlinearities more prominent. Following reference (6), the expression for the circumference

$$C(p) = C_0[1 + \alpha_0\delta (1 + \alpha_1\delta + \alpha_2\delta^2) + ...], \tag{1}$$

where $dp/p = \delta$, and C_0 is the median circumference ($\delta = 0$), and the α's are various orders of the momentum compaction. One also defines the slip factor as

$$\eta \equiv \Delta T/T / \delta, \qquad (2)$$

giving the transition gamma,

$$\gamma_t(\delta) = \gamma_{t0} \{ 1 - (\alpha_1 + 0.5 - \alpha_0/2) \delta + O(\delta^2) \}, \qquad (3)$$

the first order momentum compaction is therefore

$$\alpha_1 = -1/\gamma_{t0} \, d\gamma_t/d\delta - 1/2 + \alpha_0/2. \qquad (4)$$

Predictions of MAD (7) for the AGS are shown in Figure 1.

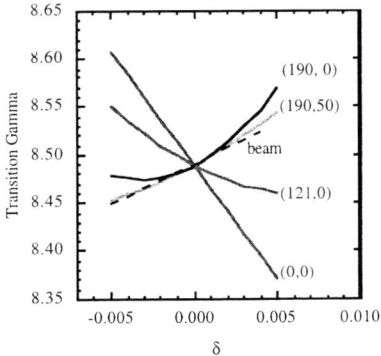

FIGURE 1. Predicted momentum dependence of γ_t as a function of sextupole and octupole settings (I_S, I_O).

OPERATION OF THE AGS

The operating mode of the AGS is shown in Figure 2. The maximum beam energy was reduced from the normal 28 GeV to around 7 GeV by shortening the acceleration period, and the γ_t jump system (8) was modified to give a short flattop period before the transition energy dropped. The beam was flattopped for 300 ms before the magnet guide field was raised slightly and then ramped down. One bunch was injected from the booster at about 0.08 sec into the acceleration cycle, when $dB/dt \sim 0.4$ T/s, and the beam was in the machine for about 500 ms total. The rf cavities were operated at ~340 kV/turn while there was beam. The bunch current was 3-5 x 10^{12} in a single bunch, and the bunch area was measured to be 1.6 - 1.5 eV-s.

Previous work had shown that the γ_t jump system caused significant changes in the nonlinear momentum compaction factor α_1, the measured maximum dispersion and the momentum aperture. Although three sextupole families are present in the ring, the locations of these sextupoles are such that these lattice changes are difficult to tune out both before and after the transition energy is moved. In order to evaluate the magnitude of these effects we remeasured the transition energy as a function of radial position by measuring beam loss as a function of phase-switch delay time and sextupole setting

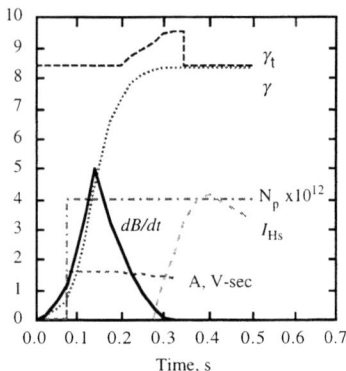

FIGURE 2, The operating mode of the AGS.

EXPERIMENTAL RESULTS

The transition energy was first measured relative to the gauss clock, which measures the instantaneous field in the dipoles as a function of time, for four values of radial displacement. The published value of α_1 was checked, by measuring the radial dependence of beam losses as a function of the time at which the phase of the rf accelerating field was flipped to compensate for the reversed direction of shear. The transition energy is obtained from the average of values with the sextupoles on, and is equal to $\gamma_t = 8.34 \pm 0.05$. The corresponding measurements of α_1 obtained using equation (4) above give the results $\alpha_1 = 7.2 \pm 1.5$ ($I_{\text{sext,H}} = 0$ A) and $\alpha_1 = 3.5 \pm 1.5$ ($I_{\text{sext,H}} = 100$ A). These measurements agree roughly with earlier published data however time constraints did not permit good statistics. There may also be a systematic error in the measurement of γ_t by this method due to the optimum time delay between the phase flip and transition time, because of the nonlinear and asymmetric behavior of the bunches near transition.

When the operating mode described in Figure 2 was implemented, the beam was found to be quite stable near transition. After the first quadrupole synchrotron oscillation all structure tended to damp out and the beam circulated stably without significant losses or beam blowup. Changes in the beam energy, to move below or above transition, or sextupole settings seemed to make little difference and the overall structure of the beam was almost unaffected by accelerator parameters. Losses were not accurately recorded, but seemed small (~5%), and no particular effort was made to minimize them.

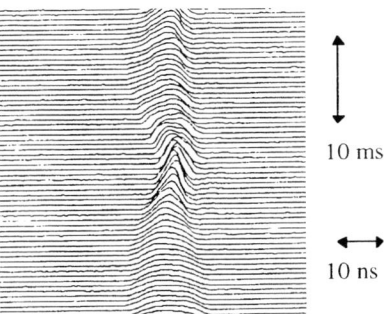

FIGURE 3. Mountain range plot of the bunch after the transition energy was dropped to the beam energy.

The results of bunching tests are shown in Figure 3 and Figure 4, where the bunch shape is plotted both before and after the γ_t was suddenly changed. The narrow final shape is noticeably asymmetric because it was impossible to entirely tune out the chromatic effects mentioned above. Runs further from transition showed a number of synchrotron oscillations and the bunch length in later minima was always longer than the first minimum. Closer to transition the bunch length contracted to a minimum and then expanded to roughly the initial value. There seemed to be little emittance blowup. Losses were a function of sextupole currents, but there was insufficient time to explore the dependence in detail. Although the minimum bunch width decreased with increasing sextupole current, the sextuple supplies produced only about half of the current predicted by MAD to optimize $\eta(\delta)$, and we were not able to produce $\alpha_1 = -1.5$, which is expected to produce the minimum bunch length.

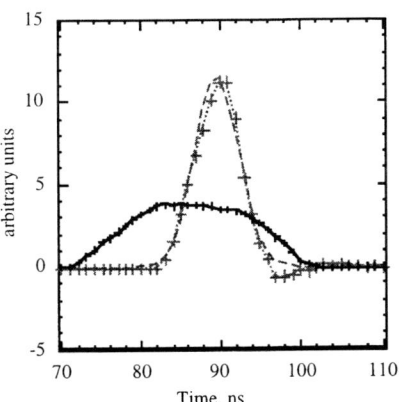

FIGURE 4. Bunch shapes before and after bunching (crosses) showing a final corrected bunch length of $\sigma = 2.66$ ns for a 1.5 V-s bunch together with a fitted Gaussian curve (dashed line).

DISCUSSION

In order to understand the mechanism of bunching we have done simulations using ESME [10]. These calculations track the longitudinal motion of particles in the bunch over the acceleration and bunching cycle with various parameters. The primary variables are $\Delta\gamma = \gamma_t - \gamma$ and α_1. Simulations show that bunching can occur over a period of ~1 ms and produce bunches with large momentum spread (full widths of $\delta = 0.024$) and narrow widths. An example of these simulations are shown in Figure 5, (showing $\Delta t = \pm 15$ns, $\delta = \pm 0.02$), which approximates the parameters shown in Figure 4. These simulations fairly closely match the structure of the data, however the accuracy of the fit seems to be a function of a number of variables such as the parameterization of the density of charge within a bunch. The shape of the shortest bunch that can be produced is sensitive to α_1 and can be used to estimate this parameter, however the bunch shapes seen do not precisely match those produced in simulations, perhaps indicating higher order terms, ($\eta \propto \delta^2$). The data seems consistent with $\alpha_1 = 0.0 \pm 1.0$. The total height of the bunch, $\delta \sim 0.024$, can circulate in the AGS which has a total momentum acceptance of $\delta \sim \pm 0.025$. Since the momentum acceptance is sufficiently large, we expect a further reduction in the bunch width is possible, however this might be accompanied with increased losses if the beams were large.

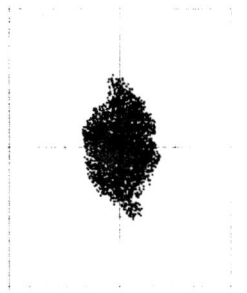

FIGURE 5. Simulation of the final phase space after γ_t is moved close to the beam energy ($\gamma = 8.40$).

CONCLUSIONS

During a short run, we were able to explore the operation of the AGS when it was flattopped close to transition. It was found that the bunch length could be reduced by a factor of ~3 by using the γ_t jump system to suddenly drop the transition energy to the beam energy, and the beam seemed fairly stable during this exercise. We also measured γ_t using: 1) the time of phase flip, 2) the synchrotron frequency, and, 3) debunching time, producing rough agreement with previous measurements and with MAD calculations for the lattice. Measurements of α_1 generally agreed with earlier data (9), but the precision of the data is limited by statistics and measurement time. Attempts to tune α_1 with sextupoles and octupoles were somewhat less successful due to power supply limitations. The stability of the beam circulating near transition was unexpected, and argues for the stability of high current isochronous storage rings.

We were unable to systematically explore possible methods of bunch manipulation due to time constraints, nevertheless, the $\sigma \sim 2.2$ ns bunch length produced by this method is sufficiently short to be useful for the proton driver of the muon collider. A further reduction by 30% would be even more desirable and this might be obtained by means of the methods tested here. Modification of the bunch shape before rotation, for example, seems useful, but was not optimized. Effects of space charge were also not studied systematically, and measurements with significant space charge effects would be highly desirable, as would the ability to move the transition energy in two directions, which could be used to produce rapid bunching. A more complete analysis of this data is being published (10).

ACKNOWLEDGEMENTS

We would like to thank the operating crew of the AGS for their cooperation performing these measurements. We would also like to acknowledge useful discussions with C. Johnstone and D. Neuffer of Fermilab, M. Yoshi and Y. Mori of KEK, and B. Autin of CERN.

REFERENCES

1. D. Neuffer, Part. Acc. 14 (1983) 75
2. $\mu^+ \mu^-$ Collider , A Feasibility Study, published in Proceedings of the Snowmass Workshop, Snowmass 96
3. D. Cline, B. Norum, R. Rossmanith, Proceedings of the European Particle Accelerator Conference, Sitges, Spain, 10-14 Jun 1996. p867
4. M. Brennan, in Proceedings of the Workshop on the Proton Driver for the Muon Collider, Argonne, Nov 14 - 15 1996. Argonne (1996).
5. Lee, Ng and Trbojevic, Phys. Rev. E48 (1993) 3040
6. J. Wei, Longitudinal Dynamics of the Non-Adiabatic Regime on Alternating- Gradient Synchrotrons, Thesis, State University of New York ad Stony Brook, Revised, (1994)
7. H. Grote, F. C. Iselin, The MAD program (methodical accelerator design) : version 8.10; user's reference manual, CERN SL 90-13 AP rev 3 . (141 p) . 1993
8. W. K. Van Asselt, L. A. Ahrens, J. M. Brennan, A. Dunber, E. Keith-Monnia, J. T. Morris, M. J. Syphers, Proceedings of the 1995 Particle Accelerator Conference, Dallas, (1995) 3022
9. J. Wei, M. Brennan, L. A. Ahraens, M. M. Blaskiewicz, D-P. Deng, W. W. MacKay. S. Peggs, T. Satogata, D. Trbojevic, A. Werner., W. K. van Asselt, Proceedings of the 1995 Particle Accelerator Conference, Dallas, (1995) 3334
10. J. MacLachlan and J. F. Ostiguy, Proceedings of the 17th IEEE Particle Accelerator Conference (PAC 97) Vancouver, Canada, 12-16 May 1997
11. C. Ankenbrandt, K-Y. Ng, J. Norem, M. Popovic, Z. Qian, L. Ahrens, M. Brennan, V. Mane, T. Roser, D. Trbojevic, W. van Asselt, Fermilab Publication, Fermilab PUB-98-006, (Submitted to Phys Rev D).

Pion Production and Targetry at $\mu^+\mu^-$ Colliders

N. V. Mokhov and A. Van Ginneken

Fermi National Accelerator Laboratory, Batavia, IL 60510[1]

Abstract. Results of simulations of pion production and power dissipation by 8 to 30 GeV proton beams in co-axial and tilted targets of liquid gallium and platinum oxide, placed in a 20 T solenoid, are reported. Pion and muon distributions are followed through the matching solenoid and decay channel.

INTRODUCTION

The proposed $\mu^+\mu^-$ collider complex [1] includes a rapid cycling (15 Hz) synchrotron which produces protons—up to 10^{14} per pulse—in the 8–30 GeV range. These protons are focused on a production target from which pions are collected and steered down a decay pipe where they produce the desired muons. Extensive simulations have been performed for pion production from 8 and 30 GeV proton beams on different target materials in a high field solenoid [1-3]. Targets of varying composition (6< Z <82), radii (0.4–3 cm) and thicknesses (0.5–3 nuclear interaction lengths (λ_I)) have been explored. This paper presents a sample of new results on pion production and target behavior with particular attention to the dependence of these results on tilt angle of the target with respect to the solenoid axis.

TARGET AND SOLENOID

The aperture of the 20 T solenoid is assumed to be 7.5 cm which results in transverse momentum acceptance ($p_\perp^{max} = |q|Ba/2$, with B the magnetic field, q the particle charge, and a the solenoid radius) of 0.22 GeV/c. The normalized phase space acceptance ($ap_\perp^{max}/m_\pi c = |q|Ba^2/2m_\pi c$) of this solenoid for pions is 0.12 m·rad. The MARS code [4], is used to simulate particle production and transport in thick targets within the solenoid field as well as to study energy deposition in the target and surrounding solenoid. The current version includes a newly developed phenomenological model for pion production in hadron-nucleus interactions [5],

---

[1] Work supported by the Universities Research Assiciation, Inc., under contract DE-AC02-76CH00300 with the U. S. Department of Energy.

that exhibits superior description of the pion production for 6 to 120 GeV/c protons compared to previous studies using ARC, MARS, DPMJET, FRITIOF and GEANT codes. In excess of 90% of all accepted muons are found to be the progeny of pions in the momentum range 0.05 - 0.8 GeV/c for 8 to 30 GeV protons. Ref. [6] suggests that tilting the incident proton beam and target with respect to the solenoid axis can increase pion yield. This idea is examined in some detail here. The tilted target geometry is shown in Fig. 1(a). Fig. 1(b) shows the dependence of energy deposition in the target and in the 2-cm innermost region of the solenoid coils, for three different targets $1.5\lambda_I$ long: (i) gallium (R=3.2 cm, beam $\sigma_{x,y}=1$ cm, $\sigma_{x',y'}=0$), (ii) Ga (R=1 cm, $\sigma_{x,y}=0.4$ cm) and (iii) Platinum oxide (R=1 cm, $\sigma_{x,y}=0.4$ cm). All three targets are co-axial with the solenoid. Power dissipation in the targets grows almost linearly with beam energy, being smallest in the 1-cm radius Ga target. Power dissipation in the inner layer of the solenoid coil also grows monotonically with energy, though at a much slower rate, and is roughly the same for the three cases. Total power deposited in the entire solenoid section around the target—and of equal length—is about 2.5 times that shown in Fig. 1(b).

π/μ PRODUCTION AND COLLECTION

The energy spectrum of the produced pions peaks at about 0.05 GeV after which it declines rapidly with energy. Most $\pi(\mu)$ producing/cooling scenarios therefore concentrate on rather low energies, typically $E_\pi < 0.5$ GeV. Tilting the target with respect to the solenoid axis helps low energy pions escape the target through their Larmor motion, with reduced secondary interactions, energy loss, and multiple scattering—but has little effect on higher energy pions. A tilted target also offers a convenient way to dispose of the remnant proton beam after target traversal. The effect on yield is demonstrated in Fig. 2(a) which shows ratios of π^+

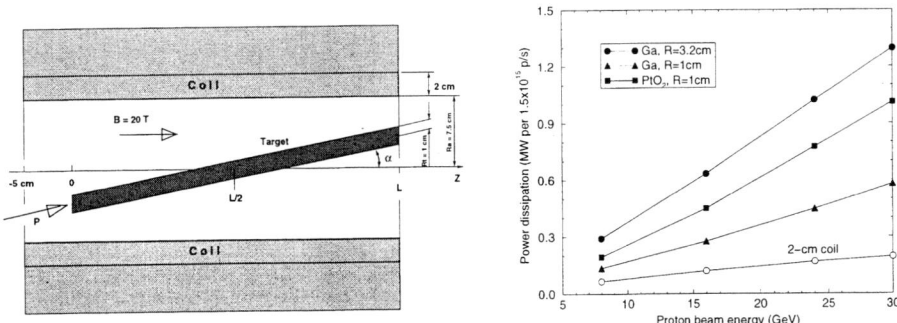

FIGURE 1. (a) Schematic view of a tilted target in solenoid; (b) Power dissipation in various targets and surrounding solenoid *vs* proton energy.

produced by a 16 GeV proton beam ($\sigma_{x,y}=1$ cm) in tilted versus untilted liquid Ga targets (L=33 cm, R=3.2 cm) as a function of pion momentum at target exit for pions traveling within 500 mrad with respect to the solenoid axis. Targets tilted up to 150 mrad show mostly increasing yields at low pion energies but with little or no effect for pions above 0.4 GeV/c. The effect is more pronounced for the thinner platinum oxide target (open jet of L=35.8 cm, R=1 cm) as illustrated in Fig. 2(b) which shows yield ratios of π^- versus pion momentum for a 16 GeV proton beam ($\sigma_{x,y}=0.4$ cm). The improved yield for PtO$_2$ vis-a-vis Ga most likely results because of (i) increased multiplicity and lower average energy of the pions in the heavier target and (ii) smaller target radius (1 vs 3.2 cm) which is beneficial via the spiraling-out mechanism. Fig. 3(a) presents yields integrated over momentum between 0.05 and 0.8 GeV/c for a set of 1.5λ_I long targets (R=1 cm, beam $\sigma_{x,y}=0.4$ cm) ranging across the periodic table and for three different proton energies. The increase in yield with mass number is quite pronounced at 30 GeV while hardly noticeable at 8 GeV. Yield is higher for larger solenoid aperture. For the 150 mrad tilted PtO$_2$ target and for 0.05< p_π <0.8 GeV/c yield grows from 0.58 to 0.75 for π^+ (and very nearly the same for π^-) as the aperture increases from 7.5 to 15 cm radius. Fig. 3(b) shows yields (0.05< p_π <0.8 GeV/c) for the 200 mrad tilted PtO$_2$ target (R=1 cm, $\sigma_{x,y}=0.4$ cm) and co-axial Ga target (R=3.2 cm, $\sigma_{x,y}=1$ cm) irradiated with a 16 GeV proton beam as a function of target thickness.

Following the target is a solenoid with decreasing field and increasing aperture. After a distance of 1–2 m both field and aperture are once again held constant at 5 T and 15 cm radius in the present simulations. The decay channel may be equipped with RF cavities to rotate the phase space prior to cooling. For results reported here no cavities are present. Pions and muons are propagated through the decay channel as in ref. [2] with decay of π and μ as well as aperture restriction taken into account. Table 1 shows the maximum yield of muons after 75 m in the decay pipe

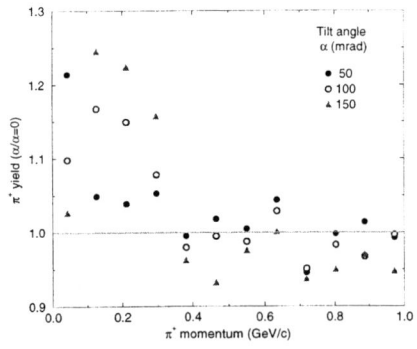 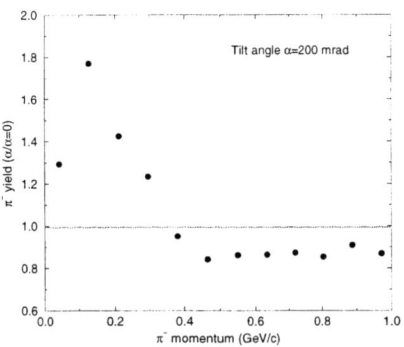

FIGURE 2. (a) Ratio of π^+ yield from Ga target tilted by 50, 100 and 150 mrad to that for untilted target vs p_π for 16 GeV protons; (b) same for π^- from PtO$_2$ target at α=200 mrad.

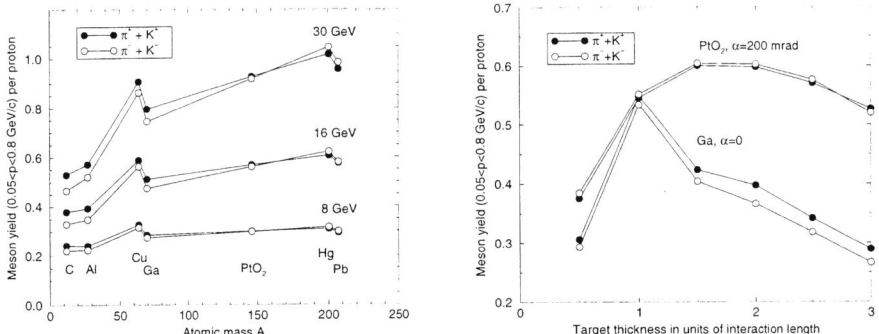

FIGURE 3. (a) Yield from $1.5\lambda_I$ 1-cm radius target irradiated with 8, 16 and 30 GeV proton beam as a function of target material; (b) Yield from 200 mrad tilted PtO_2 target ($\lambda_I=17.9$ cm) and co-axial Ga target ($\lambda_I=24.2$ cm) irradiated with a 16 GeV proton beam *vs* target thickness.

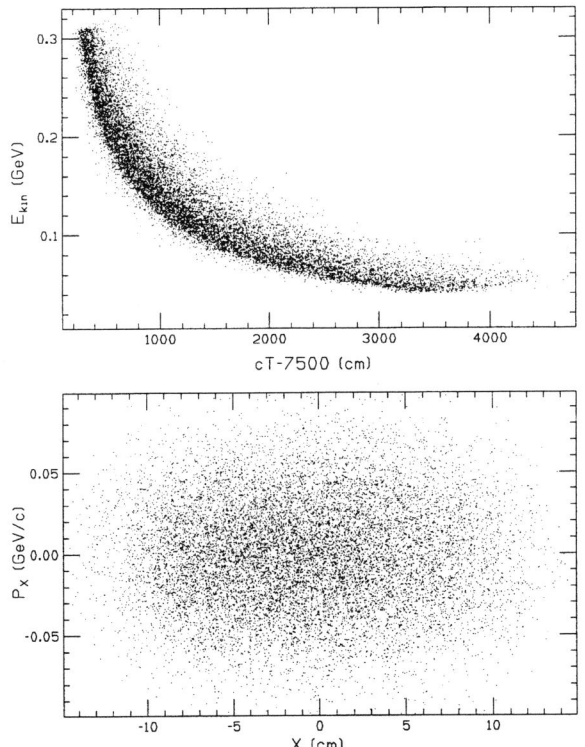

FIGURE 4. Longitudinal (top) and transverse (bottom) phase space plots of muons of both signs within $0.1 < p_z < 0.4$ GeV/c cut 75 m downstream of a 200 mrad tilted PtO_2 target.

TABLE 1. Muons in Decay Pipe after 75 m

target, tilt(mrad)		$0.1 < p_z < 0.4$		Total	
		μ^+	μ^-	μ^+	μ^-
Ga	0	0.222	0.214	0.420	0.382
Ga	200	0.230	0.232	0.394	0.373
PtO$_2$	200	0.266	0.273	0.438	0.432

for all surviving muons as well those within a $0.1 < p_z < 0.4$ GeV/c cut, i. e., those assumed to be most amenable to cooling. The gains from tilting the Ga target are considerably less here than for pions immediately after the target. Perhaps lower field/larger aperture solenoids in the target region might better accommodate the tilted targets. Fig. 4 shows longitudinal and transverse phase space plots of muons of both signs within the $(0.1 < p_z < 0.4)$-cut 75 m downstream of a 200 mrad tilted PtO$_2$ target. Phase space plots of the other cases discussed here are quite similar.

CONCLUSIONS

For low beam energy (≤ 8 GeV) any target material with $A \geq 50$ offers roughly the same pion yield. Power dissipation appears low enough to accommodate many target choices. At 16 GeV higher-A targets yield more pions but power absorption becomes the primary concern. Either liquid Ga inside a pipe or an open PtO$_2$ jet is a good choice. Yields are optimum for target about $1.5\lambda_I$ long and with radius of about 2.5 times $\sigma_{x,y}$ of the beam. Tilting the target at 150–200 mrad increases yields and allows ready disposal of the spent proton beam. The present study shows that smaller target radius, larger solenoid aperture, and higher solenoid field all help increase yield.

REFERENCES

1. $\mu^+\mu^-$ Collider: A Feasibility Study, The $\mu^+\mu^-$ Collider Collaboration, BNL–52503; Fermilab–Conf–96/092; LBNL–38946, July 1996.
2. Mokhov, N. V., Noble, R. J., and Van Ginneken, A., in *AIP Conference Proceedings* **372**, pp. 61–86 (1996).
3. Ehst, D., Mokhov, N. V., Noble, R. J., and Van Ginneken, A., presented at the 1997 IEEE Particle Accelerator Conference, Vancouver, May 1997.
4. Mokhov, N. V., Fermilab–FN–628 (1995).
5. Mokhov, N. V., and Striganov, S. I., presented at the Workshop on Physics at the First Muon Collider, Fermilab, November 1997.
6. Green, M. A., and Palmer, R. B., BNL–64747, LBL–40186 (1997).

Mesoscopic Quantum Multiplex for Channeling Bunches

Jing Shen

Institute of High Energy Physics (IHEP), Chinese Academy of Sciences
Mailing Address: P. O. Box 918(6), Beijing 100039, PRC[*]
and
China Center of Advanced Science and Technology (CCAST), P.O.B. 8730, Beijing 100080

Abstract. (1) Bogacz-Cline channeling is an interesting idea that can transform a bunch of low particle intensity to a collider of high luminosity but it was maintained as impossible to carry out because of three technical problems. (2) The first of which is discussed in this paper, and it is how to get billions particles from each bunch to enter into and channel through a single crystal channel. (3) Two basic difficulties of entrance are discussed in this paper. The first is due to the Heisenberg's uncertainty, and the second is the dimension reduction of a beam bunch in crystal from 3D to 1D. (4) To overcome these difficulties, a hybrid device named Mesoscopic Quantum Multiplex (MQM) is designed to achieve entrance and channeling. It is a quantum generalization of classical multiplex in detector readout electronics for the classical-quantum interface. It is made by nano-crystalline technology. (5) The MQM can channel the Richter-Kimura-Takada flat $e^{\pm}$ beams of NLC-JLC, and low emittance p or heavy ion beams as well as the Bogacz-Cline $\mu^{\pm}$ beams, and the Nagamine-Chu cool $\mu^{\pm}$ beams.

INTRODUCTION

S.A.Bogacz and D.B.Cline raised an argument of new conceptual design of $\mu^+\mu^-$ quantum collider using bent crystals and binary crystals at the collision point[1]. They raised a new research but their concrete scheme of collider seems not workable to most of muon collider people. However, the research seems to the author to be not the end but only the beginning. The problem is just that the physics of a new device should be researched before designing such a collider. If a penetrating analysis is made, one can find that the proposed concept of a quantum solid collider is more advanced level than the advancing technologies like the SC linac, the relativistic klystron, the wake field acceleration, the two beam FEL acceleration, etc. It is parallel to the plasma acceleration, the laser acceleration, and the plasma-laser acceleration.

SPECIAL ADVANTAGES OF CLINE'S ECONOMY

Bogacz-Cline's quantum muon collider has six important advantages:
(1) **High Luminosity-Low Intensity:** Luminosity up to 10^{32} cm^{-2}s^{-1} just by a very low intensity muon beam of 10^8 $\mu^{\pm}$ / pulse; compare this to Palmer's muon beam of $4 \times 10^{12} \mu^{\pm}$/pulse and a linear collider beam of 10^{10} $e^{\pm}$/pulse. In fact, high luminosity-low intensity is important in principle for either the muon colliders or the linear colliders.

---

[*]*Email: shenj@bepc3.ihep.ac.cn*

(2) **Low Background in Detector:** The muon decay backgrounds in detectors is a problem of detector survival; thus, low intensity $\mu^\pm$ beams are easier on detectors.
(3) **Lowest Energy Higgs Factory:** The minimum energy of Higgs factory for a proton collider is > 14 TeV, a linear collider > 500 GeV, and a muon collider just ~ 100 GeV.
(4) **Lowest Interaction between Beam and Crystal:** Interaction between muon and crystal is lower than protons, electrons and ions because of much lower synchrotron radiation emission and reaction cross section.
(5) **Lowest Scale of Ring:** We always seem to ask more questions than we have tools and budgets to answer. The dimension of instrument for basic science up to 100 km is unreasonable in current days. So developing muon machines as opposed to electron machines and proton machines as well as nanom electronics as opposed to micro electronics is natural.
(6) **Lowest Risk of R&D:** R&D of basic sciences that are connected to industrial technologies and social demands have the lowest risk because of spin-in and spin-off effects. This is the Cline's economy.

EMITTANCE QUANTUM AND FOUR LEVELS OF BUNCH

Heisenberg Emittance Quantum

Let's define the Heisenberg emittance quantum to classify four kinds of beam bunches:

$$\varepsilon_0^N \equiv \text{rad} \times \text{Compton wave length of } e \ (or \ \mu) \quad (3.1)$$
$$= 3.86159323 \times 10^{-13} \text{ m-rad (or } 3.86159323 \times 10^{-13}/207 \text{ m-rad)}$$

Macroscopic Beam Bunches

Beam bunches of all existing colliders are macroscopic including the Palmer's bunch of a muon collider[2], which is produced by conventional technologies of existing proton machines. The Palmer's bunch emittance is

$$\varepsilon_{\text{Pal}}^N = 90 \times \pi \times 10^{-6} \text{ m-rad} = 7.32 \times 10^8 \times 207 \varepsilon_0^N = 1.52 \times 10^{11} \varepsilon_0^N, \quad (3.2)$$

where the Heisenberg emittance ε_0^N is now for a muon.

Quasi-Mesoscopic Bunches

Nagamine's bunch on the other hand is made by accelerating sub-eV positive muons which are generated by a laser resonant ionization source or eV negative muons which are generated by a muon-catalyzed fusion source[3]. The emittance is

$$\varepsilon_{\text{Nag}}^N = 6 \times 10^{-11} \beta\gamma = 6 \times 10^{-11} \times 9460 \text{ m-rad} = 1.47 \times 10^6 \times 207 \varepsilon_0^N = 3.04 \times 10^8 \varepsilon_0^N, \quad (3.3)$$

where ε_0^N is for a muon. The beam bunch appears to be still a macroscopic bunch, but it could be decreased two orders of magnitude by Chu's cooling technique[4]. Hence, it reaches to be a quasi-mesoscopic bunch.

$$\varepsilon_{\text{N-C}}^N \approx 10^{-2} \varepsilon_{\text{Nag}}^N = 3.04 \times 10^6 \varepsilon_0^N. \quad (3.4)$$

Mesoscopic Bunches
Richter-Kimura-Takata's polarized flat bunches is one type of mesoscopic bunches. The first generation is that of the NLC-JLC which is a combination of high technology in X-band RF, laser, and super-lattice GaAs-AlGaAs [5,6],
$$\left(\varepsilon_{R-K-T}^N\right)_y = 3 \times 10^{-8} \text{ m-rad} = 7.77 \times 10^4 \, \varepsilon_0^N. \qquad (3.4)$$

Microscopic Bunches
Bogacz-Cline's channeled muon bunch and Huang-Chen-Ruth's dampled positron bunch [7] should be a final goal of microscopic bunch of all kinds of future colliders,
$$\varepsilon_{B-C-H-C-R}^N \approx 10^2 \varepsilon_0^N. \qquad (3.5)$$

HEISENBERG'S FORBIDDEN BUNCH CHANNELING

Nature of Bunch Channeling Problem
The most difficult problem of Bogacz-Cline's collider is to get the $10^8 - 10^{12}$ muons from each bunch to enter into and channel through a single crystal channel with a diameter of 0.4 nm. It's not only a problem of technique but also of basic physics, electronics, and photonics; even a problem of creating phononics [8].

Heisenberg's Uncertainty in Bunch Channeling
Normalized emittance in the interaction region is a kinematical constant. The normalized emittance of existing beam bunches is much larger than the Heisenberg emittance [see (3.1), (4.6), and table 1]. If a realistic bunch is focused into a channel of 0.4 nm in diameter, then the incident angle must be greater than the critical angle of channeling, so it can't enter into a crystal channel. Inversely, if the incident angles of all particles are compressed within the critical angle of channeling, then the transverse dimension of the bunch must be increased, spreading over many channels and resulting in little or no collision. The luminosity is decreased in spite of channeling, defeating the proposed scheme of Bogacz-Cline's crystal channeling for a muon collider. This is due to the Heisenberg's uncertainty principle.

Heisenberg Angle in Bunch Channeling
Let the projected angle θ denotes the Heisenberg angle in channeling
$$\theta \le \psi_c = \psi_1 f \qquad (4.1)$$
where ψ_c is critical angle of channeling. According to a Lindhard's approximation[9],
$$\psi_1 = \sqrt{\frac{2U_d}{E_{//}}} \text{ and } f = \frac{1}{\sqrt{2}} \left| \ln\left(\frac{3a^2}{\rho_2^2}+1\right) \right|^{\frac{1}{2}} = 0.911 [\text{worst case CsI}] \qquad (4.2)$$
with $U_d = \frac{Ze^2}{4\pi\varepsilon_0\varepsilon_\infty d}$ [J] $\frac{1}{1.6 \times 10^{-19}}$ [eV/J]$= \frac{Z}{d[\text{nm}]\varepsilon_\infty} \times 1.44 \text{ eV} = 60.3\text{eV [CsI]}, \qquad (4.3)$

$E_{//}$ the particle beam energy, a the Thomas Fermi shielding length

$$a = 0.8853 a_0 \Big/ \sqrt{Z_1^{2/3} + Z_2^{2/3}} = 0.119 a_0 \text{ [for CsI]}, \quad (4.4)$$

and $\rho_2 (= 0.01 a_0)$ the lattice vibration amplitude of a low temperature crystal. The other constants are: a_0 the Bohr radius, $Z_1 = 1$, and ε_∞ the high frequency dielectric constant.

Heisenberg Transverse Dimension in Bunch Channeling

$$\sigma_r = \varepsilon^N / \beta\gamma\pi\theta = \varepsilon^P / \pi\theta \quad (4.5)$$

where ε^N is the kinematically invariant, normalized emittance and ε^P the projected emittance. Substituting (4.1)-(4.4) into (4.5) yields σ_r^{min} of channeling bunch,

$$\sigma_r^{min} = \frac{\varepsilon^N}{\pi\sqrt{\beta\gamma}} \sqrt{\frac{m_0 c^2}{2U_d}} \frac{1}{f} = \frac{\varepsilon^P \sqrt{\beta\gamma}}{\pi} \sqrt{\frac{m_0 c^2}{2U_d}} \frac{1}{f},$$

$$= \frac{\varepsilon^N}{\pi} \frac{m_0 c^2}{\sqrt{2EU_d}} \frac{1}{f} = \frac{\varepsilon^P}{\pi} \sqrt{\frac{E}{2U_d}} \frac{1}{f}. \quad (4.6)$$

PRINCIPLE OF MESOSCOPIC QUANTUM MULTIPLEX

Parallel to Serial Conversion of Decreasing Dimension----Multiplex

To get the trajectories of a muon bunch pass through a single channel, let's design a device. At its input is a crystal for the parallel entrance of the macroscopic muon bunch. At the output is a single channel for the exit of the channeled muon bunch. This device then transforms the bunch dimensions from 3D to 1D.

The transformation is done in two stages. In the first stage, it is from 3D to 2D, and in the second stage from 2D into 1D.

The transformation is a quantum generalization of classical multiplex in the HEP detector readout electronics. It is a kind of parallel-to-serial conversion in which a parallel input of particles from a bunch is re-arranged to a serial output. Muons are then re-arranged one by one into a channel. This device, named as Mesoscopic Quantum Multiplex (MQM), might be invented by nano-crystalline technology.

Quantum Limit of Richter-Kimura-Takata's Bunch for NLC-JLC

Let us assign the (X,Y,Z) coordinates to a muon bunch with Z in the direction of channeling. The first stage of the MQM converts a 3D Gaussian bunch into a 2D Gaussian bunch in the X-Z plane. The multiplex transformation has been done in the Y-Z plane. This transformed-2D-bunch in the crystal channels is the quantum limit of the Richter-Kimura-Takata's flat bunch in the vacuum for the NLC-JLC.

Quantum Limit of Bogacz-Cline's Bunch for a Muon Collider

The second stage transforms a 2D bunch into a 1D bunch along the Z direction. Its 0D cross-section is a single channel in the Z direction. This is the quantum limit of Bogacz-Cline's muon bunch.

DESIGN OF MQM

Components and Materials

The first stage of the MQM is composed of two kinds of components: a 3D crystal with channel diameters of 0.3-0.4 nm is used as a 3D bunch acceptor; it is joined with a 2D component, a linear array of Carbon Bucky tubes with 3-4 nm in diameters. The lattice constant of Bucky tube wall d is 0.34 nm. The second stage of the MQM is the same but all minus one dimension.

Schematic Diagram

Let (m, n, l) denote the coordinates of an acceptor crystal with l along the direction of channels, and there are $M \times N$ parallel channels in the l direction. Also, let (j, k) denote the coordinates of a 2D (one layer of) Bucky tube array. Let

$$k // Z \text{ of the bunch, and } J=M \tag{5.1}$$

with J as the number of Bucky tubes in the 2D array. One side of the acceptor crystal is inter-faced with the bunch cross-section $\sigma_y \times \sigma_x$ and the other side with the 2D array of Bucky tubes at an angle $\chi = \angle lk$. If χ is the same as the critical angle, then the particles can channel through the crystal acceptor and into the Bucky tubes.

$$\chi = \psi_c \geq \theta = \varepsilon^N / \beta \gamma \pi \sigma_{x,y}. \tag{5.2}$$

Every Bucky tube connects to N channels of the acceptor crystal, and every microscopic channel connects to a mesoscopic dimension of I cells on the wall of a Carbon Bucky tube, where

$$I = \psi_c^{-1} \approx 10^5. \tag{5.2}$$

That means there are 10^5 gates (cells) per channel of the acceptor crystal as a mesoscopic interface on the contact face between acceptor crystal and channeling Bucky tube. The acceptor crystal can channel a 3D bunch into $M \times N$ parallel outputs and into the 2D Bucky tube array.

CHANNELING OPERATION OF MQM

At the entrance into the acceptor crystal, a particle bunch with emittance ε^N should have the entrance trajectory angle minimal and the beam size $>> \sigma_r^{min}$, so that the trajectory angles are within the critical angle ψ_c. Individual particles then can channel into and through one of the $M \times N$ channels of the crystal and enter into a Bucky tube by one of the I gates randomly. The I gates are serially lined on the wall of a Bucky tube along the k//Z direction. Finally, the M layers of N parallel outputs from the crystal channels then enter into J Bucky tubes by way of the $J \times I$ gates serially. This is micro-transformation of parallel to serial mode, which is operated in M (or J) parallel channels.

The 3D bunch on entering the channels of the acceptor crystal become a 3D parallel array. Then on entering into the 2D array of Bucky tubes through the serial gates on the side of every tube wall, the 3D parallel array transforms into a 2D array in series inside the 2D Bucky tube array. Hence, the 3D bunch is transformed into a 2D bunch, in say, the X-Z plane, which is a multiplexing in the YZ planes. The device converts a 3D bunch into a 3D array and then channels them into a 2D-array of Bucky tubes. The transformation from a 2D-bunch to a 1D-bunch is similar.

TECHNOLOGICAL DIMENSION OF MQM

The key dimension is obviously the entrance length on Carbon Bucky tube,

$$L_e \equiv \frac{\sigma_r^{min}}{\psi_c} = \frac{\varepsilon^N}{\pi\sqrt{\beta\gamma}}\sqrt{\frac{m_0c^2}{2U_d}}\frac{1}{f} \bigg/ \sqrt{\frac{2U_d}{\beta\gamma m_0c^2}}f, \quad (7.1)$$

$$= \frac{\varepsilon^N}{\pi f^2}\left(\frac{m_0c^2}{2U_d}\right) = \frac{\varepsilon^P\beta\gamma}{\pi f^2}\left(\frac{m_0c^2}{2U_d}\right). \quad (7.2)$$

Table I. The technological length of entrance contacting Bucky tube.

Bunch Type	Energy	$\beta\gamma$	ε^N [m-rad]	ψ_c [rad]	σ_r^{min} [m]	L_c [CsI]
Ric-Ki-Ta	250 GeV	489 237	3×10^{-8} y	2.7×10^{-5}	5.0×10^{-10}	26 μm
Ric-Ki-Ta	250 GeV	489 237	3×10^{-6} x	2.7×10^{-5}	5.0×10^{-8}	2.6 mm
Naga-Chu	1 TeV μ^+	9461	$6\times 10^{-11} \beta\gamma$	1.4×10^{-5}	2.8×10^{-6}	10.4 cm
Naga-Chu	1 TeV μ^-	9461	$10^{-9}\times\beta\gamma$	1.4×10^{-5}	5.3×10^{-6}	1.73 m
Palmer	50 GeV	474	$90\pi\times 10^{-6}$	6.0×10^{-5}	2.2×10^{-3}	52.1 m
Palmer	250 GeV	2400	$90\pi\times 10^{-6}$	2.7×10^{-5}	5.0×10^{-3}	52.1 m

DISCUSSION

Palmer muon collider is valuable for research on channeling. Hybrid nanom devices has many potentials to carry out the Richter-Kimura-Takata's flat bunch, and the Bogacz-Cline's channeling on the Nagamine-Chu's cooled muon bunch. The technological limitation is that the length of Bucky tube at this time is just 1 mm, but it could be replaced with crystal channels.

ACKNOWLEDGMENTS

The author thanks to D.Cline, A.Skrinsky, R.Palmer, Z.Parsa, P.Chen, K.Lee, and M.Atac for kind discussion and also to Z.P.Zheng and M.H.Ye for their encouragement.

REFERENCES

[1] S.A.Bogacz, D.B.Cline, Nucl. Phys. Proc. Suppl. 51A, 90-97, (1996).
[2] R.Palmer, et al. Nucl. Phys. Proc. Suppl.51A (1996), 61-84; PAC97 IEEE 1997.
[3] K.Nagamine, Proc. Suppl. 51A, 115-124, (1996).
[4] S.Chu, Science, 235, pp.861-866, (1991).
[5] B.Richter, Linac 86 Conf. June 6, SLAC, 1986; APS-IEEE PAC97, 1997.
[6] K.Takata and Y.Kimura, Particle Accelerator, 26, 87, (1990).
[7] Z.Huang, P.Chen, and R.D.Ruth, SLAC-PUB-6574, July, 1994.
[8] J.Shen, Nucl.Phys. Proc. Suppl. 51A (1996),98-108;143-147; in this Proceeding.
[9] J.K.Lindhard, Dan.Vidensk. Selsk. Mat. Fys. Medd., 34, No.14 (1965).

μ^-p COLLIDERS

Some Physical Problems for Future μp Colliders

Ilya F. Ginzburg
Institute of Mathematics. 630090. Novosibirsk. Russia.
E-mail: ginzburg@math.nsc.ru

Abstract. The physical problems for the μp collider which can be constructed in one complex with $\mu^+\mu^-$ machine are discussed. It is taken into account that this machine can operate only after basic $\mu^+\mu^-$ collider and LHC.

INTRODUCTION

The concept of a muon collider allows it to be built as a complex in which the basic $\mu^+\mu^-$ collider is supplemented by a μp collider with high energy and luminosity. There are two basic variants of μp colliders. From the basic $\mu^+\mu^-$ device, a $\mu^+ p$ ($\mu^- p$) collider can be formed by the replacement of the $\mu-$ ($\mu+$) source with a (specifically prepared) proton beam. Also, a μp collider can be formed from the intersection of one muon beam from $\mu^+\mu^-$ collider with one proton beam from the Tevatron ring.

In the first variant, the final proton beam energy will be the same as that of the μ^+ (μ^-) beam. Luminosity of this new collider will be the same or even larger than that of the basic $\mu^+\mu^-$ collider (since the number of protons can be larger than the number of muons). Finally, muon polarization is expected to be the same as that of the basic $\mu^+\mu^-$ collider.

Let us enumerate **characteristic features of this μp collider.**

1. Lepton–proton collision in new energy domain.

2. High luminosity.

3. Bremsstrahlung and Initial State Radiation are practically absent.

4. Large enough degree of muon polarization (I expect about 60%).

5. New type of collisions — μp instead of modern ep.

High monochromaticity of muons seems unnecessary for this collider.

The operation of the discussed collider is expected to come only after the first muon collider and LHC. Therefore, the number of really new problems of particle physics to be addressed by the collider is not too high. So, we discuss:

What really new can be studied at µp collider after the studies at a basic $\mu^+\mu^-$ collider and the LHC.

We don't expect any discovery of fundamentally new physics (new interactions beyond the SM). However, the µp collider will supplement for deeper understanding of phenomena in Modern Physics and Physics discovered at the LHC and a muon collider.

I QCD AND HADRON PHYSICS

A Modern (HERA) studies continuation

There is large list of modern studies here, which can be continued at this machine. The standard list of problems for HERA [1] can be continued here for new region of parameters. I discuss these problems only briefly with slightly detail discussion of really new opportunities.

Structure functions, etc. Large extension of the studied region in the x, Q^2 space is provided by much higher energy and luminosity, much more accuracy in the results is expected due to practical absence of Initial State Radiation.

Breaking of factorization. The cross sections for the production of heavy particles or events with high $p_\perp$ in pp or $p\bar{p}$ collision are written usually in the simple factorizable form like

$$\sigma \propto \int f(x_1,Q^2) f(x_2,Q^2) \hat{\sigma}(x_1 x_2 s) dx_1 dx_2 \qquad (1)$$

with factorization of parton densities. At small enough values of x_i this factorization should be broken (see e.g. [2]). This phenomenon is close to the effect of unitarization for BFKL Pomeron.

The expected high accuracy of data at µp collider together with data from LHC provides opportunity to see in detail this breaking.

Hard diffraction. That are the processes

$$\gamma p \to M + ... \text{ with rapidity gap}, \quad (M = \rho^0,\ \omega,\ \gamma,\ \pi^0,\ (q\bar{q}), ...) \qquad (2)$$
$$\text{at } s \gg p_{\perp M}^2 \gg \mu^2,\ \mu \sim 0.3 \text{ GeV.}$$

Here $p_\perp$ is the transverse momentum of produced system M.

The study of these processes provides opportunity to test pQCD and to investigate perturbative Pomeron and Odderon. It is stated now [3] that for the vector meson photoproduction pQCD becomes valid only at very high $p_\perp$, $\geq$ 7-10 GeV/c, where the cross section is very small to study it at HERA. This region is reachable at a µp collider due to higher luminosity and beam energy. Besides, the pQCD becomes valid at lower $p_\perp$, $\geq$ 3-4 GeV/c, in the hard Compton effect with rapidity gap; the cross section here is not so small. But, in ep collisions the extraction of this hard Compton scattering from the data is very difficult due to strong bremsstrahlung from electron. This background is greatly reduced in the μp collision. So, this process can be studied well here. One can expect to see here the regime of BFKL Pomeron (if it exists) and its unitarization (good theory of this effect based on pQCD is absent at this time).

B Structure function for axial current (Z exchange)

The axial structure functions are related to the interaction of axial current with the proton. The structure functions are determined similarly as the standard "vector" structure functions (related to the interaction of vector – electromagnetic – current with proton). They present really new objects in hadron physics and QCD. Indeed, these axial structure functions differ from vector structure function, essentially due to axial current anomaly (at $Q^2 = -(p_\mu - p'_\mu)^2 < m_q^2$, $q = c, b, t$). This difference is beyond many modern discussions but it can result in important effects. Calculations here are in progress [4]. Studies of the axial structure functions will become possible only with the high beam energies and luminosity along with variable longitudinal polarization of muons at a μp collider.

The axial structure functions will be extracted from the standard DIS experiments with longitudinally polarized muons at high Q^2:

$$\mu^{L,R} p \to \mu + ... \tag{3}$$

We consider the cross sections σ^L and σ^R for the left–hand and right–hand polarized muon as an example. The process discussed is described by diagrams with both photon and Z exchange (other diagrams contribute negligible). These photon and Z contributions, J_γ and J_Z, are related to the vector and axial currents, M_V and M_A, as

$$M_V = \frac{1}{Q^2} J^\gamma + \frac{1/4 - \sin^2 \Theta_W}{Q^2 + M_Z^2} J_V^Z; \quad M_A = \frac{1}{Q^2 + M_Z^2} J_A^Z. \tag{4}$$

With these notations we have

$$\sigma_{L,R} \propto |M_V \pm M_A|^2;$$
$$\Delta\sigma = \sigma^L - \sigma^R \propto Re(M_V^* M_A), \quad \sigma^{np} = \frac{\sigma^L + \sigma^R}{2} \propto |M_V|^2 + |M_A|^2. \tag{5}$$

The difference $\Delta\sigma$ should be not small in comparison with separate cross sections σ^L and σ^R at $Q^2/M_Z^2 \gg 0$, for example at $Q^2 ¿(1\text{-}3)\cdot 1000$ GeV^2. Therefore, the reliable accuracy in the extraction of axial contribution is possible here. Experience from the previous calculations for $e\gamma \to eH$ process shows that $\Delta\sigma$ is saturated at $Q^2 \approx 500$ GeV^2.

C Structure functions for charged current

The structure functions for the charged current are studied now in the experiments like $\mu p \to \nu + ...$ Here, the energy and virtuality of exchanged W are unknown, and we only can restore structure functions from some integral properties.

High energy and luminosity of μp collider provides opportunity to study here the process

$$\mu p \to \mu W + ... \qquad (6)$$

This process is described by diagrams shown in figure. When $(\vec{p}_{\perp\mu} + \vec{p}_{\perp W})^2 \ll$

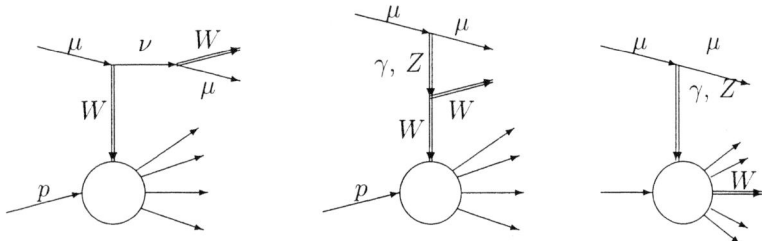

Diagrams with W boson exchange   *Diagrams with no W exchange*

s, the diagrams with W exchange dominate. One can study here the charged current interaction with hadrons with *the known energy and virtuality* of the exchanged W. Until now we have here only calculations in the parton model with very rough testing. Detailed calculations for this process are in progress.

II LEPTOQUARKS, ETC.

This collider will provide really new opportunities to see leptoquark in the fusion of muon with quark from proton (mainly, s or c.)

It is natural in the frame of SM ideas to expect that the coupling of this "light" leptoquark with quarks and leptons to be similar to that in Yukawa interaction of quark with Higgs (responsible for the quark masses)[1]:

$$g_{\ell q} \sim \left(\frac{m_\ell + m_q}{v}\right) |S_{\ell q}| \text{ with } v = 1\sqrt{/G_F\sqrt{2}} = 246 \text{ GeV}. \qquad (7)$$

Besides, it seems natural that the mixing between generations is small, and the masses of muon and s or c quarks are summed here. Therefore,for the "muoquark",

---

[1] Certainly, the less natural model with not too small $g^{\ell q}$ cannot be excluded and the search of leptoquark should be continued here.

this coupling constant is not so small,

$$g_{\mu c} \sim 0.005, \quad g_{\mu s} \sim 0.002,$$

and one can expect here visible effect (if such muoquark exists.)[2]

There is some limited list of new opportunities here related to the discovery of some new particles (SUSY, etc.). The simple example provides us *the gluino search* in the reaction

$$\gamma g \to \tilde{\gamma}\tilde{g}. \tag{8}$$

The expected high luminosity here provides high potential for the gluino discovery in this process.

III SUMMARY

A μp collider will be a simple extension of a basic $\mu^+ \mu^-$ collider providing polarized beams of high energy and intensity. Thus, it allows for studies of QCD and hadron physics extending beyond kinematic regions of HERA. Some of the possible measurements are: structure functions, breaking of factorization, axial current structure function, charged current structure function, etc. In addition, the collider may see leptoquarks if they exist and some SUSY particles.

ACKNOWLEDGEMENT

I am grateful to V. Serbo for discussions. I am thankful to A.M. Sessler, S. Chattopadhyay and M. Zolotorev for warm hospitality in LBNL before and during Conference and to Organizing Committee for the invitation. This work is supported by grants INTAS – 93-1180 ext and RFBR – 96-02-19114.

REFERENCES

1. HERA, yellow book.
2. I.F. Ginzburg, D.Yu. Ivanov, V.G. Serbo, Proc. Workshop on Physics and Experiments with Linear e^+e^- Colliders. Hawaii (1993) v.II, 600–604, World Sc. Singapore.
3. I.F. Ginzburg, D.Yu. Ivanov, Phys. Rev. **D54** (1996) 5523–5535.
4. I.F. Ginzburg, V.A. Ilyin, in preparation.

---

[2] The corresponding natural value for "electroquark" is much lower, $g_{eq} \sim 2 \cdot 10^{-5}$. It means that their discovery in similar experiments is hardly probable.

Muon-proton Colliders: Leptoquarks and Contact Interactions

Kingman Cheung

Department of Physics, University of California at Davis, Davis, CA 95616

Abstract. The muon-proton (μp) collider is an interesting option of the muon collider. Here we discuss the physics potential of the μp collider; especially, leptoquarks and contact interactions. We calculate the sensitivity reach for the second generation leptoquarks and leptogluons, R-parity violating squarks, and μq contact interactions for the μp colliders of various energies and luminosities.

INTRODUCTION

Recently, the muon collider has received a lot of attentions [1]. The muon collider of a few hundred GeV center-of-mass energy is considered a Higgs factory [2], where interactions and branching ratios of the Higgs boson can be studied in detail. It is also an excellent place to study the top-quark near the threshold region [3]. Other physics opportunities include precision studies of gauge bosons [3], search for supersymmetry and lepton-number violation, and other new physics. Muon colliders in TeV range should be feasible for studying strong electroweak symmetry breaking [4], lepton-number violation, and search for heavy exotic particles.

The R&D [1,5] of the muon collider is underway. The First Muon Collider (FMC) will have a 200 GeV muon beam on a 200 GeV anti-muon beam, which could possibly be at the Fermilab [5]. With the existing Tevatron proton beam the muon-proton collision becomes a possible option. It would be a 200 GeV ⊗ 1 TeV μp collider, where the first energy is the energy of the muon beam and the second energy is the proton beam energy. The existing lepton-proton collider is the ep collider at HERA. Lepton-proton colliders have been proved to be successful by the physics results from HERA. In this work, we shall discuss the physics potential of the μp colliders at various energies and luminosities. Other μp colliders that we are considering are 50 GeV ⊗ 1 TeV, 1 TeV ⊗ 1 TeV, and 2 TeV ⊗ 3 TeV. The center-of-mass energies and luminosities of these various designs are summarized in Table 1. The nominal yearly luminosity of the 200 GeV ⊗ 1 TeV μp collider is about 13 fb^{-1}. Luminosities for other designs are roughly scaled by one quarter power of the muon beam energy and given in Table 1.

PHYSICS POTENTIAL

The physics opportunities of μp colliders are similar to those of ep colliders, but the sensitivity reach might be very different, which depends on how precise the particles can be identified and measured in ep and μp environments. Similar to ep colliders the proton structure functions can be measured to very large Q^2 and small x in μp colliders of higher energies. At HERA, the Q^2 has been measured up to $Q^2 \simeq 10^4$ GeV^2 and x down to $x \sim 3 \times 10^{-4}$. At the 200 GeV $\otimes$ 1 TeV μp collider the Q^2 can be measured up to 10^6 GeV^2. In addition, QCD studies, search for supersymmetry and other exotic particles can be carried out at the μp colliders. Here we are particularly interested in the leptoquarks, leptogluons, R-parity violating squarks, and the contact interactions that are specific to the muon. The goal here is to estimate the sensitivity reach for these new physics at various energies and luminosities.

Leptoquarks

The second generation leptoquarks made up of a muon and a charm or strange quark are particularly interesting at the μp collider because they can be directly produced in the s-channel processes, e.g.,

$$\mu^{\pm} c \to L^0_{\mu c} \,. \tag{1}$$

It is conventional to assume no inter-generational mixing in order to prevent the dangerous flavor-changing neutral currents. The s-channel production will give spectacular enhancement in the invariant mass M of the muon and the hadronic final state, or the $x = s/M^2$ distribution.

The Lagrangian of the second generation leptoquark with the muon and charm and strange quarks is given by

$$\mathcal{L} = \lambda^0_{\mu q} \bar{q} \mu L^0_{\mu q} + \lambda^1_{\mu q} \bar{q} \gamma_\rho \mu L^{1\rho}_{\mu q} + \text{h.c.} \,, \tag{2}$$

where $q = c, s$ and the superscripts $(0, 1)$ on the leptoquark field denote the scalar and the vector leptoquarks, respectively. The production cross section of the leptoquark at the μp collider is given by

TABLE 1. The center-of-mass energies $\sqrt{s}$ and luminosities $\mathcal{L}$ of various designs of muon-proton colliders.

	$\sqrt{s}$(GeV)	$\mathcal{L}$ (fb^{-1})
30GeV $\otimes$ 820GeV	314	0.1
50GeV $\otimes$ 1TeV	447	2
200GeV $\otimes$ 1TeV	894	13
1TeV $\otimes$ 1TeV	2000	110
2TeV $\otimes$ 3TeV	4899	280

$$\sigma = \frac{\pi \lambda^2}{4s} q(x, Q^2) \times (J+1) , \qquad (3)$$

where J is the spin of the leptoquark and $q(x, Q^2)$ is the parton luminosity.

Leptogluons

A leptogluon has a spin of either 1/2 or 3/2, a lepton quantum number (in this case it is the muon), and a color quantum number (the same as gluon.) The interaction Lagrangian for a spin 1/2 leptogluon is given by

$$\mathcal{L} = g_s \frac{M_{L_{\mu g}}}{2\Lambda^2} \overline{L_{\mu g}^a} \sigma^{\mu\nu} \mu \, G_{\mu\nu}^b \, \delta_{ab} + \text{h.c.} , \qquad (4)$$

where Λ is the scale that determines the strength of the interaction. The decay width of the leptogluon into a muon and a gluon is given by

$$\Gamma(L_{\mu g} \to \mu g) = \frac{\alpha_s M_{L_{\mu g}}^5}{2\Lambda^4} . \qquad (5)$$

The leptogluon can be produced in s-channel in a μp collider and the production cross section is given by

$$\sigma = \frac{4\pi^2 \alpha_s}{s} \left(\frac{M_{L_{\mu g}}^2}{\Lambda^2}\right)^2 g(x, Q^2) , \qquad (6)$$

where $g(x, Q^2)$ is the gluon luminosity.

R-parity Violating Squarks

R-parity is in general assumed in supersymmetry, but there is no theoretical reasons why R-parity should conserve. R-parity violation is included by introducing additional terms in the superpotential:

$$\mathcal{W}_R = \lambda_{ijk} L_i L_j \overline{E_k} + \lambda'_{ijk} L_i Q_j \overline{D_k} + \lambda''_{ijk} \overline{U_i}\,\overline{D_j}\,\overline{D_k} + \mu_i L_i H_u , \qquad (7)$$

where $L, \overline{E}, Q, \overline{U}, \overline{D}, H_u$ are superfields. The relevant term in the superpotential for the direct production of the R-parity violating squark at the μp collider is $\lambda'_{ijk} L_i Q_j \overline{D_k}$. The corresponding Lagrangian is

$$\mathcal{L}_{L_i Q_j \overline{D_k}} = \lambda'_{ijk} \Big[\tilde{e}_{iL} \overline{d_{kR}} u_{jL} + \tilde{u}_{jL} \overline{d_{kR}} e_{iL} + \tilde{d}^*_{kR} \overline{(e_{iL})^c} u_{jL}$$
$$- \tilde{\nu}_{iL} \overline{d_{kR}} d_{jL} - \tilde{d}_{jL} \overline{d_{kR}} \nu_{iL} - \tilde{d}^*_{kR} \overline{(\nu_{iL})^c} d_{jL} \Big] + h.c. \qquad (8)$$

where i, j, k are the family indices, and c denotes the charge conjugate. The R-parity violating squarks can be considered special scalar leptoquarks.

The cross section for $\mu^+ p \to \tilde{t}_L \to \mu^+ X$ is given by

$$\sigma_{\tilde{t}_L} = \frac{\pi |\lambda'_{231}|^2}{4s} d\left(\frac{m_{\tilde{t}_L}^2}{s}, Q^2 = m_{\tilde{t}_L}^2\right), \tag{9}$$

where d is the down-quark luminosity. The above formula can be easily changed to represent the production of other squarks with the corresponding subscripts in λ' and the parton luminosity. If kinematically allowed the leptoquarks, leptogluons, and the R-parity violating squarks are produced in s-channel and thus give rise to spectacular enhancement in a single bin of the invariant mass M distribution or the x distribution.

Contact Interactions

The effective four-fermion contact interactions can arise from fermion compositeness or exchanges of heavy particles like heavy Z', heavy leptoquarks, or other exotic particles. The conventional effective Lagrangian of $llqq$ ($l = e, \mu$) contact interactions has the form [6]

$$L_{NC} = \sum_q \Big[\eta_{LL} \left(\overline{l_L}\gamma_\mu l_L\right)\left(\overline{q_L}\gamma^\mu q_L\right) + \eta_{RR} \left(\overline{l_R}\gamma_\mu l_R\right)\left(\overline{q_R}\gamma^\mu q_R\right)$$
$$+ \eta_{LR} \left(\overline{l_L}\gamma_\mu l_L\right)\left(\overline{q_R}\gamma^\mu q_R\right) + \eta_{RL} \left(\overline{l_R}\gamma_\mu l_R\right)\left(\overline{q_L}\gamma^\mu q_L\right)\Big], \tag{10}$$

where the eight independent coefficients $\eta_{\alpha\beta}^{lu}$ and $\eta_{\alpha\beta}^{ld}$ have dimension (TeV)$^{-2}$ and are conventionally expressed as $\eta_{\alpha\beta}^{lq} = \epsilon g^2/\Lambda_{lq}^2$, with a fixed $g^2 = 4\pi$.

We introduce the reduced amplitudes $M_{\alpha\beta}^{\mu q}$, where the subscripts label the chiralities of the initial lepton (α) and quark (β). The SM tree level reduced amplitude for $\mu q \to \mu q$ is

$$M_{\alpha\beta}^{\mu q}(\hat{t}) = -\frac{e^2 Q_q}{\hat{t}} + \frac{e^2}{\sin^2\theta_w \cos^2\theta_w} \frac{g_\alpha^\mu g_\beta^q}{\hat{t} - m_Z^2}, \qquad \alpha, \beta = L, R \tag{11}$$

where $\hat{t} = -Q^2$ is the Mandelstam variable, $g_L^f = T_{3f} - \sin^2\theta_w Q_f$ and $g_R^f = -\sin^2\theta_w Q_f$, T_{3f} and Q_f are, respectively, the third component of the SU(2) isospin and the electric charge of the fermion f in units of the proton charge, and $e^2 = 4\pi\alpha_{em}$. The new physics contributions to the reduced amplitudes $M_{\alpha\beta}$ from the $\mu\mu qq$ contact interactions of Eq. (10) are

$$\Delta M_{\alpha\beta}^{\mu q} = \eta_{\alpha\beta}^{\mu q}, \qquad \alpha, \beta = L, R. \tag{12}$$

The differential cross section are given by

$$\frac{d\sigma(\mu^+p)}{dx\,dy} = \frac{sx}{16\pi} \left\{ u(x,Q^2) \left[|M_{LR}^{\mu u}|^2 + |M_{RL}^{\mu u}|^2 + (1-y)^2 \left(|M_{LL}^{\mu u}|^2 + |M_{RR}^{\mu u}|^2 \right) \right] \right.$$
$$\left. + d(x,Q^2) \left[\left|M_{LR}^{\mu d}\right|^2 + \left|M_{RL}^{\mu d}\right|^2 + (1-y)^2 \left(\left|M_{LL}^{\mu d}\right|^2 + \left|M_{RR}^{\mu d}\right|^2 \right) \right] \right\} \quad (13)$$

$$\frac{d\sigma(\mu^-p)}{dx\,dy} = \frac{sx}{16\pi} \left\{ u(x,Q^2) \left[|M_{LL}^{\mu u}|^2 + |M_{RR}^{\mu u}|^2 + (1-y)^2 \left(|M_{LR}^{\mu u}|^2 + |M_{RL}^{\mu u}|^2 \right) \right] \right.$$
$$\left. + d(x,Q^2) \left[\left|M_{LL}^{\mu d}\right|^2 + \left|M_{RR}^{\mu d}\right|^2 + (1-y)^2 \left(\left|M_{LR}^{\mu d}\right|^2 + \left|M_{RL}^{\mu d}\right|^2 \right) \right] \right\} \quad (14)$$

The above contact interactions do not enhance the cross section in a single bin of the invariant mass distribution like the light leptoquarks do, instead, contact interactions enhance the cross section at the large Q^2 tail.

SENSITIVITY REACH

The 95% sensitivity of the contact interaction scale that can be reached by μp colliders at various center-of-mass energies and luminosities are performed in the following. We use the x-y distribution to investigate the sensitivity to the new contact interactions. We divide the x-y plane into a grid with $0.05 < x < 0.95$ and $0.05 < y < 0.95$ and 0.1 interval in both x and y directions. We calculate the number of events predicted by the standard model in each bin, call it $n_i^{\rm sm}$. We use an overall efficiency of 0.8. We assume that the observed number of events is given by the standard model. We vary one $\eta_{\alpha\beta}^{\mu q}$ at a time while keeping others zero and calculate the predicted number of events in each bin, call it $n_i^{\rm th}$. We then calculate the χ^2 using

TABLE 2. The 95% sensitivity of the $\Lambda_{\alpha\beta}^{\mu q}$, ($\alpha,\beta = L,R$; $q = u,d$) that can be reached at the various μ^+p colliders, by assuming that what will be observed is given by the SM prediction.

	30GeV ⊗ 820GeV		50GeV ⊗ 1TeV		200GeV ⊗ 1TeV		1TeV ⊗ 1TeV		2TeV ⊗ 3TeV	
$\sqrt{s}$(GeV)	314		447		894		2000		4899	
$\mathcal{L}$ (fb^{-1})	0.1		2		13		110		280	
	+	−	+	−	+	−	+	−	+	−
$\Lambda_{LL}^{\mu u}$	1.4	0.8	3.9	3.7	9.4	9.2	24	23	46	46
$\Lambda_{LR}^{\mu u}$	2.0	1.3	4.9	4.3	11	9.2	26	23	50	40
$\Lambda_{RL}^{\mu u}$	1.9	1.3	4.3	3.2	8.9	6.4	21	14	40	29
$\Lambda_{RR}^{\mu u}$	1.4	0.8	3.5	3.2	8.4	7.9	21	20	40	37
$\Lambda_{LL}^{\mu d}$	0.7	1.1	2.2	2.7	6.2	6.6	17	17	32	34
$\Lambda_{LR}^{\mu d}$	1.1	1.3	2.1	2.8	4.5	6.0	10	14	31	29
$\Lambda_{RL}^{\mu d}$	1.2	1.2	2.5	2.2	5.6	4.4	14	10	29	22
$\Lambda_{RR}^{\mu d}$	0.8	1.0	1.5	2.4	3.7	5.1	9.9	13	18	25

$$\chi^2 = \sum_i \left[2(n_i^{\text{th}} - n_i^{\text{sm}}) + 2n_i^{\text{sm}} \log\left(\frac{n_i^{\text{sm}}}{n_i^{\text{th}}}\right) \right], \quad (15)$$

where the sum is over all 9×9 bins. We know that for a larger η we will obtain a larger χ^2, which means that it is really different from the standard model beyond statistical fluctuation. Here we have 80 degree of freedom, and so for a 95% CL we set $\chi^2 = 102$. We then repeat for another η.

The sensitivity reach of $\Lambda_{\alpha\beta}^{\mu q}$ is tabulated in Table 2. The sensitivity reach depends on the sign of the contact term. The maximum reach of Λ at each center-of-mass energy roughly scales as $\Lambda \sim 10\sqrt{s}$. The effect of luminosity on Λ is rather small: Λ only scales as the 1/4th power of the luminosity.

To estimate the sensitivity reach for R-parity violating squarks we start with λ'_{231} for the top squark and the down quark luminosity. We assume the enhancement in cross section is in the mass bin of $0.9\, m_{\tilde{t}_L} < m < 1.1\, m_{\tilde{t}_L}$. We calculate the number of events predicted by the standard model in this bin, call it n^{sm}. Again, we use an overall efficiency of 0.8. Then we use the poisson statistics to estimate the nth that n^{sm} can fluctuate to at the 95% CL:

$$\sum_{n=0}^{n^{\text{th}}} \frac{(n^{\text{sm}})^n e^{-n^{\text{sm}}}}{n!} > 0.95 \quad (16)$$

where n^{th} is the first n that the above inequality is satisfied. Once the nth is obtained the λ'_{231} can be obtained using Eq. (9).

TABLE 3. 95% sensitivity on λ'_{231} for a few choices of $m_{\tilde{t}_L}$ at various $\mu^+ p$ colliders.

	30GeV ⊗ 820GeV	50GeV ⊗ 1TeV	200GeV ⊗ 1TeV	1TeV ⊗ 1TeV	2TeV ⊗ 3TeV
$\sqrt{s}$(GeV)	314	447	894	2000	4899
$\mathcal{L}$(fb^{-1})	0.1	2	13	110	280
$m_{\tilde{t}_L}$ (GeV)					
200	0.014	0.0045	0.0025	0.0015	0.0010
300	-	0.0094	0.0032	0.0018	0.0014
400	-	0.055	0.0041	0.0021	0.0017
500	-	-	0.0056	0.0024	0.0019
600	-	-	0.0086	0.0027	0.0021
700	-	-	0.016	0.0030	0.0023
800	-	-	0.045	0.0033	0.0025
900	-	-	-	0.0037	0.0026
1000	-	-	-	0.0043	0.0027
1500	-	-	-	0.012	0.0034
2000	-	-	-	-	0.0043
2500	-	-	-	-	0.0056
3000	-	-	-	-	0.0078
3500	-	-	-	-	0.013
4000	-	-	-	-	0.024
4500	-	-	-	-	0.081

The sensitivity reach of λ'_{231} is tabulated in Table 3. We have also calculated the sensitivity reach of λ'_{232} using the strange quark luminosity. We found that the reach is typically worse than that of λ'_{231}: for small $m_{\tilde{t}_L}$ the reach is about a factor of two worse while for large $m_{\tilde{t}_L}$ the reach can be ten times worse. This is because the strange quark luminosity is rather large at small x but very small at large x.

The results for the second generation leptoquarks and leptogluons are summarized in Table 4 [7]. Here only a simple criteria is defined for the discovery of the leptoquarks and leptogluons. Assuming no background and requiring five signal events for the discovery the sensitivity reach is at 99% CL.

REFERENCES

1. *Proceedings of the Symposium on Physics Potential and Development of $\mu^+\mu^-$ Colliders*, San Francisco, CA December 1995; $\mu^+\mu^-$ *Colliders: A Feasible Study*, Snowmass, CO, July 1996; and these proceedings.
2. V. Barger, M. Berger, J. Gunion, and T. Han, Phys. Rept. **286**, 1 (1997).
3. V. Barger et al., Phys. Rev. **D56**, 1714 (1997).
4. V. Barger et al., Phys. Rev. **D55**, 142 (1997).
5. *Workshop on Physics at the First Muon Collider and at the Front End of a Muon Collider*, Fermilab, Batavia IL, December 1997.
6. For a recent review, see V. Barger et al. Phys. Rev. **D57**, 391 (1998).
7. V. Barger, M. Berger, K. Cheung, J. Gunion, and T. Han, in preparation.

TABLE 4. 99% sensitivity reach on the coupling λ for the second generation leptoquarks and the new physics scale Λ for leptogluon via the resonance production $\mu p \to L$.

	30GeV ⊗ 820GeV	50GeV ⊗ 1TeV	200GeV ⊗ 1TeV	2TeV ⊗ 4TeV
$\sqrt{s}$(GeV)	314	447	894	5657
$\mathcal{L}$(fb^{-1})	0.1	2	13	280
$M^0_{\mu c}$	200 GeV	300 GeV	500 GeV	1500 GeV
$\lambda^0_{\mu c}$	0.089	0.043	0.010	0.0014
$M^0_{\mu s}$	200 GeV	300 GeV	500 GeV	1500 GeV
$\lambda^0_{\mu s}$	0.068	0.034	0.0080	0.0011
$M^1_{\mu c}$	200 GeV	300 GeV	500 GeV	1500 GeV
$\lambda^1_{\mu c}$	0.063	0.031	0.0072	0.0010
$M^1_{\mu s}$	200 GeV	300 GeV	500 GeV	1500 GeV
$\lambda^1_{\mu s}$	0.048	0.024	0.0055	0.0008
$M_{\mu g}$	200 GeV	300 GeV	500 GeV	1500 GeV
$\Lambda_{\mu g}$(TeV)	20	49	190	1700

PHYSICS SUMMARY

The Physics of Muon Colliders
A Perspective

William J. Marciano

*Brookhaven National Laboratory
Upton, New York 11973*

Abstract. The physics objectives of future muon colliders are discussed. Attractive features of various energy options are described. Comments on the utility of polarization for Z and Higgs resonance studies are given.

The standard $SU(3)_c \times SU(2)_L \times U(1)_Y$ theory of strong and electroweak interactions represents tremendous intellectual advancement. In roughly 100 years, physics has progressed from discovery of the electron (1897) to a profound understanding of the basic constituents and fundamental laws of Nature. Those laws are based on the principle of local gauge invariance and the realization that "Symmetry Dictates Dynamics". Unveiling the hidden symmetries of Nature and exploring their subtleties has become the main focus of elementary particle physics [1].

The standard model has been tested at the level of its quantum loop predictions, and passed with flying colors. Precision measurements have probed the $\pm 0.1\%$ level without uncovering any pronounced deviation. Similarly, loop induced flavor-changing neutral current effects and CP violation conform to standard model expectations. Nevertheless, the standard model exhibits a striking blemish, the source of electroweak symmetry breaking and mass generation, which almost certainly points to "new physics". Presently, an electroweak mass scale $v \simeq 250$ GeV is essentially introduced by hand via a symmetry breaking potential

$$V(\phi) = \lambda(\phi^+\phi - v^2/2)^2 \qquad (1)$$
$$\phi = \frac{1}{\sqrt{2}} \begin{pmatrix} \omega_1 + i\omega_2 \\ v + H + iz \end{pmatrix}$$

The constant field component, v, permeates space-time and quite efficiently provides particles with masses. It also leads to quark mixing and CP violation. However, couplings of other fields to ϕ and consequently their masses are largely arbitrary. It is generally believed that the true underlying source of mass generation will manifest itself near or before the TeV scale. The leading candidate for that new physics is connected with supersymmetry, but various other options exist such as strong dynamics (technicolor), additional compact spatial dimensions ($\mathcal{O}(10^{-17}\text{cm})$) [2], etc. Elucidating the true nature of electroweak symmetry breaking and mass generation is currently the primary goal of elementary particle physics.

Searches for new short-distance physics fall (roughly) into three complementary categories: High Precision, High Sensitivity, and High Energy. Of these, high energy colliders provide the most direct means for uncovering new particles and phenomena. They also allow for precision measurements and searches for rare or "forbidden" reactions. In such pursuits, the muon collider has the potential to become an extremely valuable tool. Let me outline some of the various muon collider options being considered and briefly comment on their physics potential.

Muon colliders will provide many attractive options. Most interesting is the very high energy $\mu^+\mu^-$ collider which will open new windows to never before seen short-distance phenomena [3]. Lower energy facilities may also have strong motivation. For example, Higgs scalar, Z boson or as yet unanticipated other neutral bosons can be thoroughly studied at dedicated facilities where one can sit directly on the s-channel resonance and accumulate high statistics.

The technology to produce, cool, and store intense muon beams will have other interesting applications. One can envision $\mu^-\mu^-$ or μ^-p colliders as facility options. Muon storage rings could also be used as sources for extremely intense neutrino beams [4]. Very clean low energy muon beams could similarly significantly advance searches for muon-number non-conservation. For example, the coherent conversion of muons into electrons in the field of a nucleus, $\mu^-N \to e^-N$, could be pushed to 10^{-18} sensitivity with a $10^{12}\mu^-$/sec source. Such low energy possibilities are themselves very well motivated. In addition they may provide a path for R&D towards the First Muon Collider (FMC).

What center-of-mass energies should muon collider technology strive for? Specific long term goals will become more obvious as "new physics" is uncovered. However, even now, it is clear that $\mu^+\mu^-$ facilities optimized at roughly $\sqrt{s} \sim 100$ GeV, 500 GeV–1 TeV, 3 TeV, and 10 TeV are well motivated. Since cross-sections generally fall as $1/s$, higher energy generally demands higher luminosity. Some specific studies also require outstanding beam energy resolution or beam polarization. For example, resonance studies of very narrow states or high precision threshold measurements significantly profit from the outstanding energy resolution that could be achieved at muon colliders.

Let me comment on the utility of polarization at a $\mu^+\mu^-$ collider. Polarization can be, in some cases, extremely valuable for sorting out new physics or suppressing backgrounds. For example, various supersymmetry signatures have very pronounced muon polarization dependencies. Standard model W^+W^- backgrounds can be suppressed by employing polarized muons. Resonance studies can be enhanced by employing polarization.

In the case of a muon collider, both the μ^+ and μ^- beams can be polarized [3]. One gets some degree of polarization essentially for free in the usual collection scheme of $\pi^\pm$ decays to $\mu^\pm$. Roughly ± 0.20 beam polarization is expected.

Enhancing the polarization to ±0.5 simply by muon energy cuts reduces the beam intensity by about a factor of 1/4 (for each beam), if no compensating mechanism is included. However, by increasing the $\mu^{\pm}$ production by a factor of 4 (e.g. increasing the initial proton flux by 4×), one could retain high muon currents with increased polarization. Also, at this meeting Skrinsky [5] has discussed the possibility of higher energy muon production which could give polarization ±0.70 and still yield high luminosity.

When both beams are polarized, the effective polarization [6]

$$P_{eff} = \frac{P_- - P_+}{1 - P_- P_+} \qquad (2)$$

provides a useful parametrization for s-channel physics. In that expression P_- and P_+ are the polarization of the μ^- and μ^+ respectively, with each in the range -1 (lefthanded) $\leq P_{\pm} \leq +1$ (righthanded). The magnitude of the effective polarization is illustrated in table 1 for various $P_{\pm}$ values.

TABLE 1. Effective polarization, P_{eff}, for various values of P_+ and P_-.

P_+	P_-	$P_{eff} = \frac{P_- - P_+}{1 - P_+ P_-}$
0.20	-0.20	-0.385
0.34	-0.34	-0.610
0.50	-0.50	-0.800
0.70	-0.70	-0.940
0.85	-0.85	-0.987

Polarization will be a useful commodity at the FMC. That facility is envisioned to be a $\mu^+\mu^-$ collider optimized for $\sqrt{s} \simeq 100$ GeV running. It should be capable of very high luminosity Z pole (91.186 GeV) studies [7] as well as Higgs resonance [8] running (for a narrow Higgs with $m_H \lesssim 160$ GeV). In addition, it will be useful to run somewhat above the W^+W^- threshold region ($\sqrt{s} \simeq 161$ GeV) for the purpose of precisely determining m_W and the non-abelian gauge couplings. In all of those pursuits high luminosity will be the main priority, but very good energy resolution and polarization should also prove useful.

In the case of Z pole running, a wealth of interesting studies remain to be explored. For $\mathcal{L} \simeq 10^{33}$ cm^{-2}s^{-1}, one can expect $10^9 \sim 10^{10}$ Z bosons, i.e. 2 to 3 orders of magnitude beyond LEPI. Such high statistics open up many new windows. The already successful LEPI program can be repeated with more than an order of magnitude reduction in statistical errors. With $\gtrsim 10^9$ Z decay events, predicted rare decay events such as $Z \to W^{\pm}e\nu$ or $Z \to W^{\pm}\pi^{\mp}$ should be observable. One can also search for $Z \to \tau^+\mu^-$ and other flavor-violating neutral current processes. Such a copious Z source would be desireable as a b and τ factory where rare decays of those fermions, such as $b \to s\gamma$, could be studied or high precision measurements of the tau could be cleanly made.

In the above pursuits, polarization will be extremely valuable. It can be used to increase, somewhat, the production cross-section. More important, the large $b\bar{b}$ forward-backward polarization asymmetry can be used to efficiently tag b and $\bar{b}$; thus providing a new window to oscillation and CP violation studies. Indeed, the angular distribution for $\mu^-\mu^+ \to b\bar{b}$ is given at the Z pole by [6] (for $x \equiv \cos\theta = 4p_{\mu^-} \cdot p_b/s$)

$$\frac{1}{N}\frac{dN(\mu^-\mu^+ \to b\bar{b})}{dx} = \frac{3}{8}\left(1 + x^2 + \frac{8}{3}xA_{eff}\right) \tag{3}$$

$$A_{eff} = A_{LR}^{FB}\left\{\frac{A_{LR} + P_{eff}}{1 + P_{eff}A_{LR}}\right\}$$

$$A_{LR}^{FB} = \frac{3}{4}\left(\frac{1 - \frac{4}{3}\sin^2\theta_W}{1 - \frac{4}{3}\sin^2\theta_W + \frac{8}{9}\sin^4\theta_W}\right)$$

$$A_{LR} = \frac{1 - 4\sin^2\theta_W}{1 - 4\sin^2\theta_W + 8\sin^4\theta_W}$$

For $\sin^2\theta_W = 0.2315$, $A_{LR}^{FB} = 0.70$, $A_{LR} = 0.147$ and $P_{eff} \simeq 0.8$, one finds $A_{eff} \simeq 0.6$. By restricting to forward events, one can get very good $b(\bar{b})$ tagging efficiency.

Polarization and high statistics at the Z pole can also be used to measure $\sin^2\theta_W$ with incredible precision via the left-right asymmetry. With both beams polarized and switching both polarization between LR and RL modes, the asymmetry [9]

$$\frac{N_{LR} - N_{RL}}{N_{LR} + N_{RL}} = P_{eff}A_{LR} \tag{4}$$

$$A_{LR} = \frac{2(1 - 4\sin^2\theta_W)}{1 + (1 - 4\sin^2\theta_W)^2}$$

provides $\sin^2\theta_W$ with high precision. Indeed, $\Delta\sin^2\theta_W/\sin^2\theta_W \simeq \frac{1}{10}\Delta A_{LR}/A_{LR}$. For $\gtrsim 10^8$ Z decays and $P_{eff} \simeq 0.8$, A_{LR} can be determined to better than $\pm 0.1\%$ and $\sin^2\theta_W(m_Z)$ to about ± 0.00002; i.e. ten times better than present day accuracy. Of course, P_{eff} would have to be monitored at the $\pm 0.1\%$ level. However, that may not be so difficult since muon decay provides a self-analyzing polarization measure for each beam. Also, the fractional uncertainty in P_{eff} is smaller than P_+ and P_- separately.

Above the W^+W^- threshold, the W boson mass, m_W, can be determined with very high precision. A measure of what level to aim for is provided by the fact that $m_W^2\sin^2\theta_W$ is expected to be constant in the standard model (i.e. very insensitive to the Higgs mass). A $\pm 0.01\%$ measurement of $\sin^2\theta_W$, therefore, correspond to ± 4 MeV determination of m_W. Such precision would be extremely challenging (perhaps unattainable). However, it is certainly worth striving for.

Measuring $\sin^2\theta_W$ to ± 0.00002 and m_W to ± 4 MeV would test the standard model at the $\pm 0.01\%$ level and severely constrain (or point to) "new physics". For example, each measurement could be individually used to predict the Higgs mass to about ±5%. Their comparison could be used to determine the S parameter (of Peskin and Takeuchi) via

$$S \simeq \left\{ 2 \left(\frac{m_W - 80.383 \text{ GeV}}{80.383 \text{ GeV}} \right) + \frac{\sin^2 \theta_W (m_Z)_{\overline{MS}} - 0.23121}{0.23121} \right\} \quad (5)$$

The precision mentioned above would pinpoint S to about ±0.016! It would also probe for "new physics" at about the 5 TeV level.

Polarization will also be useful for finding and studying a narrow low mass Higgs boson on resonance. For example, if $m_H \simeq 110$ GeV, the standard Higgs is expected to have a width of about $\Gamma_H \sim 3$ MeV. Employing beam resolution $\Delta E/E \simeq 3 \times 10^{-5}$ and an integrated luminosity $L = 0.05 fb^{-1}$ gives about 3000 Higgs decays [6]. For the primary decay, $H \to b\bar{b}$, signal to background is about 1 to 1. Employing polarized μ^+ and μ^- beams can enhance that ratio primarily by reducing backgrounds. Indeed, $N_S/\sqrt{N_B}$ is enhanced by

$$\frac{1 + P_+ P_-}{\sqrt{1 - P_+ P_-}} \frac{1}{\sqrt{1 - P_{eff} A_{LR}}} \quad (6)$$

Perhaps more important, high polarization can be used to reduce the scan time to locate a narrow resonance. Without polarization, and assuming $\mathcal{L} \simeq 10^{31}$ cm$^{-2}s^{-1}$, one finds the scan time for finding a 110 GeV Higgs scalar is about 3 years [10], which is unacceptably long. (Outstanding energy resolution will lead to a luminosity reduction.) If higher luminosity $\mathcal{L}$ along with two beam polarization is employed, the scan time is reduced by about

$$\frac{(1 + P_+ P_-)^2}{1 - P_+ P_-} \frac{\mathcal{L}}{(10^{31} \text{ cm}^{-2}s^{-1})} \quad (7)$$

Using information about the forward-backward background asymmetry can also enhance signal/$\sqrt{\text{background}}$ and reduce scan time [7].

Once the Higgs boson resonance is found, it becomes important to precisely measure the Higgs width and its various branching ratios $H \to b\bar{b}, c\bar{c}, \tau^+\tau^-$ etc. One can, in principle, use that information to differentiate non-SUSY and SUSY Higgs. Also, the Higgs couplings to quarks provides, in principle, a precise measure of quark masses.

If a Higgs resonance factory can be realized with high luminosity ($5 \times 10^{31} \sim 10^{32}$ cm$^{-2}s^{-1}$), outstanding energy resolution ($\Delta E/E \sim 3 \times 10^{-5}$), and some degree of polarization $P_{eff} \simeq 0.5$–0.94, it becomes a must do facility. It will pinpoint the Higgs mass, elucidate its decay properties ($H \to b\bar{b}, c\bar{c}, \tau\bar{\tau}, WW^*, ZZ^* \ldots$) and search for "new physics".

Beyond the FMC, one can envision a $\mu^+\mu^-$ facility in the range $\sqrt{s} \sim 160$–500 GeV. It would extend the W^+W^- studies beyond threshold, explore the $t\bar{t}$ threshold (near $\sqrt{s} \sim 350$ GeV), and likely explore ZH associated production (depending on m_H). The W^+W^- physics would precisely measure trilinear gauge boson couplings and search for deviations from strong underlying dynamics. It would also continue the study of m_W and Γ_W. Top quark studies at threshold would pinpoint m_t, α_s and V_{tb} with very good precision. High statistics ZH running (at $\sqrt{s} \simeq m_Z + \sqrt{2}m_H$) would complement the Higgs factory with precision measurement of the ZZH coupling. If the scale of SUSY is not too high, such a facility will cross various thresholds $\tilde{\chi}^+\tilde{\chi}^-$ (charginos), $\tilde{\ell}^+\tilde{\ell}^-$ (sleptons) etc. In all those studies, high luminosity $\gtrsim 10^{33}$ cm^{-2}s^{-1} is of crucial importance. Polarization will be useful for sorting out properties of "new physics" and reducing backgrounds.

Beyond $\sqrt{s} \simeq 500$ GeV, new physics discoveries (probably at the LHC) will be needed as guidance. A facility in the $\sqrt{s} \sim 2\text{--}4$ TeV is likely to be highly motivated. If SUSY is discovered at the LHC, such a muon collider will be effective in producing heavier SUSY states, producing the entire Higgs structure, and measuring the more than 120 parameters of SUSY. If strong new dynamics is hinted at by the LHC, a 3 TeV $\mu^+\mu^-$ collider will be ideal for unfolding strong W^+W^- and ZZ interactions. One then expects new meson and baryon resonances (lots of spectroscopy) to manifest themselves. Exploring their properties via $\mu^+\mu^-$ collisions will be our primary access to such phenomena. In addition Z' bosons, compositeness, leptoquarks, etc. may start to appear. A $\sqrt{s} \sim 3$ TeV $\mu^+\mu^-$ collider can probe compositeness scales of $\mathcal{O}(100 \text{ TeV})$.

Are there needed collider facilities beyond $\sqrt{s} \sim 3$ TeV? It seems likely that higher energy will be called for, but right now it is hard to motivate a specific energy. Some say $\sqrt{s} \sim 10$ TeV seems to be the likely place to aim. At that energy the true unique value of a $\mu^+\mu^-$ collider manifests itself, since it is hard to imagine e^+e^- linear colliders reaching such scales. A 10 TeV facility will follow-up SUSY discovery by searching for the source of SUSY breaking. Perhaps new dynamics or new (small) space dimensions open up. Large violations of conservation law could occur. If physics is rich at ~ 1 TeV, the motivation to push to higher energy will be great. Of course, an important issue at high energies is the neutrino radiation problem. We will have to be creative in developing technology that will retain high luminosity while maintaining neutrino radiation safety.

The muon collider movement is gaining momentum. No show stoppers have been found. The physics case for various energy facilities has been elucidated and the high energy community is genuinely excited by the novelty and promise of $\mu^+\mu^-$ colliders. If we can focus that excitement and confront the technological challenges, the First Muon Collider will be realized and will pave the way for follow-up generations of higher energy facilities. A good idea is eventually recognized and realized.

REFERENCES

1. W. Marciano in "Future High Energy Colliders" AIP Conference 397, edited by Z. Parsa (1997) p. 11.

2. G. Chapline and R. Slansky, *Nucl. Phys.* **B209**, 461 (1982); I. Antoniadis, *Phys. Lett.* **B246**, 377 (1990).
3. "Muon Collider Feasibility Study", BNL Report BNL52503 (1996).
4. B. King and S. Geer in these proceedings.
5. A. Skrinsky in these proceedings.
6. B. Kamal, W. Marciano and Z. Parsa in these proceedings.
7. M. Demarteau and T. Han, hep-ph/9801407 (1998).
8. D. Cline, in "Future High Energy Colliders", AIP Conference 397, edited by Z. Parsa (1997) p. 203.
9. A. Czarnecki and W. Marciano, hep-ph/9801394 (1998).
10. J. Gunion in these proceedings.

APPENDIX

e^+e^- and ep Options for the Very Large Hadron Collider

J. Norem

Fermilab, Batavia IL 60610 / Argonne, Argonne IL 60439

Abstract Although the linear collider is ultimately capable of higher energies, a circular e^+e^- collider installed in the large tunnels of a Very Large Hadron Collider (VLHC) has attractive features, including very light magnet system and unchallenging vacuum requirements. An ep collider, built either in the 3 TeV booster or the large tunnel, could extend the HERA program beyond $\sqrt{s} \sim 1$ TeV. Both machines could perhaps use the same rf system, first in the booster tunnel and then as part of the large collider.

INTRODUCTION

The possibility of installing a large electron/positron collider in the tunnels of a Very Large Hadron Collider was first considered at the 1996 Snowmass meeting (1), and studied more fully at the PAC97 conference (2). The ep machine has been studied fairly recently and less completely. These machines are being considered as part of the VLHC effort, and recent progress can be found on the WWW (3).

While the NLC and the Muon collider have received the primary design effort, it seems desirable to also consider the potential advantages of these machines, until the physics goals are more clearly defined by new results.

THE e^+e^- COLLIDER

Using the constraints that the total synchrotron beam loss would be 100 MW and the total circumference was 531 km, a parameter list was developed which would describe the facility. Interesting features of this machine are: 1) the luminosity, which is limited by radiated beam power and magnet aperture, has a maximum at a center of mass roughly equal to the $\bar{t}t$ threshold, decreasing at energies above or below this energy, (it is interesting to note that the energy at which the maximum luminosity occurs is proportional to $R^{0.2}$, where R is the radius of the machine, so the lepton collider parameters are comparatively independent of the VLHC parameters), 2) the energy resolution at this energy is comparatively good, $\sigma_E = 0.26$ GeV, 3) the required field for the dipole magnets is very low, $B \sim 100$ G at full field and $B \sim 10$ G at injection, requiring a good shielding against stray fields including the earth's field, 4) because the machine radius is so large, it would be difficult to evenly deposit the synchrotron power on a vacuum chamber wall, thus it seems desirable to use localized beam absorbers which would be pumped, at intervals of about 100 m. The pumping requirements on these absorbers would only be about 200 L/s at each unit. The power deposition from a 50 - 100 MW synchrotron load would be significant and would be distributed over a wide area. A complete design of this system will include a method of sinking this heat locally along the length of the tunnel without cooling towers.

Although the bending field is low, measurements on a prototype magnet indicated that it should be possible to use the yokes as a shield without additional active or passive components. The low field itself results in very light magnet yokes and very low excitation currents which make the magnet very light, 40 kg/m, and perhaps inexpensive.

The important parameters of the machine would be shown in the following table, taken from Reference 2.

TABLE 1 Parameters of a circular $\bar{t}t$ factory.

Beam energy	180	GeV
Circumference	531	km
Luminosity Goal	$9.15 \cdot 10^{31}$	cm^{-2} s^{-1}
Beam-beam tune shift, $\xi_x = \xi_y$	0.03	
Total current /beam	37	mA
Number of bunches	512	
Synchrotron loss	1.3	GeV/turn
RF voltage	1.6	GV
Beam Aperture, $A_x : A_y$	53 : 38	mm
Total Generator Power	102	MW

THE *ep* COLLIDER

This machine would be an extension of the DESY/HERA program, and the facility itself would be located in the 34 km tunnel planned for the VLHC booster synchrotron. The design constraints on this machine are 1) that the luminosity be at the 1 fb^{-1} scale, 2) that leptons polarization be possible, and 3) collisions with both electrons and positrons be possible (4). A possible parameter list is shown in Table 1, which includes the goals of the design effort. It is possible to reach $\sqrt{s}$ = 7 TeV with 50 TeV protons in the large tunnel.

TABLE 2 Parameters of a large *ep* collider.

Proton beam energy	3	TeV
Electron beam energy	80	GeV
Circumference	34	km
Luminosity goal	1	fb^{-1}/y
$\sqrt{s}$	1	TeV
Total Generator Power	50	MW

A preliminary look seems to indicate that it may be possible to satisfy the requirements, however many technical questions remain.

AN ELECTRON INJECTOR CHAIN

Very preliminary work is underway to look at a possible electron injector chain utilizing the Booster and Main Injector (MI). Because of synchrotron radiation losses, it seems desirable to run the Booster below 4.5 GeV and the MI below 12 GeV. In order to produce positrons, it would be necessary to build a new electron linac and positron accumulator, modeled after those at CERN or the Argonne APS.

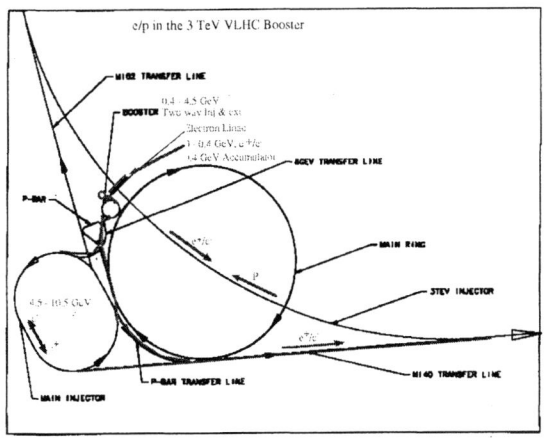

FIGURE 1, An injector chain option

CONCLUSIONS

The cost of circular colliders depends on a variety of parameters, and these very large examples are comparatively far from normal design experience, thus it seems possible that innovative designs could significantly improve the performance/cost ratio over initial expectations. The e^+e^- and ep colliders could considerably extend the physics capabilities of the VLHC. If the machines could share rf systems and other components, they could be built more economically. More detailed studies examining these options are underway at Fermilab and Argonne.

REFERENCES

1. J. Norem and E. Keil, *New Directions for High-Energy Physics*, Snowmass 96 (1997), 325
2. J. Norem, J. Jagger, S. Sharma, E. Keil, G. W. Foster, E. Malamud, E. Chojnacki, D. Winn, proceedings of the 1997 Particle Accelerator Conference, Vancouver (1998)
3. http://www-ap.fnal.gov/VLHC/electrons/index.html
4. D. Krakauer and J. Norem, in *Very Large Hadron Collider Information Packet*, S. Mishra, ed. Fermilab TM (1998)

Suppressing Beam Loss and De-Channeling Control via Modulating Potential by Phononics

Jing Shen[*]

Institute of High Energy Physics (IHEP), Chinese Academy of Sciences
Mailing Address: *P. O. Box 918(6), Beijing 100039, PRC*
and
China Center of Advanced Science and Technology (CCAST), P.O.B. 8730, Beijing 100080

Abstract: Low channeling efficiency is the second reason that crystal collider or accelerator applications are maintained as impossible. However, channeling efficiency recently has made significant improvements. So, in this paper two problems are discussed. (1) Various materials have been investigated on channeling, de-channeling and scattering to design quantum devices for channeling based on potential advancements in the technology of fusion, electronics, and photonics. (2) A phonon mode device PMD has been designed by nanom supper lattice technology. The technology is based on the fact that channeling potential is deformed by phonons at 0.01 nm or less with high probability and at greater than 0.02 nm with low probability. However, de-channeling radii of LiH, Si, GaAs are approximately 0.03 nm. Hence, phonons influence channeling slightly but decrease scattering significantly. Thus, PMD might help advance the Bogacz-Cline channeling, and would make the Richter-Kimura-Takata beam and the Nagamine-Chu beam unnecessary.

INTRODUCTION

The problem of entrance in Bogacz-Cline channeling is discussed in the author's proceeding paper "Mesoscopic Quantum Multiplex"[1]. In this paper, the second problem, de-channeling and beam loss, is discussed. At FNAL in 1984, a bent silicon crystal was used to replace a magnet in a secondary charged particle beam, raising the maximum momentum which could be transmitted from the magnetic septum limit of 225 GeV to the full primary beam level of 400 GeV. Later at Superkhov IHEP in 1989, the channeled particles were deflected 80 mrad with an extraction efficiency of 5×10^{-5} [2]. This extremely low efficiency made a crystal storage ring of Bogacz-Cline channeling improbable. In this paper a new consideration is described using the new experimental results.

INCREASING OF CHANNELING EFFICIENCY

Although the efficiency of channeling seems too low for constructing a realistic collider, the delicate steering and deflection (channeling) of particle beam inside crystals is increasingly studied for high energy physics application as well as being an interesting physics topic in itself. Recently, the efficiency of using crystal channeling technique to steer and deflect particle beams by bent crystal has improved significantly. At CERN in 1990, SPS proton beam of 450 GeV was bent 37.4 mrad which is equivalent to a 6 m magnet with an 11 Tesla field, and its efficiency was 15%

---

[*] *Email: shenj@bepc3.ihep.ac.cn*

or 66 m-Tesla bending magnet[3]. At FNAL in 1996, a tightly focussed beam of 900 GeV from the Tevatron was made incident on a specially made crystal with a vertical bend of 0.64 mrad. The beam at the exit was with a tail below the spot showing an improved efficiency in beam extraction by the crystal[4]. At CERN in 1997, a 33 TeV beam of lead ion (400 GeV/charge) is steered through 4 mrad using a crystal with an efficiency of 15%[5]. At Protvino (also known as Superkhov) IHEP in 1988, a 7 mm long silicon crystal with a 1.75 mrad bend was used to extract a beam of 70 GeV. The channeled beam was up to 2×10^{11} protons per spill of 1.1 sec duration. Its efficiency was 20%[6]. The three major experiments at IHEP, CERN, FNAL were comparable in results. The efficiency was increased 4,000 times at IHEP in the last 8 years. And it was boosted due to an increased number of proton encounters with a radically shortened crystal called a "multi-pass" extraction. A theory predicts that increasing the "multiplicity" factor in extraction by shortening the crystal in this way would greatly improve the efficiency, up to 70-90% efficiency for crystals with optimal lengths in the crystal extraction and collimation projects at TeV range accelerators[6]. New results are the first proof of this prediction.

CHANNELING EQUATION

Beam straggling [7] due to scattering in the medium also results in de-channeling. In the Nagamine-Chu's scheme [8,9], a cooled μ^- beam is more difficult to produce than a cooled μ^+ beam, and the quality of the μ^-'s is usually worse than the μ^+'s. The probability of de-channeling and straggling of the μ^-'s are higher than the μ^+'s. So, we focus in this paper on Nagamine's μ^- beam. Its beam has $E_\perp = 2$ eV. Channeling equation is

$$\left[-\frac{h^2}{2m_0}\nabla^2 - \gamma e\alpha \cdot A(r) - \gamma\rho_3 m_0 c^2\right]\psi(r) = E_\perp^R \psi(r) \tag{3.1}$$

Neglecting spin, creation and annihilation, then in bunch coordinate eqn. (3.1) becomes

$$\left[-\frac{h^2}{2m_0}\nabla^2 - \gamma U_1(r)\right]u(r) = E_\perp^R u(r) \tag{3.2}$$

The Lindhard's approximation [10] in the laboratory coordinates gives

$$\gamma U_1(r) \rightarrow U_{Lin}(r) = \frac{E_{//}\psi_1^2}{2}\ln\left[\left(\frac{C^2 a^2}{r^2}+1\right)\frac{1}{B}\right], \; E_\perp^R \rightarrow E_\perp \tag{3.3}$$

where $\psi_1^2 = 2U_d/E_{//}$, (3.4)

$a = 0.8853 a_0 / \sqrt{Z_1^{\frac{2}{3}} + Z_2^{\frac{2}{3}}}$ the Thomas-Fermi shielding length, (3.5)

and $B = \frac{C^2 a^2}{r_0^2} + 1 = C^2 a^2 \pi N d + 1$, $C^2 = 3$. (3.6)

When $r_{min} \ll r_0$ [see (4.2) and (4.3)], the single atom row approximation of (3.3) is

$$U_{Lin}(r) \approx U_d \ln\left[\frac{C^2 a^2}{r^2} + 1\right] \tag{3.7}$$

where $U_d = \dfrac{Ze^2}{4\pi\varepsilon_0 \varepsilon_\infty d}$ [J] $\dfrac{1}{1.6 \times 10^{19}}$ [eV/J] $= \dfrac{E_\parallel \psi_1^2}{2}$. \hfill (3.8)

MINIMUN CHANNELING RADIUS

De-channeling occurs when the channeling radius is smaller than the average radius of ground state of channeling the minimum radius r_{min}.

$$U_{Lin}(r_{min}) = E_\perp, \quad r_{min} = \int u^*_{l=0}(r) r u_{l=0}(r) dr \tag{4.1}$$

$$r_{min} = Ca\left[B\exp\left(\frac{2E_\perp}{E_\parallel \psi_1^2}\right) - 1\right]^{-\frac{1}{2}} = Ca\left[B\exp\left(\frac{E_\perp}{U_d}\right) - 1\right]^{-\frac{1}{2}} \tag{4.2}$$

where r_0 the channeling radius with $\pi r_0^2 = \dfrac{1}{Nd}$, $N = \dfrac{\rho}{A} N_0$, $r_0 = \sqrt{\dfrac{A}{\pi \rho N_0 d}}$ \hfill (4.3)

$a_0 = 0.529 \times 10^{-8}$ cm the Bohr radius, $N_0 = 6.03 \times 10^{23}$ the Avagadro's number, d the lattice constant, and ρ the density. Substituting (3.5), (3.6), and (3.8) into (4.2), one obtains the de-channeling radius.

DE-CHANNELING RATE

From (4.2) and (4.3), we obtain the de-channeling rate

$$\eta = \frac{r_{min}^2}{r_0^2} = \frac{3a^2}{r_0^2}\left[\left(\frac{3a^2}{r_0^2} + 1\right)\exp\frac{E_\perp}{U_d} - 1\right]^{-1} \tag{5.1}$$

$$= 5.8 N_0 a_0^2 \left(1 + Z^{\frac{3}{2}}\right) A^{-1} \rho d \left\{\left[5.8 N_0 a_0^2\left(1 + Z^{\frac{3}{2}}\right) A^{-1} \rho d + 1\right]\exp\frac{4\pi\varepsilon_0 \varepsilon_\infty E_\perp d}{Ze^2} - 1\right\}^{-1} \tag{5.2}$$

It's a function of crystal parameters. When $r < r_{min}$, muons are de-channeled and scattered by lattice kernels. The critical angle of de-channeling α is as following:

$$\tan\frac{\alpha}{2} = \frac{Ze^2}{4\pi\varepsilon_0 r_{min}} \frac{1}{E_\parallel} = \frac{d}{2r_{min}} \psi_1,$$

$$\alpha = \psi_1 \frac{d}{r_{min}} = 11.4\psi_1 > \psi_1 f, \quad f = \frac{1}{\sqrt{2}}\left|\ln\left(\frac{C^2 a^2}{\rho_2^2} + 1\right)\right|^{\frac{1}{2}} \tag{5.3}$$

where $\rho_2 (= 0.01 a_0)$ is the rms vibration amplitude of a low temperature lattice.

Table I. The basic estimation of various performance of application.

	$\langle Z \rangle$	$\langle A \rangle$	ρ g/c.c	d nm	a nm	r_0 nm	ε_∞	U_d eV	ψ_1^{50GeV} μ rd	r_{min} nm	η %
H	1	1	.076	0.27	.033	.161	2	2.67	10.5	.048	9.0
LiH	2	4	0.53	0.41	.029	.099	3.61	1.95	8.8	.032	10.1
C Diamond	6	12	3.52	0.36	.023	.071	5.5	4.36	13.2	.039	29.8
C Bucky Tube	6	12	Single Tube	3.60	.023	1.80	5.5?	0.44	0.42	.004	.001
Si	14	28	2.33	0.54	.018	.108	11.8	3.16	11.2	.030	7.9
GaAs	32	72	3.50	0.64	.014	.130	10.9	6.61	16.3	.039	9.0
NaI	32	75	3.67	0.53	.014	.143	2.91	29.9	34.6	.063	19.4
CsI	54	130	4.51	0.42	.012	.190	3.07	60.3	49.1	.105	30.5
TlI	67	166	7.29	0.40	.011	.173	4.21	54.6	46.7	.085	24.5

Si crystal and C Bucky tubes are best, LiH and GaAs are better than others, Metal H is good and not yet available. On the other hand, Bogacz-Cline channeling collision has intensity redundancy of $10^{2\to4}$. It has enough leeway of allowing beam loss to use sub-atomic modulation to control scattering and de-channeling.

POTENTIAL CONFIGURATION IN 3 CHANNEL REGIONS

Channeling Region

$r_0 > r > r_{min}$: muons feel Lindhard's channeling potential. See (3.5), (3.7), and (3.8).

De-channeling and Straggling Region

$r_C < r < r_{min}$: muons feel Coulomb potential

$$U_C(r) = \frac{Ze^2}{4\pi\varepsilon_0\varepsilon_\infty r}[J]\frac{1}{1.6\times10^{19}}[eV/J] \qquad (6.1)$$

It is a Moliere scattering when $\theta < \theta_0$, it is a Rutherford scattering when $\theta > \theta_0$.

$$\theta_0 = \frac{13.6\,\text{MeV}}{\beta cp}\sqrt{\frac{x}{X_0}}\left[1+0.038\ln\frac{x}{X_0}\right], \quad X_0 = \frac{716.4\,\text{g}\cdot\text{cm}^{-2}\,A}{Z(Z+1)\ln(287/\sqrt{Z})} \qquad (6.2)$$

where x/X_0 is the thickness of medium in the unit of radiation length.

Scattering Region

$r < r_C$: muons feel Rutherford-Mott-Rosenbluth-Feynman potential. It's a kind of Hofstadter-Friedman-Kendall-Taylor scattering.[11]

$$U_{RMRF}(r) = U_C(r)\frac{\cos\frac{\theta}{2}}{\sin^2\frac{\theta}{2}}\left(\frac{E'}{mc^2}\right)\sqrt{\frac{E}{E'}}\cos\frac{\theta}{2}\left(\frac{F_E^2+F_M^2}{1+\tau}+2\tau F_M^2\tan^2\frac{\theta}{2}\right)^{\frac{1}{2}} \qquad (6.3)$$

$$= U_C(r)u(E,E',\theta,q,F_E,F_M), \quad \text{where} \quad \tau = q^2/4M^2c^2 \qquad (6.4)$$

DE-CHANNELING RADIUS IN INCOHERENT HEATING

Crystal at Low Temperature

All kinds of scattering beam loss results after de-channeling. From (3.7)

$$U_{Lin}(r) = \frac{Ze^2}{4\pi\varepsilon_0\varepsilon_\infty d}\ln\left(\frac{C^2a^2}{r^2}+1\right) \to \frac{Ze^2}{2\pi\varepsilon_0\varepsilon_\infty d}\ln\left(\frac{Ca}{r}\right) \quad (7.1)$$

when $r \to r_{min} << Ca \le 10^{-1}$ nm---compare with ρ_2 ($\approx 10^{-2}a_0 \le 10^{-3}$ nm), (7.2)

$$U_{Lin}(0) = \infty \quad (7.3)$$

The de-channeling radius r_{min} of ordinary crystal is obtained by

$$U_{Lin}(r_{min}) = E_\perp \quad (7.4)$$

with $r_{min}^2 = \dfrac{3a^2}{3\pi Na^2 d + 1}\exp\left(1 - \dfrac{E_\perp}{E_{//}}\dfrac{2}{\psi_1^2}\right).$ (7.5)

When $a << d$ and $\dfrac{2E_\perp}{\psi_1^2 E_{//}} << 1$, then $r_{min}^2 \approx 3ea^2$ or $r_{min} \approx 2.86a$, (7.6)

which is the case of Nagamine's bunch.

Hot Crystal Case

The temperature of channeling crystal reaches several hundreds Celsius[6]. So the lattice kernel gets heated and emits phonons. Hence, the potential of channeling is modulated by the lattice vibration the phonons as following:

$$U^{ther}(r) = \frac{2}{\rho_2^2}\int_0^\infty e^{-\frac{r'^2}{\rho_2^2}} U(r - r') r' dr' \quad (7.7)$$

$$U_{Lin}^{ther}(r) = \frac{Ze^2}{4\pi\varepsilon_0\varepsilon_\infty d}\left[2\ln\frac{Ca}{r} - \text{Ei}\left(\frac{r^2}{\rho_2^2}\right)\right]$$

$$\approx \frac{E_{//}\psi_1^2}{2}\left[2\ln\frac{Ca}{\rho_2} + \Gamma - \frac{r^2}{\rho_2^2} + \frac{1}{2}\frac{r^4}{\rho_2^4} - \cdots\right] \quad (7.9)$$

where $\Gamma = 0.577215$ is the Euler's number, ρ_2 the rms amplitude of lattice vibration which < 0.02 nm at 900° K for Si. De-channeling radius r_{min}^{ther} of vibrating lattice is from (7.4)

$$U_{Lin}^{ther}(r_{min}^{ther}) = E_\perp, \text{ and } (r_{min}^{ther})^2 = \rho_2^2\left\{1 \pm \sqrt{1 + 4\frac{E_\perp}{E_{//}}\frac{1}{\psi_1^2} - 2\ln\frac{Ca}{\rho_2} - \Gamma}\right\} \quad (7.10)$$

where $\rho_2^2 = C^2a^2 \exp\left(-\dfrac{x}{2}\right)$ (7.11)

Hence, the roots of min and max $\{(r_{min}^{ther})^2\}$ for the Nagamine's beam of (7.6) are

$$x = 1.113, 0.536, 0.195, -2.772 \quad (7.12)$$

Then, $(r_{min}^{ther})^2_{max,min} = 3a^2 \exp\left(-\dfrac{x}{2}\right)(1 \pm \sqrt{1 - 2\Gamma - x})$, for all $\le 3ea^2 = r_{min}^2$ (7.13)

If $E_\perp \geq U_{Lin}^{ther}(0) \neq \infty$, muons are de-channeled and scattered by lattice in $r < r_{min}^{ther}$, but if $r_{min}^{ther} \gg \rho_2$ and $r > r_{min}^{ther}$, heat vibration decreases de-channeling rate. That means heated crystals can channel muons even better.

DE-CHANNELING RADIUS IN COHERENTLY EXCITED SUPER LATTICE

The super lattice is excited coherently using a laser light. The Lindhard's U_{Lin} of channeling is expressed in the quantized form of Haung-Zhu's quantum dipole oscillator model [13] as a phonon-muon potential

$$U_{SuLa}(r) = U_{Lin}(r) + U_{H-Z}(r) \tag{8.1}$$

$$U_{H-Z}(r) = \frac{U_d}{d^2} \frac{1}{\sqrt{N^3}} \frac{1}{\sqrt{\mu|q_s|}} \sum_{q,i} Q(q,i) \sum_i a_s(q,i) \exp(iq_i \cdot r) \tag{8.2}$$

where $Q(q,i) = \sqrt{\frac{h}{2\omega(q,i)}} \left[b_i^{0-}(-q,i) + b_i^{0+}(-q,i) \right]$ (8.3)

and $q_s = q + \frac{2\pi}{L} s z^0, |s| = \frac{1}{2}(m+m')$, (8.4)

L is the super lattice period, μ the mass, and N^3 the volume cell number, $m+m'$ the phonon quantum number, q the phonon wave vector, i the longitudinal mode index, $b_i^{0-,+}$ the annihilation and creation operators respectively of phonon intensity, and $\omega(q,i)$ the eigen-frequency which depends on super lattice parameters. It is similar to (4.1), (7.4), and (7.10). Let

$$\langle \alpha | U_{Lin}(r_{min}) + U_{HZ}(r_{min}) | \beta \rangle = E_\perp, \text{ then one can obtain } \langle \alpha | r_{min}^{SuLa} | \beta \rangle, \tag{8.5}$$

which is the de-channeling radius. It depends on super lattice parameters which is a controllable parameter.

PHONON MODULATING SCATTERING CROSS-SECTION

The scattering potential modulated by phonon is given by

$$U_{RMRF}^{PhMo}(r) = \sum_l u(l) \cdot \frac{\partial}{\partial t} U_{RMRF}(r - r_l) = \sum_l u(l) \cdot \frac{\partial}{\partial t} U_C(r - r_l) u(E, E', \theta, q, F_E, F_M) \tag{9.1}$$

Substituting (6.1), (6.3) into (9.1) one obtains the scattering potential modulated by phonon. Hence, modulated scattering cross section is obtained by

$$\sigma_{RMRF}^{PhMo}(r) = 4\pi^2 m^2 h^2 \frac{P'}{P_0} \langle p' \alpha' | U_{RMRF}^{PhMo}(r) | p \alpha \rangle \tag{9.2}$$

where P' and P_0 are the final and initial momenta, $|\alpha'\rangle$ and $|\alpha\rangle$ are the final and initial states of observable set of quantities. Let's define the ratio of scattering and de-channeling

$$\xi \equiv \sigma_{sc} / 2\pi r_{min}^2 \tag{9.3}$$

In the ultra relativistic approximation of the lepton-hadron, in the range of 50 GeV to 1 TeV, the scattering cross-section σ_{sc} should be σ_{Ros}, the Rosenbluth's scattering cross-section. If lattice vibrate thermally or coherently, then $U_C(r_{min})$ is replaced by $U_{Lin}^{ther}(r_{min})$ or $U_{RMRF}^{PhoMo}(r_{min})$ respectively. Let

$$\xi^{PhoMo} \equiv \frac{\sigma_{Ros}}{2\pi r_{min}^2} = \frac{U_{RMRF}^{PhoMo}(r_{min})^2}{4c^2 p^2}. \quad (9.4)$$

When $r \leq r_{min}$, and if $U_{RMRF}^{PhoMo}(r)^2 \ll U_{Lin}^{ther}(r)^2 \ll U^2(r)$, then (9.5)

$$\xi^{PhoMo} \ll \xi^{ther} \ll \xi^{cool} \quad (9.6)$$

Hence, controllable lattice vibration can decrease scattering and increase de-channeling.

DISCUSSION

C Bucky tube, supper lattice seems hopeful to be the quantum devices to control de-channeling and scattering. It's also a new R&D field of clean nuclear energy [14] and new information device of electronics, photonics, and phononics.

The plentiful quantum effects of H, LiH crystal, GaAs-AlGaAs super-lattice, C Bucky tube, nm Si, SiC, NaI, CsI, TiI, etc. might be selected for the interaction regions of muon colliders and linear colliders. These materials have been researching in fusion fields also.

ACKNOWLEDGMENTS

The author thanks D.Cline, A.Skrinsky, R.Palmer, Z.Parsa, P.Chen, M.Atac, and K.Lee for kind discussions and also Z.P.Zheng and M.H.Ye for their encouragement.

REFERENCES

[1] J. Shen, This Proceeding, 1998.
[2] Protvino (Surpukhov) IHEP, CERN Courier, 29, Dec. 24, (1989).
[3] CERN, CERN Courier, 30, May, 5, (1990).
[4] FNAL, CERN Courier, 36, Jan. 3, (1996).
[5] CERN, CERN Courier, 37, Jan-Feb. 4, (1997).
[6] Protvino (Surpukhov) IHEP, CERN Courier, 38, Mar. 12, (1998).
[7] R.Palmer, et al. Nucl.Phys.Proc.Suppl.51A, 61-84, (1996).
[8] K.Nagamine, Nucl.Phys.Proc.Suppl.51A (1996), 115-124; Talk on 97 MuMu.
[9] S.Chu, Science, 235 (1991) pp.861-866; Lecture of Nobel prize, 1997.
[10] J.K.Lindhard, Dan.Vidensk.Selsk.Mat.Fys.Medd., 34, No.14 (1965).
[11] Lectures of Nobel prize, Hofstadter,1961; Friedman-Kendall-Taylor,1990.
[12] C.Erginsoy, Phys.Rev.Lett.,15, 360, (1965).
[13] K.Huang, B.F.Zhu, Phys.Rev.B 38, 2183; 13377, (1988).
[14] W.D.Philips,P.L.Gould, and D.Paul, Lett. in Sciences, 239, pp.877-883, (1988).

Lane Fuzzy Collision in Channel with Potential Deformation by Photon-Phonon-Electron Excitation and Sub-atomic Control

Jing Shen*

Institute of High Energy Physics (IHEP), Chinese Academy of Sciences
Mailing Address: P. O. Box 918(6), Beijing 100039, PRC
and
China Center of Advanced Science and Technology (CCAST), P.O.B. 8730, Beijing 100080

Abstract: Collision between μ^+ and the μ^- beams in the crystal are forbidden due to the two beams having different "lanes" in a channel. A laser pulse of ps-fs shocks lattice kernel vibration and dilates lattice electron distribution. It deforms the Lindhard's potential which is then expressed in a quantized form as the Huang-Zhu's potential[1]. The dynamic lanes can be made to overlap in a channel to allow collision without ductile fracture. This raises a new technology of sub-atomic information & control, which has been raised by T.D. Lee.

INTRODUCTION

After the discussions on "entrance" and "de-channeling"[1], we discuss in this paper the channeling collision. Collision between the μ^+ and the μ^- beams in the crystal are forbidden due to the two beams having different "lanes" in a channel. Actually, it isn't that the μ^+'s and the μ^-'s are in separate lanes. Because the μ^+'s see a positive periodic Lindhard's potential well, with the maximum ($+\infty$) at the kernel lattice sites and the minimum (0) at the electron lattice sites along the center axis of a channel, they run in the "central axis lane" with a Bogacz-Cline's micron betatron oscillation[2]. However, the μ^-'s in the same channel sees a negative periodic Lindhard's potential well with the maximum (0) at the electron lattice sites along the center axis of a channel and the minimum ($-\infty$) at kernel lattice sites, they run in the "lattice kernel lane" with different Bogacz-Cline's micron betatron oscillation. Hence, in a channel there is a "double yellow line" between the μ^+ lane and the μ^- lane of channeling "free way". The μ^+'s and the μ^-'s travel face to face but could not collide. In the case of mesoscopic beams like the Nagamine-Chu's muon beams or the Richter-Kimura-Takada's electron-positron beams, lane crossing and collision are quantum forbidden. This is the third difficult problem of the Bogacz-Cline's collider.

LANE OVERLAP BY DEFORMING LINDHARD'S POTENTIAL

To cope with this problem, let's make a coherent lattice kernel vibration shock of THz and a resonant lattice electron dilating excitation. Both excited by a ps-fs laser. The lattice displacement due to linear phonons is much smaller than lattice constant.

---

* Email shenj@bepc3.ihep.ac.cn
CP441, *Physics Potential and Development of mu-mu Colliders:* Fourth International Conference
edited by David B. Cline
© 1998 The American Institute of Physics 1-56396-723-5/98/$15.00

It has a limit of the bond separation. Thus, the geometrical displacement of the lane is very small. However, the kernel vibration and the electron dilation excited by the laser light results in a non-linear Lindhard's potential with large deformations. It makes lanes to "cross", and thus, the channeled particles are allowed to collide.

COLLISION PARAMETER AS 2D LINDHARD'S ATOMIC RADIUS

The collision parameters are averaged transverse positions of the μ^+'s and the μ^-'s in a channel. They might be considered as the radii of 2D atoms resulting from projection onto a transverse plane of the muon position along the entire channel and can be determined from the quantum equation of the Lindhard's approximation.

$$\left[-\frac{\hbar^2}{2m_0}\nabla^2 - U_{Lin}(r)\right]u(r) = E_\perp u(r,\varphi) \tag{3.1}$$

$$U_{Lin}(r) \approx \frac{\pm Ze^2}{4\pi\varepsilon_0\varepsilon_\infty d}\ln\left[\frac{C^2 a^2}{r^2}+1\right] \tag{3.2}$$

where $\pm$ represents respectively the positive and the negative charged particles. Let r be in units of Bohr radius a_0 ($=4\pi\varepsilon_0\hbar^2/m_0 e^2$) and $E_\perp$ be replaced with E in units of $\hbar^2/2\mu a_0^2$. In cylindrical coordinates, (3.1) becomes

$$u(r,\varphi) = R(r)\Phi(\varphi), \text{ and } P(r) = \sqrt{r}R(r) \tag{3.3}$$

$$-\frac{d^2 P(r)}{dr^2} + \left[\frac{(1/4 - m^2)}{r^2} - U_{Lin}(r)\right]P(r) = EP(r) \tag{3.4}$$

$$-\left[\frac{1}{Rr}\frac{d}{dr}\left(r\frac{dR}{dr}\right) + \frac{1}{\Phi}\frac{d^2\Phi}{d\varphi^2}\right] - U_{Lin} = E \tag{3.5}$$

$$\frac{d^2\Phi}{d\varphi^2} = -m^2\Phi \tag{3.6}$$

Thus, the radius of channeling particle is equivalent to a 2D Lindhard's atom. Its radius

$$(a_{Lin})_n^\pm = \sqrt{\langle P_n(r)|r^2|P_n(r)\rangle} \approx Ca\left[\exp\left(\pm d/(a_{Lin})_n^\pm\right) - 1\right]^{\frac{1}{2}} \tag{3.7}$$

Hence, the collision parameters, the lane positions of $\pm$ beams in channel, $(a_{Lin})_n^\pm$ are solutions of the non-linear equation (3.7), where d is a parameter that can be modulated by the laser light.

PHOTO-PHONON DEFORMED LINDHARD'S POTENTIAL

If the duration of laser < 0.3 ps, then the diffusion of thermal excitation could be neglected. The longitudinal wave solution of thermal strain in the lattice is

$$\eta_{ij} = \frac{u_i(l_j)}{d} = \frac{3(1-R')Q\alpha B_{ij}}{2AC_p\varsigma\rho v_L^2}\exp\left[-(z-v_L t)/\varsigma\right]\text{Sgn}\left[(z-v_L t)/\varsigma\right] \tag{4.1}$$

where $u_i(l_j)$ is displacement of lattice kernel, B_{ij} the volume modulus [FL^{-2}], C_p the heat capacity [Q/K°], Q the heat quantity, α the linear coefficient of expansion [L/K°], d the lattice constant, a and A the laser illumination radius and area, ς the absorption length, ρ the density, v_L the velocity of acoustic wave, and N_{photon} the incident photon number[3]. Thus, the deformation of the Lindhard's potential of beam resulting from thermal strain is

$$\partial U_{Lin}^{PhMo}(r) = \sum_l u(l) \cdot \frac{\partial}{\partial r} U_{Lin}(r) = \sum_l \eta_{ij} d \frac{\partial}{\partial r} U_{Lin}(r) \qquad (4.2)$$

Acoustic intensity in lattice is

$$I_{ac} \propto \frac{\Delta n}{n} = 3 \times 10^{-24} N_{photon} \frac{1}{\pi a^2} \frac{1}{v_L \tau_{phonon}} \qquad (4.3)$$

PHOTO-ELECTRONIC DILATION MODULATING POTENTIAL

If the quantum energy of laser is equal to the energy gap of crystal,

$$\omega_{phon} = \omega_{Las} = 2\pi \times 8.55 \, \text{THz (GaAs)}, \qquad (5.1)$$

and the excitation time is

$$\tau_{las} < 1/\omega_{phon} = \tau_{phon} \leq 10^{-12} \, \text{s}. \qquad (5.2)$$

Thus, the coherent excitation of electrons is stimulated. Finally, the thermal to electron dilation ratio is

$$\delta = \varsigma / \langle v_{phon} \rangle \tau_{phon} \qquad (5.3)$$

EXCITING FUZZY OF "DOUBLE YELLOW LINE"

The stress the energy density of excited electrons and phonons

$$\frac{U_{cell}}{d^3} = \sigma_{ij} = \sigma_{ij}^{elec} + \sigma_{ij}^{phon} \qquad (6.1)$$

$$\sigma_{ij}^{elec} = \sum_{k'} \delta n_{elec}(k') \frac{\partial E(k')}{\partial \eta_{ij}} \qquad (6.2)$$

$$\sigma_{ij}^{phon} = \sum_{k'} \delta n_{phon}(k') h \frac{\partial \omega(k')}{\partial \eta_{ij}} \qquad (6.3)$$

Where δ is (5.3), σ_{ij} the stress (6.2), η_{ij} the strain (4.1), n_{elec} and n_{phon} the densities of electron and phonon, $E(k')$ the electron energy, $h\omega(k')$ the phonon energy, and k' the phonon vector. The expansion of lattice cell is

$$\partial_e d = \left[\frac{E(k')}{\sigma_{ij}^{elec}}\right]^{\frac{1}{3}}, \quad \partial_p d = \left[\frac{h\omega(k')}{\sigma_{ij}^{phon}}\right]^{\frac{1}{3}} \qquad (6.4)$$

which modifies the radii of Lindhard's atom. Substituting (6.1-4) into (3.7), one gets

$$\partial_e (a_{Lin})_n^+ = Ca \left\{ \exp\left[\sqrt[3]{E(k')/\sigma_{ij}} / (a_{Lin})_n^+ \right] - 1 \right\}^{\frac{1}{2}} \qquad (6.5)$$

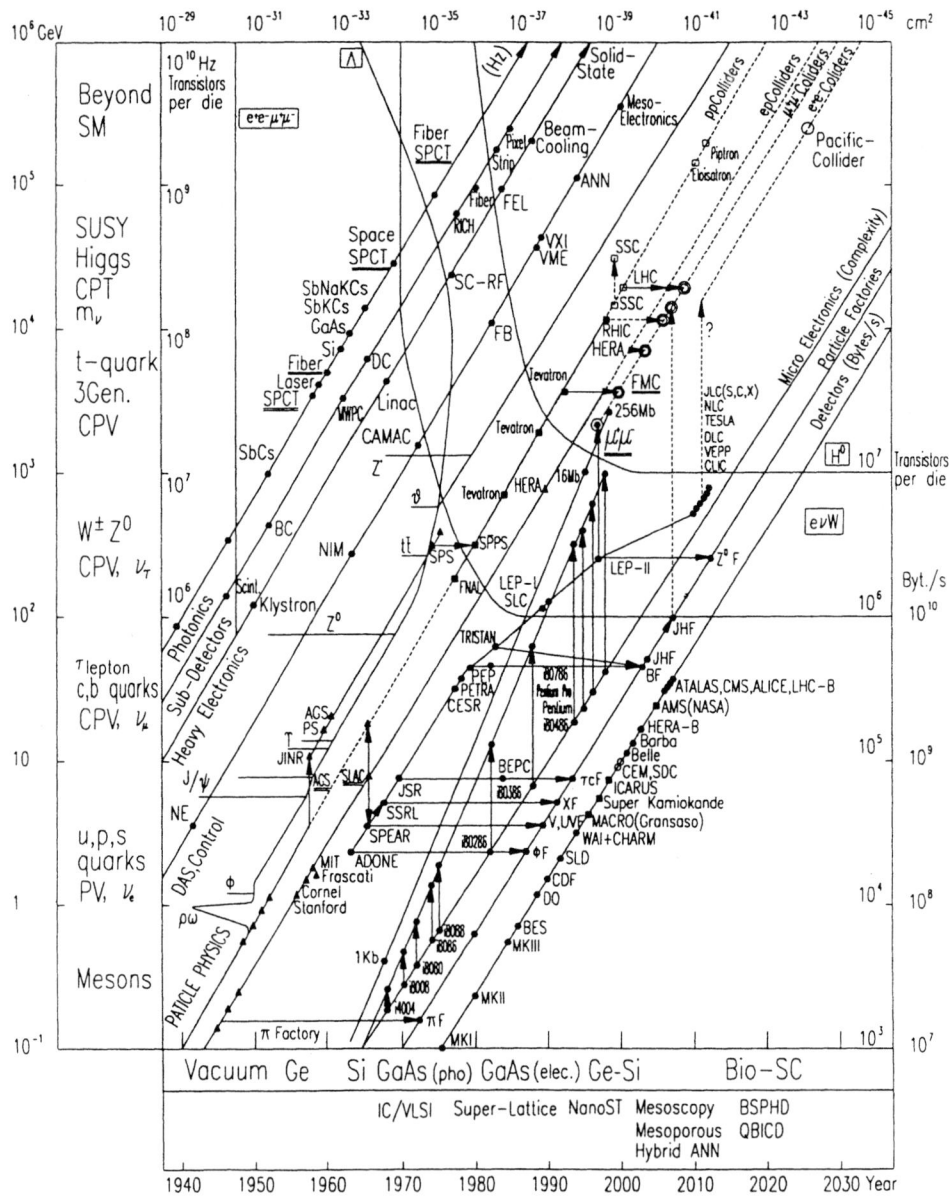

State Space of DPPD-TEEET-CDDC

It makes the μ^-'s run away from the kernel axis, and the μ^+'s close to the kernel axis.

$$\partial_p(a_{Lin})_n^- = Ca\{\exp[-\sqrt[3]{\hbar\omega(k')/\sigma_{ij}}/(a_{Lin})_n^+] - 1\}^{\frac{1}{2}} \quad (6.6)$$

Thus, the $\pm\mu$'s trajectories in a channel can be controlled to collide by the laser light.

SUB-ATOMIC CONTROL AND INFORMATION

Recently the nanom technology appears a leap on the chart of Moore law. (See figure) Internet, parallel computer, ANN developed rapidly, but still lacks information. In the early 1970s, T.D. Lee turned away from the conventional matters and explored abnormal state matters and urged A.M. Sessler to upgrade heavy ion collider (now RHIC)[4-6]. During that time, the author worked on Quantum Bio-Information Code Dynamics for diagnosis of carcinogenesis by scintillating imaging.[7-9]. In 1980s, T.D.Lee developed a parallel computer for calculating lattice QCD[10]. D.H. Weingarten developed one also at IBM[11,12]. In 1994 T.D. Lee talked on quark-gluon plasma state as extremely high entropy density for information[13]. In 1995, Bogacz, Cline, and Shen considered quantum colliders, and Cline researched CPV effects in the origin of life[14]. Thus, particle physics is coming into information and life sciences as well as astrophysics and cosmology.

ACKNOWLEDGMENTS

The author thanks D.Cline, A.Skrinsky, R.Palmer, Z.Parsa, P.Chen, M.Atac, and K.Lee for kind discussions and also Z.P.Zheng and M.H.Ye for their encouragement.

REFERENCES

[1] Jing Shen, Other two papers in this Proc. 1998.
[2] S.A.Bogacz, and D.B.Cline, Nucl. Phys. Proc. Suppl. 51A, 90-97, (1996).
[3] C.Thomsen, et al. Phys.Rev.B34 (1986), 4129; Opt. Comm 60, 55, (1986).
[4] T.D.Lee and G.C,Wick, Phys. Rev. D 9, 2291, (1974).
[5] T.D.Lee and M.Margulies, Columbia Univ. Preprints CO-2271-33, 1974.
[6] T.D.Lee, Rev. Mod. Phys. 47, 267, (1975).
[7] Jing Shen, Quantum Bio-Information Code Dynamics, 2nd CIE Conf. 1978.
[8] Jing Shen, Proc. 1979 CIE Conf. of founding SNENDT, pp.203-242, 1980.
[9] Jing Shen, Proc. Symp. Signal and Information Processing, CIE, p.187,1981.
[10] T.D.Lee,et. al. Phys.Lett.B,122,217,(1983); J.Math.Phys.35,11,p.5600,(1994).
[11] F.Butler, H.Chen, J.Sexton, A.Vaccarino, D.Weingarten, P.R.L.70, 2847, (1993).
[12] D.Weingarten, Scientific American 274, No.2, (1996).
[13] T.D.Lee, Lecture at Chinese Science and Technology Univ. Beijing, June, 1994.
[14] D.B.Cline, UCLA-PPH0070-3/95; -PPH-0073-12/95; -PPH-0072-4/96,1996.

Program for the 4th International Conference on Physics Potential and Development of $\mu^+\mu^-$ Colliders

December 10, Wednesday
10:00 am Registration

I) *Opening Session* [Chair, *D. Cline (UCLA)*]

1:30 pm	Overview of Physics at Muon Colliders	V. Barger (U Wisc-M)
2:05	Review of Physics at the FMC and the Collider Front End Workshop	S. Geer (FNAL)
2:40	Quantitative Higgs Physics at a Muon Collider in Comparison with LHC and NLC	J. Gunion (UC Davis)
3:05	Review of Recent LEP II Results	Y. Pan (U Wisc-M)
3:30	Physics Potential of CMS/LHC	S. Dasu (U Wisc-M)
3:55	Hadron Colliders, the Top Quark and the Higgs Sector	C. Quigg (Princeton/FNAL)
4:20	Comparison of Lepton and Hadron Colliders for the Discovery of Supersymmetry	H. Baer (Florida S U)
4:45	Higgs Searches at LEP II	P. Bambade (Orsay, France)

5:10 Break

II) *Low Energy Physics with Cold $\mu^\pm$ Beams* [Chair, *Z. Parsa (BNL)*]

5:20	Low Energy Physics with High Intensity Muon Sources	W. Marciano (BNL)
5:45	MECO Experiment, to Measure μ–e Conversion at BNL	W. Molzon (UC Irvine)
6:05	A $\mu \rightarrow e + \gamma$ Experiment at 10^{-14} Sensitivity	H. C. Walter (PSI)

6:25 pm Reception Buffet

December 11, Thursday
7:00 am Continental breakfast

IIIa) *Physics Goals of Future Colliders* [Chair, *V. Barger (U Wisc-M)*]

8:30 am	Strongly Interacting Electroweak Physics: A Comparative Study for Future Colliders	T. Han (UC Davis)
8:55	Threshold Cross-Section Measurements	M. Berger (Ind U)
9:20	Top Physics at a Polarized Muon Collider	S. Parke (FNAL)

IIIb) *Physics Goals of Future Colliders* [Chair, G. Kane (U Mich)]
9:45   Sorting out SUSY Particles at a Muon Collider   J. Gunion (UC Davis)

10:00   s-Particle Masses from Kinematical End-Points   J.D. Lykken (FNAL)

10:25   s-Channel sneutrino Production in   J. Feng (UC Berkeley/LBNL)
R-Parity Violating SUSY Theories

10:50   Coffee

IIIc) *Physics Goals of Future Colliders* [Chair, D. Neuffer (FNAL)]
11:05   CP Violation at Muon Colliders   W.Y. Keung (U Ill/FNAL)

11:30   New Particles and Interactions at High-Energy   S. Godfrey (Carleton U, Canada)
Muon Colliders

11:55   Physics of Like-Sign Muon Colliders   C. Heusch (UC Santa Cruz/SLAC)

12:20   Testing non-GUT Standard Model Extensions   S. Rajpoot (CSULB)
at a $\mu^-\mu^-$ ($\mu^+\mu^+$) Collider

12:35   Lunch

IVa) $\mu^+\mu^-$ *Collider Studies* [Chair, A. Sessler (LBNL)]
1:45   Muon Collider Designs and Cooling Simulations   R. Palmer (BNL)

2:20   Higgs Factory 50-GeV × 50-GeV Muon Collider   A. Garren (UCLA/LBNL)

2:40   Dynamic Aperture and Chromatic Correction for   C. Johnstone (FNAL)
50-GeV × 50-GeV Muon Collider: Background Reduction and Beam Scraping

3:00   Lattice Design for 50-GeV × 50-GeV   K.Y. Ng (FNAL)
Muon Collider

3:15   Design of the JHF 50-GeV Proton Synchroton   Y. Mori (KEK / Tanashi)
and a New Type of RF System

3:50   Crystal Room

IVb) $\mu^+\mu^-$ *Collider Studies* [Chair, A. Tollestrup (FNAL)]
4:00   Polarization   R. Rossmanith (DESY)

4:20   Towards Ultimate Luminosity with   A. Skrinsky
Polarized Muon Collider   (INP, Novosibirsk)

4:40   New $\mu^\pm$ Cooling for $\mu^\pm$ Colliders and   N. Nagamine (KEK / Riken)
Possible Realizaton at JHF

5:05   Muon Dynamics in a Toroidal Sector Magnet   J. C. Gallardo (BNL)

5:20	Ionization Cooling Studies I	R. Fernow (BNL)
5:35	Ionization Cooling Studies II	D. Neuffer (FNAL)
5:50	3-D Cooling Scheme	F. Mills (FNAL)
6:05	Some Electric, Electronic, Engineering, and Economic Problems of Quantum Muon Colliders and Quantum Muon Detectors	J. Shen (IHEP, China)

6:20 Cocktails

7:00 pm Banquet

December 12, Friday
7:00 am Continental breakfast

V) Physics at Higgs Factory [Chair, *A. Skrinsky (INP, Novosibirsk)*]

8:30 am	Overview of a Higgs Factory Concept	D. Cline (UCLA)
8:55	Estimates of and Constraints on Higgs Boson Masses	G. Kane (U Mich)
9:30	Beam Decays	I. Stumer (BNL)
9:45	Beam Halo	N. Mokhov (FNAL)
10:00	Tracking Concepts of High-Luminosity Muon Colliders	M. Atac (UCLA/FNAL)
10:25	Calibrating the Energy of a 50-GeV × 50-GeV Muon Collider Using Spin Precession	Raja / A. Tollestrop (FNAL)

10:45 Coffee

VI) Precision Electroweak Data Studies [Chair, *S. Geer (FNAL)*]

11:00	Precision Electroweak Data: Present Status and Future Prospects	P. Renton (Oxford, UK)
11:25	Z Pole Physics at a Muon Collider	T. Han (UC Davis)

11:50 Lunch

VII) μ–p Colliders [Chair, *Z. Parsa (BNL)*]

1:30	The Physics Potential of μ–p Collider and Comparison with Other Colliders	I. Ginzburg (Inst. Math, Novosibirsk)
1:55	Possible Luminosity Monitoring at $\mu^+\mu^-$ and μ–p Collider	I. Ginzburg (Inst. Math, Novosibirsk)

2:10	μ–p Colliders: Leptoquarks and Contact Interactions	K. Cheung (U Texas-Austin)
2:35	*Coffee*	
2:50	Pion Production and Targetry	N. Mokhov (FNAL)
3:00	Bunching Experiments in the AGS	J. Norem (FNAL / Argonne)
3:10	Ring Cooler for Muon Collider	V. Balbekov (FNAL)
3:20	Neutrino Physics at Muon Colliders	B. King (BNL)
3:30	*Break*	

VIII) ***Summary Session*** [Chair, *B. Palmer(BNL)*]

3:40	R&D Program	J. Wurtele (UC Berkeley/LBNL)
3:55	Advanced Accelerator Physics Research	D. Sutter (DOE)
4:35	Physics with $\mu^+\mu^-$ Colliders	W. Marciano (BNL)

5:15 pm End

December 13, Saturday
7:00 am Continental Breakfast

Muon Collaboration Meeting [Organizer, *J. Gallardo (BNL)*]

9 am - Lunch

Speakers and Conference Organization	J. Wurtele (UC Berkeley / LBNL)
Superconducting Magnets for a Muon Collider	R. Scanlan (LBNL)
Pion Production and Targetry at $\mu^+\mu^-$ Colliders	N. Mokhov (FNAL)
The AGS E910 Experiment: Measurement of π Production	H. Kirk (BNL)
Liquid Jet Targets	K. McDonald (Princeton)
ICOOL Simulations of Alternation Solenoid Transverse Cooling	R. Palmer (BNL)
ICOOL: Recent Developments	R. Fernow (BNL)
Theory of Emittance Exchange	R. Palmer (BNL)

Lunch Break

A New Scheme of Linear Accelerator	Y. Zhao (BNL)
Temperature Distribution on Beryllium Windows in $\mu^+\mu^-$ Cooling RF Cavity	D. Li (LBNL)
Linear Orbit Recirculators	F. Mills (FNAL)
Cooling Experiment	S. Geer (FNAL)
An Update on a Cooling Scenario	J. Norem (FNAL / ANL)
Longitudinal Impedance Tuner Using High Permeability Material	Y. Mori (KEK)
Vertex Tagging	B. King (BNL)
Matching and Emittance Exchange Section for Ionization Cooling with Li Lenses	V. Balbekov (FNAL)
General Discussion	

End

LIST OF ATTENDEES

4th International Conference Sponsored by UCLA on Physics Potential & Development of μ+μ- Colliders

Fairmont Hotel, San Francisco, California
December 10-12, 1997

ARISAKA, Katsushi: UCLA, Physics & Astron. Dept., Box 951547, Los Angeles, CA 90095-1547, USA; e-mail: arisaka@physics.ucla.edu; (310) 825-4925, F: (310) 206-1091
ATAC, Muzaffer: FNAL, PPD/CMS, P.O. Box 500, MS 205, Batavia, IL 60510, USA; e-mail: matac@fnal.gov; (630) 840-3960, F: (630) 840-2194
AUTIN, Bruno R.: CERN, PS/DI Div., CH-1211, Geneva 23, SWITZERLAND; e-mail: autin@ps.msm.cern.ch; 41-22-767-2525, F: 41-22-767-9145
BAER, Howard A.: Florida State U., Physics Dept., 511-KEN, Tallahassee, FL 32306-3016, USA; e-mail: baer@hep.fsu.edu; (904) 644-3523, F: (904) 644-6735
BALBEKOV, Valeri: FNAL, Beam Physics, P.O. Box 500, MS 220, Batavia, IL 60510, USA; e-mail: balbekov@fnal.gov; (630) 840-6318, F: (630) 840-6039
BAMBADE, Philip S.: Laboratoire de l'Accelerateur Lineaire, Dept. of Physics, Batiment 200, Centre d'Orsay, F-91405 Orsay Cedex, FRANCE; e-mail: bambade@lal.in2p3.fr; 33-164-46-8370, F: 33-169-07-9404
BARGER, Vernon D.: U. Wisc-Mad, Physics Dept., 1150 University Ave., Madison, WI 53706, USA; e-mail: barger@pheno.physics.wisc.edu; (608) 262-8908, -4906, F: (608) 262-8628
BARKLOW, Timothy: SLAC, MS 96, P.O. Box 4349, Stanford, CA 94309, USA; e-mail: timb@slac.stanford.edu; (650) 926-3199, (650) 926-3199
BERGER, Mike: Indiana U., Physics Dept., Swain West 117, Bloomington, IN 47405, USA; e-mail: berger@indiana.edu; (812) 855-2609, F: (812) 855-5533
BROCK, Raymond: Michigan State U., Dept. of Physics and Astron., 207 Physics and Astron., East Lansing, MI 48824, USA; e-mail: brock@chip.pa.msu.edu; (517) 353-8662, F: (517) 353-4500
CHEUNG, Kingman: UCD, Physics Dept., Davis, CA 95616, USA; e-mail: cheung@bethe.ucdavis.edu; (916) 752-0820, F: (916) 752-4717
CLINE, David B.: UCLA, Physics & Astron. Dept., Box 951547, Los Angeles, CA 90095-1547, USA; e-mail: dcline@physics.ucla.edu; (310) 825-1673, -4649, F: (310) 206-1091
DAHL, Per F.: LBNL, AFRD, c/o 9 Commodore Drive, A211, Emeryville, CA 94608-1952, USA; e-mail: pfdahl@aol.com; (510) 601-7276
DASU, Sridhara: U. Wisc-Mad, Physics Dept., 1150 University Ave., Madison, WI 53706, USA; e-mail: dasu@slac.stanford.edu; (650) 926-3196, F: (650) 926-2923
FENG, Jonathan L.: LBNL, Theoretical Physics Group, 1 Cyclotron Rd., MS 50A-5101, Berkeley, CA 94720, USA; e-mail: feng@thsrv.lbl.gov; (510) 486-4739, F: (510) 486-6808
FERNOW, Richard C.: BNL, Physics Dept., 901A, PO Box 5000, Upton, NY 11973-5000, USA; e-mail: fernow@bnl.gov; (516) 344-3741, F: (516) 344-3248
FUKUI, Yasuo: KEK/FNAL, PPD Div., c/o P.O. Box 500, MS 223, Batavia, IL 60510, USA; e-mail: fukui@fnal.gov; (630) 840-8473, F: (630) 840-2968
GALLARDO, Juan C.: BNL, Physics Dept., Bldg. 910A, PO Box 5000, Upton, NY 11973-5000, USA; e-mail: jcg@bnlarm.bnl.gov; (516) 344-3523, F: (516) 344-3248

GARREN, Alper: LBNL, Accel. & Fusion Research Div., 1 Cyclotron Rd., Berkeley, CA 94720, USA; e-mail: garren@lbl.gov; (510) 486-6574, F: (510) 486-7981
GEER, Stephen: FNAL, 2H PPD-EPP-Group, P.O Box 500, MS 318, Batavia, IL 60510, USA; e-mail: sgeer@fnal.gov; (630) 840-2395, F: (630) 840-3867
GEORGE, Joan: UCLA, Physics & Astron. Dept., Box 951547, Los Angeles, CA 90095-1547, USA; e-mail: joang@physics.ucla.edu; (310) 825-4649, F: (310) 206-1091
GILLESPIE, George H.: G.H. Gillespie Associates, Inc., 10855 Sorrento Valley Rd., Ste. 201, San Diego, CA 92121, USA; e-mail: ghga@ghga.com; (619) 677-0076, F: (619) 677-0079
GINZBURG, Ilya F.: Inst. of Mathematics, Dept of Theoretical Physics, Universitetsky Prosp., 58 ap. 3, Novosibirsk, 630090, RUSSIA; e-mail: ginzburg@math.nsk.su; 73-83-235-1620, F: 73-83-235-0652
GODFREY, Stephen: Carleton U., Physics Dept., 3372 HP, 1125 Colonel By Drive, Ottawa, Ontario K1S 5B6, CANADA; e-mail: godfrey@physics.carleton.ca; (613) 520-2600, x4386, F: (613) 520-4061
GUNION, John F.: UCD, Physics Dept., High Energy Div., Davis, CA 95616, USA; e-mail: jfgucd@ucdhep.ucdavis.edu; (916) 752-1134
HAN, Tao: U.: Wisc-Mad, Physics Dept., 1150 University Ave., Madison, WI 53706, USA; e-mail: than@ucdhep.ucdavis.edu; (916) 752-9855, F: (916) 752-4717
HEUSCH, Clemens A.: UCSC, IPP, Santa Cruz, CA 95064, USA; e-mail: heusch@slac.stanford.edu OR clemens.heusch@cern.ch; (650) 926-2540, F: (650) 926-4905
JOHNSTONE, Carol J.: FNAL, Beams Physics, P.O. Box 500, MS 220, Batavia, IL 60510, USA; e-mail: johnstone@adcalc.fnal.gov; (630) 840-3794, F: (630) 840-6039
KAHN, Stephen: BNL, Bldg. 902A, P.O. Box 5000, Upton, NY 11973-5000, USA; e-mail: kahn1@bnl.gov; (516) 344-2282, F: (516) 344-2170
KANE, Gordon L.: U. Michigan, Dept. of Physics, Randall Physics Lab., Ann Arbor, MI 48109-1120, USA; e-mail: gkane@umich.edu; (313) 764-4451, F: (313) 763-2213
KEUNG, Wai-Yee: U. Illinois, Chicago, Physics Dept., Rm 2236, 845 W. Taylor St., M/C 273, Chicago, IL. 60607-7059, USA; e-mail: keung@uic.edu; (312) 413-2778, F: (312) 996-9016
KING, Bruce: BNL, Physics Dept., P.O. Box 5000, Bldg. 901A, Upton, NY 11973-5000, USA; e-mail: bking@sun2.bnl.gov; (516) 344-3845, -7987, F: (516) 344-3248
KIRK, Harold G.: BNL, Physics Dept., PO Box 5000, Bldg. 901A, Upton, NY 11973-5000, USA; e-mail: hkirk@bnl.gov; (516) 344-3780, F: (516) 344-3248
KOLONKO, James J.: UCLA, Physics & Astron. Dept., Box 951547, Los Angeles, CA 90095-1547, USA; e-mail: kolonko@physics.ucla.edu; (310) 206-4548, F: (310) 206-1091
KUNO, Yoshitaka: KEK, Inst. of Part. & Nucl. Studies, 1-1 Oho, Tsukuba-shi, Ibaraki-ken 305 , JAPAN; e-mail: yoshitaka.kuno@kek.jp; 81-298-64-5419, F: 81-298-64-7831
LEE, Kevin: UCLA, Dept. of Physics & Astron., Box 951547, Los Angeles, CA 90095-1547, USA; (310) 825-1214, F: (310) 206-1091
LI, Derun: LBNL, Accel. & Fusion/Ctr. for Beam Physics, 1 Cyclotron Rd., MS 71-259, Berkeley, CA 94720, USA; e-mail: dli@lbl.gov; (510) 486-5053, F: (510) 486-7981
LU, K. U.: CSULB, Dept. of Mathematics, High Energy Div., Long Beach, CA 90840, USA; e-mail: klu6654348@aol.com; (562) 985-4731, F: (562) 985-8227
LYKKEN, Joseph: FNAL, Theory Dept., P.O. Box 500, MS 106, Batavia, IL 60510, USA; e-mail: lykken@fnal.gov; (630) 840-4689, F: (630) 840-5435

MARCIANO, William J.: BNL, Dept. of Physics, Building 510A, Upton, NY 11973-5000, USA; e-mail: marciano@bnl.gov; (516) 344-3151
McDONALD, Kirk T.: Princeton U., Physics Dept., P.O. Box 708, Princeton, NJ 08544, USA; e-mail: mcdonald@puphep.princeton.edu; (609) 258-6608, F: (609) 258-6360
MERY, Pierre: L'Universite de la Mediterranee, Doyen de la Faculte des Sciences de Luminy, 163 Ave. de Luminy, Case 901, 13288 Marseille Cedex 9, FRANCE; e-mail: mery@cpt.univ-mrs.fr; 33-912-69-010, F: 33-912-69-200
MILLS, Frederick E.: FNAL, Beams Div., P.O. Box 500, Batavia, IL 60510, USA; e-mail: fredmills@aol.com; (630) 840-4440, -0423 OR (520) 825-7321
MOKHOV, Nikolai: FNAL, Beam Physics, P.O. Box 500, MS 220, Batavia, IL 60510, USA; e-mail: mokhov@fnal.gov; (630) 840-4409, F: (630) 840-6039
MOLZON, William: UCI, Dept. of Physics and Astron., 4129 Physical Sciences 2, Irvine, CA 92697-4575, USA; e-mail: wmolzon@uci.edu; (714) 824-5987, F: (714) 824-2176
MORETTI, Alfred: FNAL, Beam Div., P.O. Box 500, MS 307, Batavia, IL 60510, USA; e-mail: moretti@fnal.gov; (630) 840-4843, F: (630) 840-4552
MORI, Yoshiharu: KEK-Tanashi, 3-2-1 Midori-Cho, Tanashi-Shi, Tokyo 788, JAPAN; e-mail: moriy@kekvax.kek.jp; 81-424-69-9577, F: 81-424-68-5543
NAGAMINE, Kanetada: KEK, Meson Science Lab., 1-1 Oho, Tsukuba-shi, Ibaraki-ken 305, JAPAN; e-mail: nagamine@mslaxp.kek.jp; 81-298-64-5603, F: 81-298-64-5623
NEUFFER, David: FNAL, Beams - Muon Colliders, P.O Box 500, MS 345, Batavia, IL 60510, USA; e-mail: neuffer@adcon.fnal.gov; (630) 840-2640
NG, King-Yuen: FNAL, Beams Physics, P.O. Box 500, Baltavia, IL 60510, USA; e-mail: ng@fnal.gov; (630) 840-4597, F: (630) 840-6039
NOJIRI, Mihoko: Kyoto U., YITP, Oiwake-cho, Kitashirakawa, Sakyoku, Kyoto, JAPAN; e-mail: nojirim@theory.kek.jp; 81-298-64-5394, F: 81-298-64-5755
NOREM, James H.: FNAL/ANL, HEP Div., c/o P.O. Box 500, MS 307, Batavia, IL 60510, USA; e-mail: norem@hep.anl.gov; (630) 840-8909, F: (708) 252-5076
PALMER, Robert B.: BNL, Director's Office, Bldg. 901A, P.O. Box 5000, Upton, NY 11973-5000, USA; e-mail: palmer@bnl.gov; (516) 344-3248, F: (516) 344-2842
PAN, Yibin: CERN, Physics Dept/PPE Div., CH-1211, Geneva 23, SWITZERLAND; e-mail: pan@wisconsin.cern.ch; 41-22-767-3764, F: 41-22-767-8370
PARKE, Stephen J.: FNAL, Theoretical Physics Dept., P.O. Box 500, MS 106, Batavia, IL 60510, USA; e-mail: parke@fnal.gov; (630) 840-4517, F: (630) 840-5435
PARSA, Zohreh: BNL, Physics Dept., P.O. Box 5000, Bldg. 901A, Upton, NY 11973-5000, USA; e-mail: parsa@bnl.gov; (516) 344-2085, F: (516) 344-3248
PETERSON, Jack M.: LBNL., Ctr. for Beam Physics, AFRD, MS 71-259, 1 Cyclotron Rd., Berkeley, CA 94720, USA; e-mail: cbp@lbl.gov; (510) 486-6570, F: (510) 486-7981
QIAN, Zubao: FNAL, BD/RFI, P.O Box 500, MS 307, Batavia, IL 60510, USA; e-mail: zubao@fnal.gov; (630) 840-4551, F: (630) 840-4552
QUIGG, Chris: FNAL, Theoretical Physics, P.O. Box 500, MS 106, Batavia, IL 60510, USA; e-mail: quigg@fnal.gov; (630) 840-3578, F: (630) 840-5435
RAJA, Rajendran: FNAL, D Construction Dept., P.O. Box 500, MS 357, Batavia, IL 60510, USA; e-mail: raja@fnal.gov; (630) 840-4092, F: (630) 840-8481
RAJPOOT, Subhash: CSULB, Physics & Astron. Dept., 1250 Bellflower Blvd., Long Beach, CA 90840, USA; e-mail: rajpoot@csulb.edu; (562) 985-4847, F: (562) 985-7924
RENTON, Peter B.: Oxford U., Nuclear Physics Lab., Keble Rd., Oxford OX1 3RH, UK; e-mail: p.renton1@physics.oxford.ac.uk; 440-186-527-3327, F: 440-186-527-3418

RICHTER, Burton: SLAC, P.O. Box 4349, MS 80, Stanford, CA 94309, USA; e-mail: brichter@slac.stanford,edu; (650) 926-2601, F: (650) 926-4500

ROWSON, Peter C.: SLAC, SLD, P.O. Box 4349, MS 96, Stanford, CA 94309, USA; e-mail: rowson@slac.stanford.edu; (650) 926-2635, F: (650) 926-2923

SCANLAN, Ronald M.: LBNL, Superconducting Magnet Program, 1 Cyclotron Rd., MS 46-161, Berkeley, CA 94720, USA; e-mail: rmscanlan@lbl.gov; (510) 486-7241, F: (510) 486-5310

SESSLER, Andrew M.: LBNL, Collider Physics, Ctr. for Beam Physics, 1 Cyclotron Rd., MS 71-259, Berkeley, CA 94720 , USA; e-mail: andrew_sessler@macmail.lbl.gov; (510) 486-4992, F: (510) 486-6485

SHEN, Jing: Inst. of High Energy Physics, PO Box 918(6), Beijing 100039, PEOPLE'S REPUBLIC OF CHINA; e-mail: shenj@bepc3.ihep.ac.cn; 86-10-68210011, x3590

SILVA, John: ESD, Inc., 32258 Devonshire St., Union City, CA 94587, USA; e-mail: john@silva.com; (510) 783-0455, F: (510) 783-4718

SKRINSKY, Alexander N.: Russian Academy of Science, The Budker Inst. of Nuclear Physics Siberian Branch, 630090 Novosibirsk-90, Ac. Lavrentiev Av. 11, RUSSIA; e-mail: skrinsky@inp.nsk.su OR skrinsky@vxcern.cern.ch; 3832-35-60-31, F: 3832-35-21-63

STUMER, Iuliu: BNL, Dept. of Physics, P.O. Box 5000, 510-A, Upton, NY 11973-5000, USA; e-mail: stumer@bnl.gov; (516) 344-3944, F: (516) 344-5568

SUMMERS, Donald J.: U. Mississippi, Physics Dept., 108 Lewis, University, MS 38677, USA; e-mail: summers@umsphy.phy.olemiss.edu; (601) 232-5045, F: (601) 232-7032

SUTTER, David F.: USDOE, Div. of High Energy Physics/Technology R&D, 19901 Germantown Rd., Germantown, MD 20874-1290, USA; e-mail: hep-tech@oer.doe.gov; (301) 903-5228, F: (301) 903-2597

TOLLESTRUP, Alvin: FNAL, CDF Group, P.O. Box 500, MS 318, Batavia, IL 60510-0500, USA; e-mail: alvin@fnal.gov; (630) 840-4331, F: (630) 840-2968

WALTER, Hans Chrisitian: Paul Scherrer Inst., Nucl. & Part. Physics, PSI-West, CH-5232 Villigen, SWITZERLAND; e-mail: chris.walter@psi.ch; 41-56-310-3139, F: 41-56-310-3644

WAN, Weishi: FNAL, Beam Physics Dept., P.O. Box 500, MS 220, Batavia, IL 60510, USA; e-mail: wan@fnal.gov; (630) 840-6289, F: (630) 840-6039

WUERTELE, Jonathan S.: UCB/LBNL, Dept. of Physics, 366 Le Conte Hall , Berkeley, CA 94720 , USA; e-mail: wurtele@physics.berkeley.edu; (510) 486-6572

ZHAO, Yongxiang: BNL, Physics Dept., P.O. Box 5000, Bldg. 901A, Upton, NY 11973-5000, USA; e-mail: yzhao@muon.cap.bnl.gov; (516) 344-4565, F: (516) 344-3248

Author Index

A
Ahrens, L. A., 314
Ankenbrandt, C., 314

B
Balbekov, V. I., 310
Bambade, P., 31
Barger, V., 3
Berger, M. S., 79
Brennan, M., 314

C
Cheung, K., 338
Cline, D. B., 159

D
Dasu, S., 85
Drozhdin, A., 209, 242

F
Feng, J. L., 92
Fernow, R. C., 282

G
Gallardo, J. C., 282
Garren, A., 209, 242
Geer, S., 18
Ginzburg, I. F., 265, 333
Godfrey, S., 98
Gunion, J. F., 44

H
Heusch, C. A., 116

J
Johnstone, C., 209, 242

K
Kamal, B., 174
King, B. J., 132

L
Lykken, J. D., 106

M
Mane, V., 314
Marciano, W. J., 174, 347
Mokhov, N., 209, 242, 320
Molzon, W. R., 146

N
Nagamine, K., 295
Neuffer, D., 270
Ng, K.-Y., 220, 314
Norem, J., 314, 357

P
Palmer, R. B., 183, 282
Parke, S., 72
Parsa, Z., 174, 289
Popvic, M., 314

Q
Qian, Z., 314
Quigg, C., 57

R

Raja, R., 228
Rajpoot, S., 126
Renton, P. B., 168
Roser, T., 314

S

Shen, J., 325, 360, 367
Skrinsky, A., 249

T

Tollestrup, A., 228
Trbojevic, D., 220, 314

V

van Asselt, W., 314
Van Ginneken, A., 310, 320

W

Walter, H.-K., 143
Wan, W., 209, 242